The World
of Physics
Mysteries Magic & Myth

The World
of Physics

Mysteries Magic & Myth

John W. Jewett, Jr.

California State Polytechnic University, Pomona

BROOKS/COLE

THOMSON LEARNING

Australia • Canada • Mexico • Singapore • Spain • United Kingdom • United States

For more information about our products,
contact us at:
Thomson Learning Academic Resource Center
1-800-423-0563

For permission to use material from this text,
contact us by:
Phone: 1-800-730-2214
Fax: 1-800-731-2215
Web: www.thomsonrights.com

Asia
Thomson Learning
60 Albert Complex, #15-01
Alpert Complex
Singapore 189969

Australia
Nelson Thomson Learning
102 Dodds Street
South Street
South Melbourne, Victoria 3205
Australia

Canada
Nelson Thomson Learning
1120 Birchmount Road
Toronto, Ontario M1K 5G4
Canada

Europe/Middle East/South Africa
Thomson Learning
Berkshire House
168-173 High Holborn
London WC1 V7AA
United Kingdom

Latin America
Thomson Learning
Seneca, 53
Colonia Polanco
11560 Mexico D.F.
Mexico

Spain
Paraninfo Thomson Learning
Calle/Magallanes, 25
28015 Madrid, Spain

To Lisa Jewett, my wife and best friend, who was always there to support me and to drive me on when giving up seemed to be the best choice.

Contents

Preface

This is a book for *students*—students in formalized classrooms at the high school and university levels, teachers who see themselves as life-long learners, and those who are forever students and are always striving for more understanding of the world around us. The information in this book is intended to allow for exploration into the applications of physics in everyday life and to provoke the reader to examine his or her own understanding of physical concepts.

The chapter organization is similar to the standard order in many introductory physics textbooks. This is not intended to represent advocacy for this particular schedule of teaching or learning but is followed more for the ease of the reader in finding information about a particular topic. The material is presented in three formats within each chapter—*Mysteries*, *Magic*, and *Myth*. These "M's" are described below, followed by suggested uses for various readers.

Mysteries

These are questions about everyday life that can be understood by means of applications of the principles of physics. Some refer to common observations (e.g., "Why is the sky blue?", Mystery #3, Chapter 25) while others represent questions that may not have crossed the reader's mind (e.g., "Why does it take a different time to fly east to west than it does to fly west to east between the same two points?", Mystery #1, Chapter 1). Each of the Mysteries can be "solved" with an appropriate understanding and application of physical principles.

Magic

The Magic "tricks" in this book are demonstrations and activities that can be performed in and out of the classroom. In general, they belong to the classification of *discrepant events*, in that the outcome is different from that which one might expect. Most of the activities can be performed with simple apparatus available from the hardware store or supermarket. A small number require simple laboratory equipment such as a laser or an oscilloscope, but none requires sophisticated equipment. Some require a trek to an appropriate location, such as a quiet pond (The Dark Water, Chapter 28), and others require waiting for appropriate phenomena, such as the rainbow (The Broken Rainbow, Chapter 26), to occur.

Myth

The Myths in this book are presented as statements of fact, but the reader needs to be aware that *the statements are generally untrue.* The Myths are common misunderstandings, misconceptions and mistakes from our every-day culture. Some are well-established, but errant, beliefs such as the "weightlessness" of astronauts (Myth #2, Chapter 6) or "centrifugal force" (Myth #1, Chapter 9). Others are misuses of physics in the print, broadcast and movie media, such as the sound of space explosions (Myth #1, Chapter 18) or the use of a diverging lens to light a fire as described in a novel (Myth #3, Chapter 27). Others fall into a variety of other categories.

The book also contains the following features—

Concepts   A brief outline of the physics concepts appropriate for each chapter is included. This will assist in focusing the reader's thoughts on some of the principles to be used in addressing the Mysteries, Magic and Myth in the chapter. This is not a textbook, so these sections are by no means complete expositions of the topic of the chapter. The author refers the reader to a physics textbook for more complete descriptions of the physical principles.

Discussions   A discussion section is found at the end of each chapter, in which comments on each "M" are available. Note that this is not called an "Answer" section. The important goal in this book, as should be the case in all scientific endeavors, is the motivation to perform effective thinking, not to obtain the right "answer". The discussion section contains scientific explanations of phenomena that are well understood, but there are also examples of cases where science has not been able to establish the definitive "answer". Please refrain from the temptation to jump immediately to the back of the chapter after reading an "M". Important learning (and enjoyment) will take place by pondering these items for a while and developing and testing various hypotheses.

The discussion section also contains many references to other printed material related to the phenomenon being discussed. Most of the references are to physics teaching journals (e.g., *The Physics Teacher, American Journal of Physics, Physics Education*) or to popular science journals (e.g., *Scientific American*). Some more advanced articles are referenced for the more mathematically adventuresome. Many books are also referenced, both physics textbooks and popular treatments of scientific ideas.

Bibliography   At the back of the book, a bibliography lists a very large number of books and journal articles which can be consulted for more details on physics in everyday life.

Suggested Uses

For students in formalized classrooms—

Physics is an endeavor that requires clear and logical thinking. It is also an endeavor in which many misconceptions abound and there are many dead ends in logical thinking awaiting the unwary student. The items in this book allow an opportunity for the student to apply the knowledge from the classroom to real-world situations. Many of the items will be familiar from everyday life, helping you to keep in mind that physics is a relevant science. Thinking about these "M's" will test your understanding of the concepts and your sense of logical process. The Myths will help you to avoid some of the common misconceptions that occur in the physics classroom. The references in the bibliography will allow you to extend your knowledge beyond the necessarily compact discussions of the "M's" that appear in this book.

The primary emphasis in this book is on conceptual understanding rather than mathematical understanding. While many physics courses require you to perform mathematical calculations and develop your problem-solving skills, this is difficult to do without a solid conceptual understanding of the principles. As a result of this emphasis, the amount of mathematics in the book is less than that in your physics textbook.

As your physics course progresses, keep up with the topics in your course by following along in this book. Use the Mysteries and Myths to test your understanding of the principles. Perform the Magic demonstrations and activities at home. Perhaps your instructor may perform some of these in the classroom. If there are questions that you have about any of the items in this book, ask your instructor for help.

For those students who use this book in association with a science methods class, in preparation for becoming teachers, there is a wealth of teaching ideas contained in these pages. Challenge your future students with these, and help them to become the critical thinkers that we will need to solve our problems.

For teachers in formalized classrooms—

The Mysteries can be used in a variety of ways. They can simply be presented to the class for classwide discussion. They can be presented as assignments to small groups of students for cooperative work in formulating a hypothesis. They can be given as homework assignments for creative and critical thinking. They can be assigned as research projects for library work. They can be presented on overheads for student thought while roll is taken at the beginning of class.

The Magic demonstrations can be performed before the appropriate material is covered, as a "teaser" or a motivator, or can be performed along with the material as visual applications of the concepts. Those that can be done, or need to be done, outside of the classroom can be given as special assignments or can even form the basis of a field trip.

Each of the Myths can be "debunked" by an appropriate application of the physical principles. These can be used in the classroom in a variety of ways. They can be grouped with other, true statements for students to determine which is the incorrect statement. Small groups can work on creating clear descriptions of why the Myth is untrue. The debunking of new Myths can be given as a homework assignment to help students develop critical thinking skills and communication skills.

Although the material is aimed at a high school or introductory university level, many of the activities and concepts are adaptable to the middle school and elementary level.

For the informed layperson—

Physics is a fascinating science and is all around you all the time. This book contains a treasure of examples of how physics interacts with you in your everyday life. It will help you to appreciate the physics that surrounds you. There are some references here to social issues, such as the Greenhouse Effect and solar energy, and these will help make you a better informed citizen. But the primary goal is *fun*. Thinking is fun and these pages will give you plenty of opportunity to challenge your thinking skills. Share these "M's" with your spouse and children and help them to develop enhanced critical thinking. But, most importantly, for *all readers*—enjoy!

Acknowledgements

A project such as this is not possible without the assistance of a great many people. First of all, let me make it perfectly clear that I am not claiming to have generated all of the ideas in this book on my own. This publication is the result of my own ideas combined with those of my previous teachers, previous and present colleagues, the writings of other scientists and the input of many members of the non-scientific public.

I would like to express specific thanks to some individuals who have been especially helpful in bringing this volume to its existence. Thanks go to Soumya Chakravarti, Harvey Leff, John Mallinckrodt and Peter Siegel of California State Polytechnic University, Pomona and Roger Nanes of California State University, Fullerton for their efforts in reading several early drafts of this work and providing important suggestions for improvement. The high school physics teachers of Science IMPACT (Institute for Modern Pedagogy and Creative Teaching) receive warm appreciation for their encouragement to prepare the "M's" in book form after being bombarded with them during three summer institutes. Special thanks go to California State Polytechnic University, Pomona, for awarding me with sabbatical time during which the original draft of this project was prepared.

I have benefited greatly from the reviewers of the draft manuscript who provided me with numerous suggestions for improvement. Special thanks go to the following reviewers who provided in-depth analysis of the book: Beverley A. P. Taylor of Miami University (Hamilton, Ohio) and Edmund F. Nevirauskas, Needham High School (Needham, Massachusetts).

Deep gratitude is extended to the following members of the staff at Saunders College Publishing, without whom this book would not have been possible: Susan Pashos, for her professional support, and Marc Sherman, for his help in the book production phase of the project. I am especially indebted to John Vondeling for his many contributions to the support of my writing efforts.

Finally, I am deeply indebted to my dear wife, Lisa Jewett, who endured many lonely evenings while I pounded on the keyboard as well as many lonely conversations during which I responded to her discussions with my voice but was frantically searching in my mind for a clearer way to explain an "M".

John W. Jewett Jr.
Pomona, California
October 1999

Chapter 1
Vectors, Measurement and Other Mathematical Preliminaries

𝕸𝖞𝖘𝖙𝖊𝖗𝖎𝖊𝖘:

1.) Listed below are data from a recent airline schedule for flights between New York and Los Angeles. The departure and arrival times are combined with the time zone correction to determine the actual flight time.

From	To	Depart	Arrive	Difference	Time Zone Correction	Flight Time
LA	NY	7:45 AM	3:55 PM	8 hrs 10 min	- 3 hours	**5 hrs 10 min**
NY	LA	8:10 AM	10:50 AM	2 hrs 40 min	+ 3 hours	**5 hrs 40 min**

Why does it take a different time to fly east to west than it does to fly west to east <u>between the same two points</u>?

2.) How does a sailboat or a windsurfer travel *into* the wind?

Mysteries:

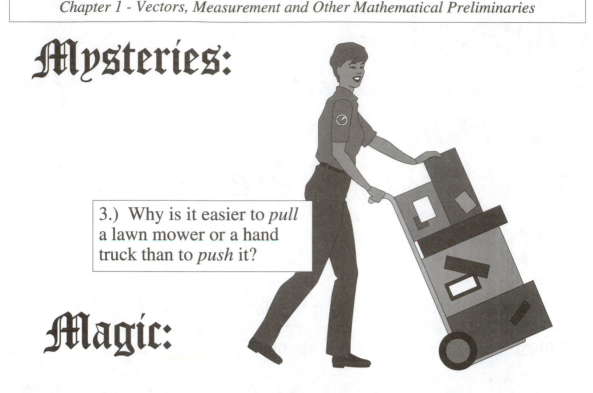

3.) Why is it easier to *pull* a lawn mower or a hand truck than to *push* it?

Magic:

A Real Humdinger

Physicists use a number of devices for making measurements. One device for measuring the <u>time</u> between recurring events is a *stroboscope*. This is a flashing light which illuminates a system exhibiting some kind of periodic motion. If the flash rate of the stroboscope matches the frequency of the system (or is an integral submultiple ($^1/_2, ^1/_3, ^1/_4$, etc.) of the frequency), then the system's motion will appear to have stopped, since it will always be illuminated at the same point in its motion and the rest of the motion takes place in the dark.

An example of a built-in human stroboscope is as follows. Set up a computer so that the monitor displays a large area of a single bright color, preferably white. Now, hum at the screen. That's right - <u>hum</u>. Watch carefully and you will see horizontal gray lines on the screen. As you change the frequency of your humming, you can make the lines move upward or downward. But don't ask anyone else if they see the lines. They will only see them if they are humming, too!

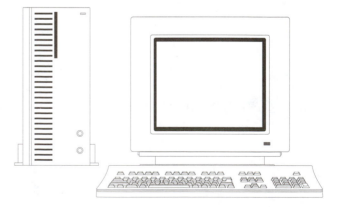

𝕸𝖆𝖌𝖎𝖈:

Go North to Go West?

This activity will require a globe and a toy car. The toy car must have a wheel with some sort of marking so that the number of rotations can be counted. We will roll the car along the globe and use the number of rotations of the wheel as a measure of distance traveled on the globe.

We will imagine taking a trip from Baja California to Shanghai, China. Start off with the wheel whose rotations you will be counting on the globe at Baja California at the location of the 30° latitude line. Roll the car west along the 30° latitude line, counting the wheel rotations until the wheel arrives at Shanghai. Record the number of rotations.

For our second trip, we will follow a curved path along a northern route and count the rotations again. Start again at Baja California on the 30° latitude, but roll the car along the California coast. At about San Francisco, the coast will take a turn toward the north. Don't follow the coast, but continue on your original heading, into the ocean, curving slightly toward the west and aiming at the end of the Aleutian Islands off Alaska (Some islands in this area are named *Amlia*, *Atka* and *Adak*.). By the time you reach these islands, you should be heading due west. Continue curving gently, aiming to roll through the Sea of Japan and "parallel" to the Japanese Islands. Finally, pass through South Korea and end at Shanghai. Record the number of rotations.

How do the number of rotations compare?

𝔐𝔶𝔱𝔥:

1.) Consider the graph to the right which shows interest rates on a certificate of deposit. The graph is taken from an actual flyer distributed by Bank A. At first glance, it would appear that Bank A gives the depositor *many times* the interest of Bank F.

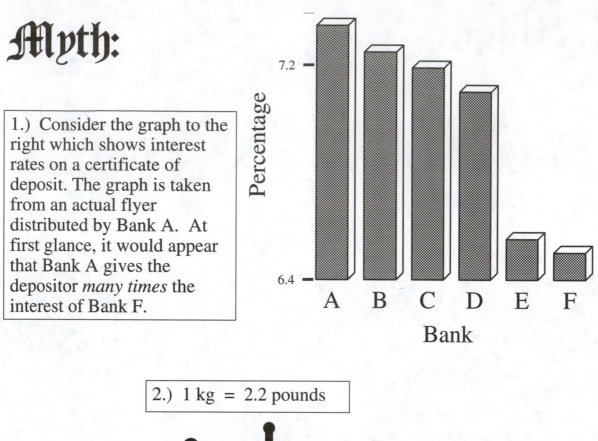

2.) 1 kg = 2.2 pounds

3.) A toy which is commercially available is a ball with an off-center weight so that the center of mass is not at the center the ball. On the package of the toy is the statement, "The weight adds a third force (*centrifugal motion*) to the two normal forces (*gravity and acceleration*) that determine a moving object's direction."

Say

What???

𝔐𝔶𝔱𝔥:

4.) In Olympic ski racing, skiers take turns racing down the course and their times are recorded. The one with the shortest time wins.

5.) For those readers who still have phonograph turntables, check your instruction manual. See if you can find a phrase such as the following: "The tracking force of the stylus should be about 3 grams".

𝔐𝔶𝔱𝔥:

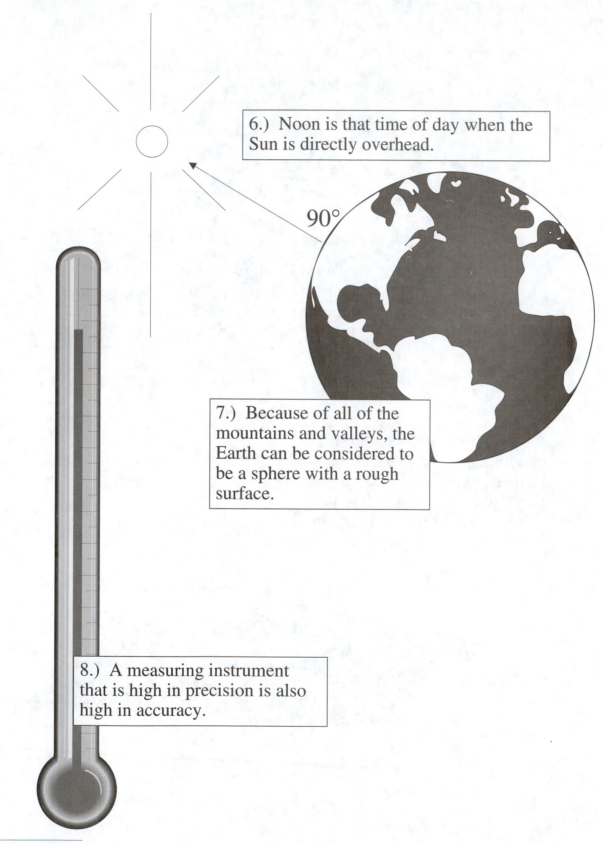

6.) Noon is that time of day when the Sun is directly overhead.

90°

7.) Because of all of the mountains and valleys, the Earth can be considered to be a sphere with a rough surface.

8.) A measuring instrument that is high in precision is also high in accuracy.

Concepts of Vectors, Measurement and Other Mathematical Preliminaries

Vectors can be used to represent physical variables that have both a magnitude and a direction. Examples include velocity, force and electric field. Variables that have only a magnitude are called *scalars*, such as mass, temperature and energy.

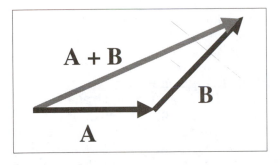

It is often necessary to add vectors. For the addition of two vectors, this can be done graphically by drawing one vector and then placing the tail of the second vector at the head of the first, such as in the diagram at the left, in which two vectors, **A** and **B**, are being added. The gray vector in the diagram is then drawn from the tail of the first vector (**A**) to the head of the second (**B**). This vector represents the sum of the vectors **A** and **B**.

An important skill when using vectors is the resolution of a vector into its components. This is replacing a vector with two other vectors such that when the two are added together, the original vector results. There are an infinite number of such pairs, but we choose that one pair that happen to lie parallel and perpendicular to an *x*-axis that we choose for convenience. This pair is indicated in the diagram to the right, showing that a vector (gray), of magnitude *A*, making an angle θ with respect to the *x*-axis, can be replaced with the sum of two vectors (black) with magnitudes:

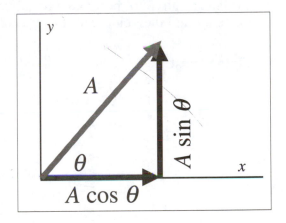

$A \cos \theta$ (parallel to the *x*-axis - the *x-component*)
$A \sin \theta$ (parallel to the *y*-axis - the *y-component*)

Physics, as a science, requires measurements of various aspects of the universe. In order to discuss these measurements with other scientists, it is important that there be precise definitions and a common language so that communication is effective and accurate. Associated with a measurement is a unit - meters for lengths, seconds for time, etc. A measurement is not complete unless the unit is included, so that another scientist can interpret the measurement.

Every measurement has some uncertainty associated with it - there is no such thing as a perfectly accurate measurement. This uncertainty arises from the measuring instrument, the person doing the measurement, the method of making the measurement and from characteristics of the system being measured. When a number of measurements are combined mathematically, the uncertainties will propagate through the calculation, so that the uncertainty in the final result is larger than the individual uncertainties in the measurements.

Physics also requires mathematical manipulation of data. Possible skills necessary for this manipulation include graphing, statistics, error analysis, estimation, etc.

Discussions; Chapter 1 - Vectors, Measurement and Other Mathematical Preliminaries

𝕸𝖞𝖘𝖙𝖊𝖗𝖎𝖊𝖘:

1.) Why does it take a different time to fly east to west than it does to fly west to east <u>between the same two points</u>?

The difference in flight times arises because there is usually a prevailing wind blowing with a component from West to East across the United States. When an airplane is flying toward the East, the prevailing wind helps the airplane, while, when flying West, the airplane has to fight the headwind. Thus, the westward velocity of the airplane with respect to the ground is smaller than the corresponding eastward velocity, and it takes longer to fly West than to fly East. In terms of vectors, we can find the velocity of the airplane relative to the ground by adding vectorially the velocity of the airplane relative to the air and the velocity of the prevailing wind:

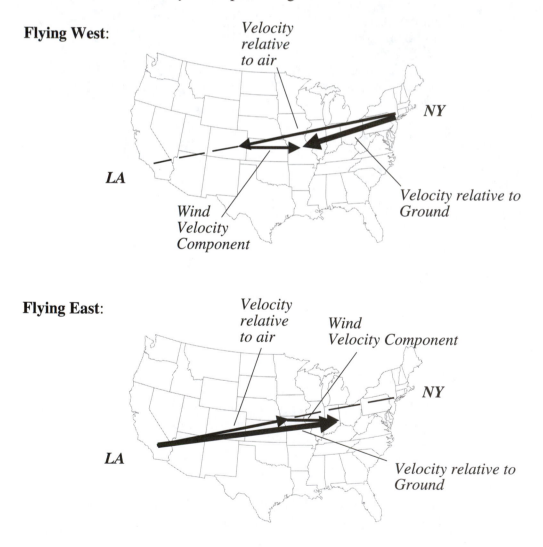

Flying West:

Velocity relative to air

NY

LA

Wind Velocity Component

Velocity relative to Ground

Flying East:

Velocity relative to air

Wind Velocity Component

NY

LA

Velocity relative to Ground

We have set up the diagrams on the previous page with the air velocity vectors having the same magnitude for the planes traveling in either direction. To that vector, we have added a vector representing that component of the prevailing wind velocity that is directed toward the East. The resultant of these two vectors is the velocity vector of the airplane with respect to the ground. Notice that the magnitude of the West to East vector is <u>larger</u> than that for the trip from East to West. Thus, the <u>speed</u> of the airplane in going toward the East is larger than that going to the West, resulting in the shorter flight time that we see in the original data. It should be pointed out that these are <u>not</u> the actual vectors for the velocities associated with an airplane flight. We arbitrarily set the air velocity vector magnitudes equal and the vectors directed from one city to the other for ease in comparison of the vectors for velocity relative to the ground. Once we add the wind velocity vectors in each case, we see that the final ground velocity vectors will result in our arriving south of the destination city in each case. Thus, in reality, the pilot of the aircraft must alter the direction of the aircraft toward the north to compensate for this effect. Corrected flight velocity vectors, resulting in a ground velocity vector aimed at the destination city, are as shown below:

Flying West:

Flying East:

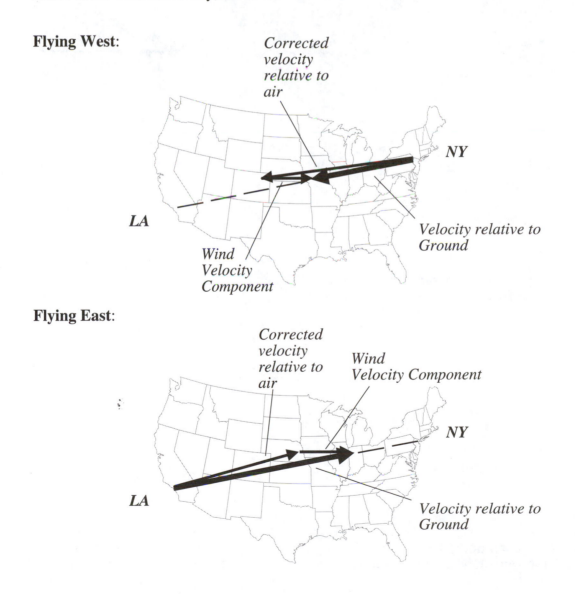

2.) How does a sailboat or a windsurfer travel *into* the wind?

Sailing into the wind in a sailboat is called *tacking*. The boat does not sail <u>directly</u> into the wind. Rather, the boat zigzags back and forth across the wind direction so that the <u>net</u> motion is in the direction from which the wind is coming. The dynamics of sailboat tacking is complicated. A simplified version will be presented here that still carries the flavor of the effect. The diagram below shows the vector diagram of the force exerted on the sail of a tacking sailboat as a parcel of air hits the sail and "reflects" off. We assume frictionless air, so that the force of the air on the sail has only a component <u>perpendicular</u> to the sail.

Force on Sail

Sail

Wind Velocity

Velocity of the "reflected" parcel of air

Now, we resolve the force on the sail into vector components parallel to and perpendicular to the keel of the sailboat:

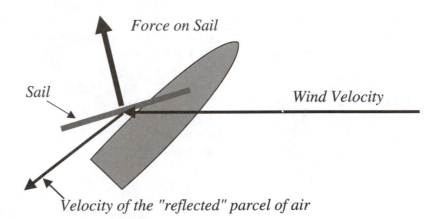

Force Component Perpendicular to Boat

Force on Sail

Force Component Parallel to Boat

Wind Velocity

Velocity of the "reflected" parcel of air

Now, the important thing to notice is that the force component parallel to the boat is in the *forward* direction! Thus, the boat sails "into" the wind. After sailing for some time under the conditions shown in the diagram, the boat is turned across the direction of the wind and sails in a new direction for some time. This <u>zigzagging</u> continues until the desired destination is reached. The diagram below shows the boat and sail position for a series of two zigs (we will define the above discussion as appropriate for a "zig"!) and two zags.

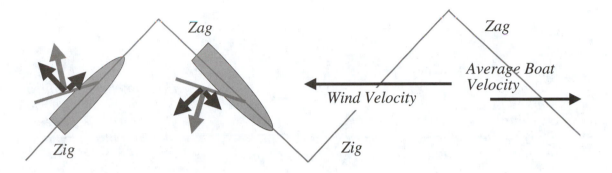

3.) Why is it easier to *pull* a lawn mower or a hand truck than to *push* it?

Before discussing the hand truck, let us discuss the more easily understood situation of a crate sliding across a floor. The diagram below shows two possibilities for applying a force to cause this sliding - one with an upward component and another with a downward component.

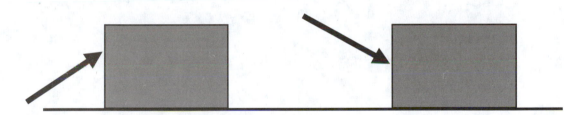

On the left, applying a force with an upward component will cause a reduction of the normal force. In turn, as discussed in Chapter 4, this will result in a decrease in the friction force between the crate and the floor. Then, as a result, less horizontal force is needed to move the crate. On the other hand, if we consider the situation on the right, applying a force with a downward component will increase the normal force, thus increasing the friction force and making it <u>harder</u> to move the crate along the floor.

Now, let's apply this same idea to the hand truck. Consider the two vector diagrams shown on the next page for pushing and pulling a hand truck in the direction toward the right side of the page. All of the forces shown are those on the hand truck as a system. The gravitational force on the hand truck is not included, since it does not enter into the argument and would unnecessarily clutter the diagrams further.

In the pushing case, as the operator applies a horizontal force, the body of the hand truck is compressed. This compression results in the application of a force with an upward component on the handle of the hand truck. Thus, the operator must apply an additional <u>downward</u> force component to establish equilibrium for the handle. This downward force on the hand truck results in a larger upward normal force from the floor which, in turn, results in a larger frictional force.

The situation is the reverse in the pulling case. Here, the horizontal force component from the operator causes the body of the hand truck to be in tension. Thus, the body pulls down on the handle and the operator must provide an upward component of force to keep the handle in equilibrium. Thus, there is a net upward force component on the hand truck, which reduces the normal force and, therefore, the friction, making it easier to roll.

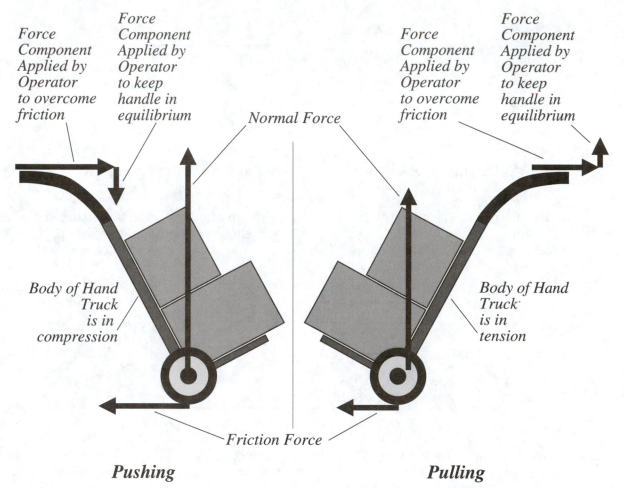

Force Component Applied by Operator to overcome friction

Force Component Applied by Operator to keep handle in equilibrium

Normal Force

Force Component Applied by Operator to overcome friction

Force Component Applied by Operator to keep handle in equilibrium

Body of Hand Truck is in compression

Body of Hand Truck is in tension

Friction Force

Pushing

Pulling

This effect is smaller for a hand truck than for the crate, since the rolling friction is fairly small to begin with. It is more evident in a lawn mower, where rolling over rough grass creates a larger friction force. For a discussion of this idea with regard to pushing or pulling a wheelbarrow over a curb, see G. Faucher, "Pushing or Pulling a Wheelbarrow", *The Physics Teacher*, **27**, 379 (1989).

𝔐𝔞𝔤𝔦𝔠:

A Real Humdinger

As you hum, your head vibrates. As a result, your eyes also vibrate and the light from an object which you are viewing will oscillate on and off the fovea. If your humming frequency is equal to, or a multiple of, the scan rate on the computer monitor, then your eyes will change direction slightly each time the scan reaches a certain level on the screen and that area will appear darker. You make the gray areas appear to move up or down by changing your humming frequency just as a rotating fan can be made to appear to move forward or backward by changing the frequency of a stroboscope. For studies on the effect of humming on viewing a rotating striped disk, see W. A. H. Rushton, "Effect of Humming on Vision", *Nature*, **216**, 1173 (1967). For a comment on a similar effect with a television, see P. C. Williams and T. P. Williams, "Effect of Humming on Watching Television", *Nature*, **239**, 407 (1972).

Go North to Go West?

Your distance measured along the northern route should be about 5 - 10% <u>shorter</u> than the route along the 30° latitude line. This demonstration points out the difference between *Euclidean geometry*, which is the geometry of the flat plane and *non-Euclidean geometry*, the geometry of curved surfaces. It may seem intuitive to travel directly west to reach Shanghai, but this distance is actually <u>farther</u> than taking a northern route. The northern route described in the demonstration is a segment of a Great Circle, which is a circle whose center coincides with the center of the Earth. Airplanes travel Great Circle routes rather than direct routes, since fuel is saved by traveling the shorter distance.

A second demonstration or student activity can be performed as follows: measure the two distances described with a flexible measuring tape, such as that used in sewing. You will have trouble measuring the direct westward route, because you cannot get the tape to lie flat along the latitude line. You will have to bend it to the side somewhat and approximate the measurement. The northern route will be easier to measure - the tape will lie flat, since its radius of curvature coincides with that of the globe.

𝔐𝔶𝔱𝔥:

1.) Consider the graph below which shows interest rates on a certificate of deposit. The graph is taken from an actual flyer distributed by Bank A. At first glance, it would appear that Bank A gives the depositor *many times* the interest of Bank F.

The graph is misleading in that the vertical axis does not start at zero. This is not mathematically, scientifically or socially illegal, but it may lead the unsuspecting consumer to an incorrect conclusion. If we redraw the graph with the vertical axis starting at zero, it appears as shown at the right.

Thus, Bank A is still the winner, but this graph more realistically represents the fact that the differences among the six banks is small.

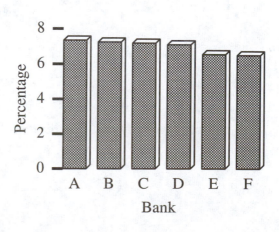

2.) 1 kg = 2.2 pounds

The problem with the statement is in the interpretation of the equal sign. The unit on the left is a unit of *mass*. On the right is a unit of *weight*. Since these are different physical variables, they cannot be related by an equal sign. Despite the fact that this conversion statement can be found in print, it is incorrect as an equality. The statement should be interpreted as follows: "An object whose mass is 1 kg will have a weight of 2.2 pounds when it is located on the surface of the Earth". If the object is anywhere else in the universe, however, its mass will continue to be 1 kg while its weight will depend on its location.

There are many other examples of confusing units in measurement. Another example of the above error occurs on soup cans where the "net weight" is listed as 10 $^3/_4$ oz (305 grams). While the first value is a weight, the second value, in parentheses, is a mass. To make matters worse, consider the label on a typical soda can: 12 FL OZ (355 mL). This may appear at first glance to be another similar situation, but a fluid ounce is <u>not</u> a weight, it is a *volume*, despite the use of the word "ounce" in its name! This type of confusion appears elsewhere, for example, in the use of electromotive *force*, when we are talking about an *electrical potential difference*! See Myth #5, Chapter 20 for more discussion of this particular example. Be very careful to keep the language straight, despite possible pitfalls such as these!

3.) A toy which is commercially available is a ball with an off-center weight so that the center of mass is not at the center the ball. On the package of the toy is the statement, "The weight adds a third force (*centrifugal motion*) to the two normal forces (*gravity and acceleration*) that determine a moving object's direction."

This is a terrible mixture of terms that was clearly not checked by a physicist before manufacturing and marketing. We will split the statement into sections and make comments:

Phrase in statement:	*Comments:*
"The weight adds a third force (*centrifugal motion*)..."	<u>Force</u> is not equal to <u>motion</u>.
	In addition, if centrifugal *force* were meant, this is not a "real" force (See Chapter 9, Myth #1).
	What's more, how can a weight, which is a downward force, "add" a force in the "centrifugal" direction, which is constantly changing as the ball bounces?
"...to the two normal forces (*gravity and acceleration*)"	"Normal" force has a particular meaning in physics, other than the "ordinary" forces suggested in this phrase.
	Also, acceleration is not a force, normal or otherwise.

What about other forces that determine the ball's motion - friction with the air and friction with the surfaces from which it bounces, the actual normal force from the surfaces, and others?

"forces (...) that determine a moving object's direction."

Forces do not determine the direction of motion, they determine the change in motion, as described by Newton's Second Law. For example, for a projectile moving in two dimensions, the force is down (gravity), but the direction of motion is never straight down.

What's more, quantities other than forces determine the direction of motion - the initial velocity, for example.

4.) In Olympic ski racing, skiers take turns racing down the course and their times are recorded. The one with the shortest time wins.

The problem here is one of uncertainty. Suppose that the times for two skiers on a given course are as follows:

Skier 1	3:47.85 seconds
Skier 2	3.47.83 seconds

According to the procedure, Skier 2 is the winner. But what are the uncertainties for these times? Think about the areas where uncertainty can arise. How is the timing mechanism started? Is it done in exactly the same way for each skier? And how is the timing mechanism stopped? Can the timing mechanism be stopped earlier if the skier is leaning over at the end? Is there any human error involved in starting and stopping the timing mechanism? No measurement is perfectly accurate, yet the times of ski runs are reported without "error bars". If the uncertainties in the above times were included, and the uncertainties were larger than 0.02 seconds, perhaps it would not be so clear which skier was the winner.

5.) For those readers who still have phonograph turntables, check your instruction manual. See if you can find a phrase such as the following: "The tracking force of the stylus should be about 3 grams".

The problem here is with units, a problem which occurs often in everyday usage of scientific terms. If the tracking force is a force, then it cannot be measured in grams, which is a unit of mass.

6.) Noon is that time of day when the Sun is directly overhead.

If there were no tilt of the Earth's axis relative to the plane of its orbit, then the Sun would appear directly overhead at noon only to observers on the Equator. Due to the 23.5° tilt that actually exists, however, the range of observers who can see the Sun directly overhead at noon at some time during the year is from latitude 23.5° north of the equator to 23.5° south of the Equator. These limits define the Tropics of Cancer and Capricorn, respectively. For the Tropic of Cancer, the Sun is directly overhead at noon at the summer solstice in June, while this event occurs in December for the Tropic of Capricorn.

The first diagram on the right shows a <u>hypothetical</u> Earth with no tilt. Notice that Tropic of Cancer observers (on the left side of the Earth in the diagram) must look south of their local vertical to see the Sun, while Tropic of Capricorn observers must look north of their vertical. Observers on the equator look directly upward, along the local vertical to see the Sun. It is noon for all observers, but only the equatorial observers see the Sun directly overhead.

The second diagram to the right shows the actual tilt of the Earth and its position at noon in the Southern Pacific on the December solstice. Notice that observers on the tropic of Capricorn look directly overhead to see the Sun at noon. Observers on the equator must now look toward the south to see the Sun, while observers in the Northern Hemisphere on or above the Tropic of Cancer (not indicated on this diagram) must look quite a bit south to see the Sun. Observers in the Southern Hemisphere south of the Tropic of Capricorn now see the Sun higher in the sky than in the case of the non-tilted Earth. This is the origin of the <u>seasons</u> - it is summer in the Southern Hemisphere in this diagram and winter in the Northern Hemisphere. In Myth #3, Chapter 14, there is more information about the seasons.

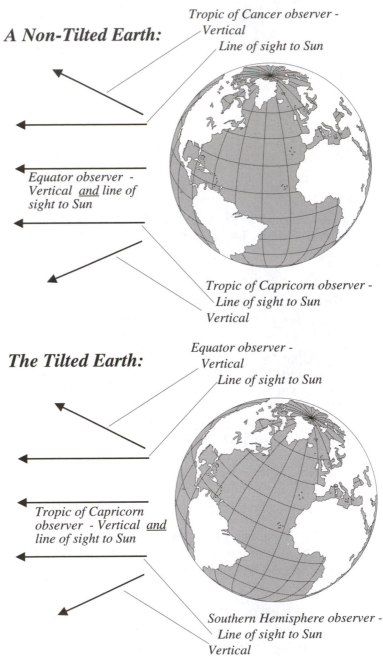

A Non-Tilted Earth:

Tropic of Cancer observer - Vertical
Line of sight to Sun

Equator observer - Vertical <u>and</u> line of sight to Sun

Tropic of Capricorn observer - Line of sight to Sun
Vertical

The Tilted Earth:

Equator observer - Vertical
Line of sight to Sun

Tropic of Capricorn observer - Vertical <u>and</u> line of sight to Sun

Southern Hemisphere observer - Line of sight to Sun
Vertical

As far as the United States is concerned, the only state farther south than the Tropic of Cancer is Hawaii, so <u>no one</u> in the mainland U.S. sees the Sun directly overhead at noon on <u>any</u> day of the year.

7.) Because of all of the mountains and valleys, the Earth can be considered to be a sphere with a rough surface.

This statement involves a value judgment - what do we mean by a rough sphere, or, alternatively, a smooth sphere? The question has to do with <u>scaling</u>. Although the surface of the Earth certainly looks rough to us, is it rough on a scale of the radius of the Earth? The highest mountain on Earth is Mount Everest, whose peak is 8.9 km above sea level. Let us express this as a percentage of the radius of the Earth:

$$\frac{\text{height of Mt. Everest}}{\text{radius of Earth}} \times 100 = \frac{8.9 \times 10^3 \text{ m}}{6.4 \times 10^6 \text{ m}} \times 100 = 0.14\%$$

Now, let us consider this amount of "roughness" on a sphere of a more imaginable size. A basketball has a radius of about 12 cm. If we let this represent the size of the Earth, then Mount Everest would be a bump on the basketball of height given by,

height of bump = (0.14%)(radius of basketball) = (0.14%)(12 cm) = 0.017 cm = 0.17 mm

Thus, the height of Mount Everest is proportionately less than the height of the standard "pebbles" that cover the surface of a standard basketball. Is a basketball a smooth sphere? If you think so, then the Earth is a smooth sphere also. Actually, of course, even if there were no mountains or valleys, the Earth would still not be a smooth <u>sphere</u>, due to the equatorial bulge caused by its rotation - See Myth #1, Chapter 11.

Another startling example of scaling is that of a map of Lake Erie. If the 386 km length of Lake Erie is drawn as a map on a standard piece of paper, so that the length of the lake is 25 cm, then we have a scale factor of :

$$386 \text{ km} \quad \leftrightarrow \quad 25 \text{ cm} \quad \Rightarrow \quad 15.4 \text{ km} \quad \leftrightarrow \quad 1 \text{ cm}$$

Now, the thickness of the piece of paper is about 0.01 cm. This corresponds to a scaled distance of 0.154 km = 154 m. Since the deepest part of Lake Erie is only 64 m, Lake Erie is not even half as deep as the single piece of paper on which it is printed!

8.) A measuring instrument that is high in precision is also high in accuracy.

Precision and accuracy are two words that are commonly used interchangeably. They do not, however, mean the same thing. The term *precision* refers to the amount of uncertainty in a measurement. If a large number of measurements are taken on a sample, and the measurements show very little deviation from the average, then the measurement is relatively precise. The meaning of *accuracy*, on the other hand, is related to the agreement between the measurement and other standard measurements.

As an example, consider making measurements of the length of an object with a meter stick on which someone has cut off the first two centimeters. We will make measurements of the length of an object by lining up the cut off end of the stick with one end of the object and noting where the other end of the object is on the meter stick. By careful measuring, the measured lengths can be *precise*, in that many measurements will agree quite closely. But the measurements are not *accurate*, since they will all give values that are about two centimeters longer than the correct length.

Chapter 2
Translational Kinematics

Mysteries:

Los Angeles   39 Miles

1.) In the metric system, units are related by nice numbers like 10, 100, 1000, etc. In the British system of units, which is in wide use in the United States, the relationships are not so simple. For example, why is a mile equal to *5280 feet*?

2.) Another example of a strange unit comes to us from horse racing. Where did the word *furlong* come from?

3.) How does an auto-focus camera know how far away the subject is?

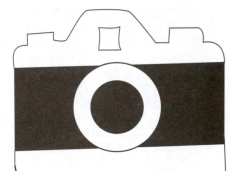

4.) How does a phonograph stylus resolve the two channels of stereo sound from a <u>single</u> groove on the record?

Magic:

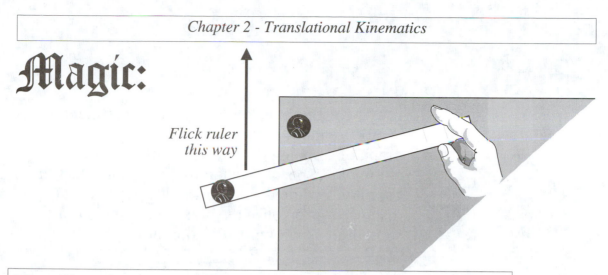

Flick ruler this way

The Simultaneous Pennies

Place a penny on the edge of a table. Place a ruler next to the penny and another penny on the top of the part of the ruler that hangs off the edge of the table, as shown in a view from above in the diagram. Flick the ruler quickly toward the first penny so that it projects the penny horizontally off the table. At the same time, the ruler will slide out from under the second penny, which will fall straight down. Listen for the sounds of the pennies hitting the floor. What can you say about the sounds?

Myth:

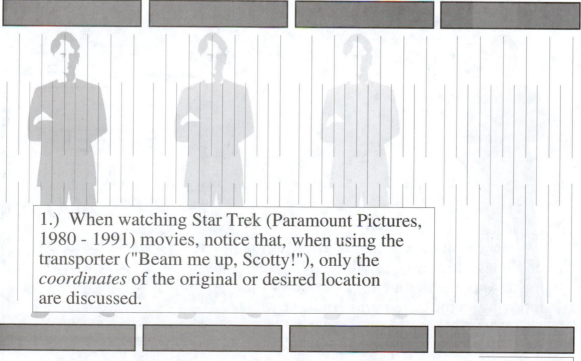

1.) When watching Star Trek (Paramount Pictures, 1980 - 1991) movies, notice that, when using the transporter ("Beam me up, Scotty!"), only the *coordinates* of the original or desired location are discussed.

𝕸𝖞𝖙𝖍:

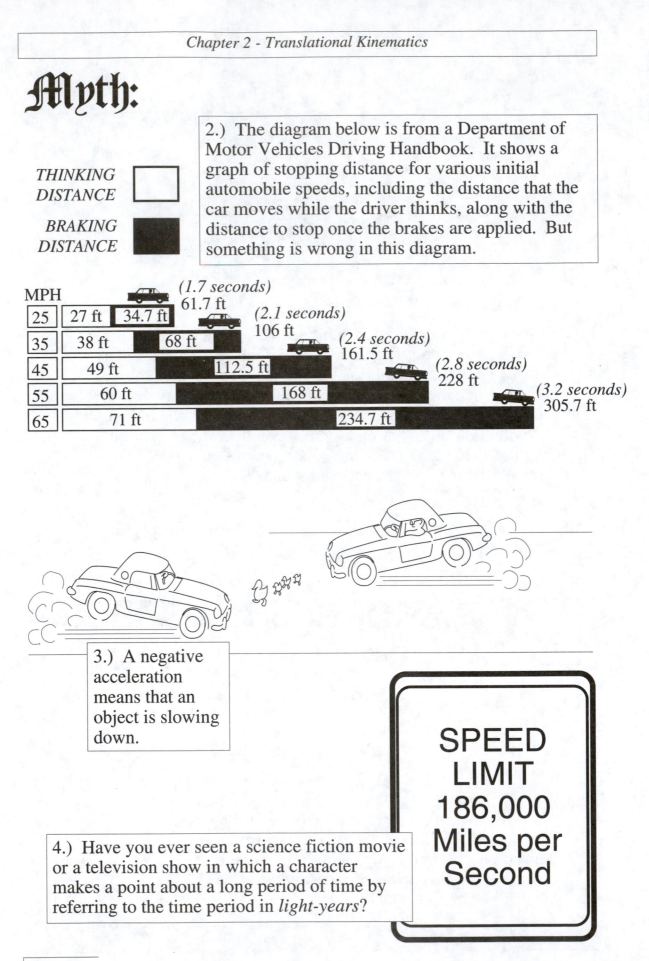

THINKING DISTANCE ☐

BRAKING DISTANCE ■

2.) The diagram below is from a Department of Motor Vehicles Driving Handbook. It shows a graph of stopping distance for various initial automobile speeds, including the distance that the car moves while the driver thinks, along with the distance to stop once the brakes are applied. But something is wrong in this diagram.

MPH

25	27 ft	34.7 ft

(1.7 seconds) 61.7 ft

35	38 ft	68 ft

(2.1 seconds) 106 ft

45	49 ft	112.5 ft

(2.4 seconds) 161.5 ft

55	60 ft	168 ft

(2.8 seconds) 228 ft

65	71 ft	234.7 ft

(3.2 seconds) 305.7 ft

3.) A negative acceleration means that an object is slowing down.

4.) Have you ever seen a science fiction movie or a television show in which a character makes a point about a long period of time by referring to the time period in *light-years*?

SPEED LIMIT 186,000 Miles per Second

𝔐𝔶𝔱𝔥:

5.) At baseball games, everyone sings "Take Me Out to the Ball Game" together.

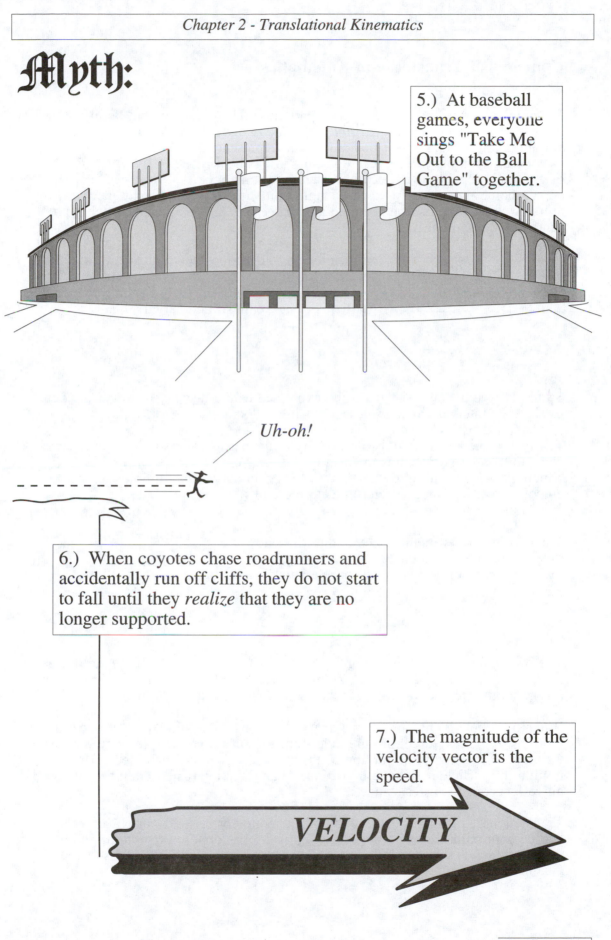

Uh-oh!

6.) When coyotes chase roadrunners and accidentally run off cliffs, they do not start to fall until they *realize* that they are no longer supported.

7.) The magnitude of the velocity vector is the speed.

VELOCITY

Concepts of Translational Kinematics

Kinematics is the study of motion of particles. The variables that describe the motion of a particle are its *position*, its *velocity* and its *acceleration*. All three variables are vectors. The position, often represented by **x** (**bold** symbols will represent vectors in this book), of a particle represents its location relative to some reference point and is measured in a length unit such as meters in the metric system. A change in position is called a *displacement* and is a vector, often represented by the symbol Δ**x**.

> *Position - where is it? (x)*

> *Velocity - how fast is the position changing? (v)*

The velocity, **v**, describes how quickly in time the position of the particle is changing. The <u>average</u> velocity is a measure of the total displacement of a particle during a finite time interval:

$$\mathbf{v}_{average} = \bar{\mathbf{v}} = \frac{\Delta \mathbf{x}}{\Delta t}$$

Similarly, the average <u>speed</u> can be expressed as the <u>distance</u> traveled divided by the time interval. Speed is a scalar, while velocity is a vector.

The <u>instantaneous</u> velocity is the velocity of the particle at a single point in time and can be approximated as the average velocity over a very short time interval which includes the desired instant of time. A similar definition holds for instantaneous speed. A speedometer in a car measures instantaneous speed.

> *Acceleration - how fast is its velocity changing? (a)*

The acceleration of the particle, **a**, describes how quickly its velocity is changing. Just as with velocity, we can define an <u>average</u> acceleration over a time interval as follows:

$$\mathbf{a}_{average} = \bar{\mathbf{a}} = \frac{\Delta \mathbf{v}}{\Delta t}$$

The <u>instantaneous</u> acceleration of the particle is the acceleration of the particle at a single point in time and can be approximated as the average acceleration over a very short time interval which includes the desired instant of time.

When a particle is moving in two or more perpendicular directions at once, the motions in the two directions are independent. Thus, for example, for a particle falling due to gravity, its motion in the vertical direction is unaffected by any motion the particle may have in the horizontal direction (This statement ignores any effects due to air resistance.).

Translational kinematics is the study of the motion of particles, or objects which can be approximated as particles, <u>through</u> space, as opposed to *rotational kinematics*, which is the study of spinning objects, and which will be investigated in Chapter 10.

Discussions; Chapter 2 - Translational Kinematics

𝕸𝖞𝖘𝖙𝖊𝖗𝖎𝖊𝖘:

1.) In the metric system, units are related by nice numbers like 10, 100, 1000, etc. In the British system of units, which is in wide use in the United States, the relationships are not so simple. For example, why is a mile equal to *5280 feet*?

2.) Another example of a strange unit comes to us from horse racing. Where did the word *furlong* come from?

These two mysteries will be discussed together, since their origin is related. The word *mile* comes from the Latin *mille*, for one thousand. The original mile was the distance of 1000 paces by Roman soldiers. A typical pace was five feet, so the Roman mile was about 5000 feet. Later on, English farmlands were marked by fences indicating the length of a standard furrow. This length was a *furrow-long*, which became abbreviated to *furlong,* and is equivalent to 220 present-day yards. Upon adopting the mile from the Romans, the English found that it was between seven and eight furlongs. Keeping the furlong as a standard unit and wanting the mile to be related to the furlong by a "nice" number, the English increased the Roman mile to exactly eight furlongs, which made it 1760 yards or 5280 feet.

3.) How does an auto-focus camera know how far away the subject is?

An auto-focus camera is designed to emit an ultrasonic sound wave toward the subject. It is also designed to receive the reflected signal and measure the time between the emitted and reflected signals. By dividing this time by the speed of sound, the distance to the subject can be calculated and the focus adjustment made.

The auto-focus camera can be fooled if a picture is taken through a window. The ultrasonic signal will bounce off the window, which is then where the camera will focus. The scene on the other side of the window will be out of focus on the film.

It is an interesting exercise to calculate the error in the focus distance due to the changing speed of sound as the air temperature changes. Let us assume that the camera is designed to be used at room temperature, $T = 23°C$. The speed of sound as a function of temperature over a range of temperatures around room temperature can be approximated as,

$$v = 331 \text{ m·s}^{-1} + (0.6 \text{ m·s}^{-1}\text{·}°\text{C}^{-1}) \, T_C$$

So, for our design temperature of 23°C,

$$v = 331 \text{ m·s}^{-1} + (0.6 \text{ m·s}^{-1}\text{·}°\text{C}^{-1}) \, (23°\text{C}) = 345 \text{ m·s}^{-1}$$

Let us assume that the subject is 5 m from the camera, so that the total distance traveled by the ultrasonic beam is 10 m. This gives a transit time of:

$$\Delta t = \frac{\Delta x}{v} = \frac{10 \text{ m}}{345 \text{ m} \cdot \text{s}^{-1}} = 0.0290 \text{ s}$$

Now, let the temperature fall to 0°C, at which the speed of sound is 331 m·s⁻¹. Since the sound moves more slowly now, the transit time will be longer, as we can calculate:

$$\Delta t = \frac{\Delta x}{v} = \frac{10 \text{ m}}{331 \text{ m} \cdot \text{s}^{-1}} = 0.0302 \text{ s}$$

Now, the camera will detect this transit time but electronically make the assumption that the speed of sound is still the design speed, 345 m·s⁻¹. Thus, it will determine the distance to the subject as,

$$\Delta x = \frac{1}{2} v_{design} \Delta t_{measured} = \frac{1}{2}(345 \text{ m} \cdot \text{s}^{-1})(0.0302 \text{ s}) = 5.2 \text{ m}$$

The error in measurement, 5.2 m - 5 m = 0.2 m is likely to be less than the depth of field (see Mystery #10, Chapter 27), so temperature does not appear to be a problem for the focusing capabilities of the camera.

4.) How does a phonograph stylus resolve the two channels of stereo sound from a <u>single</u> groove on the record?

This phenomenon depends on the independence of perpendicular motions, as discussed in the Concepts section. The two channels of information are recorded separately on the two sides of the record groove, which are slanted at 45° to the record surface so as to be at an angle of 90° from each other. Thus, the two channels cause the stylus to vibrate in <u>perpendicular directions</u>. Since these motions are independent, the information remains separate. Of course, no system is perfectly free from flaws, so there is some "crosstalk" between the two channels. "Stereo separation" is a hi-fi specification which quantifies the amount of unwanted information "leaking" into the wrong channel.

𝔐𝔞𝔤𝔦𝔠:

The Simultaneous Pennies

When the demonstration is performed, you should not hear two separate sounds of pennies hitting the floor, but only one sound. As mentioned in the Concepts section, when an object moves in two perpendicular directions at the same time, the motions in the two directions are independent. The instructions in this demonstration create a situation where a penny is projected horizontally at the same time that another penny is dropped from rest from the same height. As far as the vertical motion is concerned, each penny is dropped from rest. The fact that one has an additional horizontal velocity does not affect the vertical motion. Thus, both pennies hit the floor at the same time.

𝔐𝔶𝔱𝔥:

1.) When watching Star Trek (Paramount Pictures, 1980 - 1991) movies, notice that, when using the transporter ("Beam me up, Scotty!"), only the *coordinates* of the original or desired location are discussed.

Let us imagine the situation where the *Enterprise* is orbiting a planet and the crew is beamed to the surface of the planet below. The coordinates discussed before the teleportation correspond to our kinematic variable, the *position*. But the surface of the planet is moving with a different *velocity* than the *Enterprise*. If the crew members are beamed to a position on the planet with the velocity that they have on the *Enterprise*, they will arrive on the surface with a rather large velocity <u>relative to the surface</u>. This will result in a painful surprise as they materialize!

2.) The diagram below is from a Department of Motor Vehicles Driving Handbook. It shows a graph of stopping distance for various initial automobile speeds, including the distance that the car moves while the driver thinks, along with the distance to stop once the brakes are applied. But something is wrong in this diagram.

The problem is with the *time intervals* listed above the bars. The person responsible for these time intervals at the Department of Motor Vehicles simply divided the total stopping distance by the initial velocity. But this calculation ignores the fact that the velocity is not constant. The correct calculation would require a time to be determined for the "thinking" portion which is then added to a second time interval determined for the braking period. The correct times are longer than those indicated on the original graph. For more information, see J. Jewett, "DMV Physics - The Case of the Disappearing Time Intervals", *The Physics Teacher*, **29**, 563 (1991).

3.) A negative acceleration means that an object is slowing down.

This is a common misconception and is even stated in some physics books. Acceleration is a vector and a negative sign means that the acceleration vector points in the negative direction. But the velocity vector could be pointing in <u>either</u> direction. If the velocity is positive and the acceleration is negative, then, indeed, the object is slowing down. This is case for an elevator moving upward and arriving at its destination floor. The velocity is up and the acceleration is down - the elevator slows down. But now the elevator begins its return to the ground floor. It starts moving down and accelerates downward. Here both the velocity and acceleration are negative, and the object is speeding up.

4.) Have you ever seen a science fiction movie or a television show in which a character makes a point about a long period of time by referring to the time period in *light-years*?

A *light-year* is a unit of distance - that which light travels in a time of one year (about 9.45×10^{15} m). This is another example of the type discussed in Myth #2 of Chapter 1. Even though the word "year" appears in the unit, it represents a <u>length</u>, not a time.

5.) At baseball games, everyone sings "Take Me Out to the Ball Game" together.

The mythical aspect of this statement is in the last word - "together". In a stadium, where there are large distances involved, singing in unison with an <u>audible</u> cuing signal is virtually impossible. If there were a <u>visible</u> cuing signal, such as a conductor on the pitcher's mound leading everyone, then unison might be possible. But usually, the song is sung along with a recorded version emanating from loudspeakers. Those spectators seated close to the loudspeakers will sing right along with the recorded sound. Due to the relatively low speed of sound, however, the sound from the loudspeaker will take a measurable amount of time to reach the spectators across the stadium. They will sing right along with the sound <u>that they hear</u>, which will be delayed from those singers on the other side. Then the sound from the far singers will take more time to reach the singers near the loudspeakers, so that, to the near singers, the singing of the far singers will sound significantly behind what the loudspeaker is playing. This was particularly evident in the past, when a single set of loudspeakers was used, often near the location of the scoreboard. Many stadia today have loudspeakers distributed throughout the stadium, which significantly reduces the effect.

6.) When coyotes chase roadrunners and accidentally run off cliffs, they do not start to fall until they *realize* that they are no longer supported.

This model might be called the *realization theory of gravity*. Despite the fact that millions of children have seen this representation of gravity in Saturday morning cartoons, the falling starts as soon as the normal force from the ground has been removed, regardless of the coyote's state of realization. M. McCloskey ("Cartoon Physics", *Psychology Today*, **18(4)**, 52 (1984)) uses this idea and others to explore students' misconceptions about the way the world works. The falling of the coyote is similar to medieval ideas about projectiles, according to this study. For ideas on the use of cartoons in physics classes, see R. Akridge, "Cartoon Physics", *The Physics Teacher*, **28**, 336 (1990). C. Chiaverina ("Reviewing the Troops", *The Physics Teacher*, **27**, 268 (1989)) shows a number of animated cartoons on the last day of physics class before a vacation and his students find dozens of physics errors.

7.) The magnitude of the velocity vector is the speed.

This statement is only true for *instantaneous* velocity and speed. It is not generally true for *average* velocity and speed. As an example, consider a particle moving at constant speed in a circular path of radius r.   If it moves halfway around the circle in a time Δt, then the magnitude of the <u>average velocity</u> is,

$$|v_{avg}| = \frac{|\text{ displacement }|}{\text{time}} = \frac{2r}{\Delta t}$$

Now, the <u>speed</u> is given by,

$$s = \frac{\text{distance}}{\text{time}} = \frac{\pi r}{\Delta t}$$

which is a different number.   Only when we allow the time interval to shrink to zero (or when the particle is moving in a straight line) will the distance and displacement be the same, so that the speed is the magnitude of the velocity vector.

Chapter 3
Newton's Laws

𝕸ysteries:

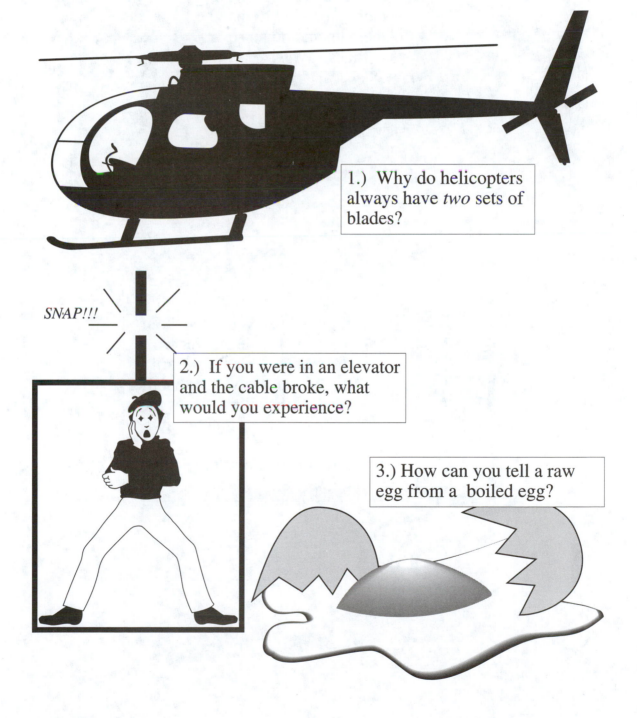

1.) Why do helicopters always have *two* sets of blades?

SNAP!!!

2.) If you were in an elevator and the cable broke, what would you experience?

3.) How can you tell a raw egg from a boiled egg?

Mysteries:

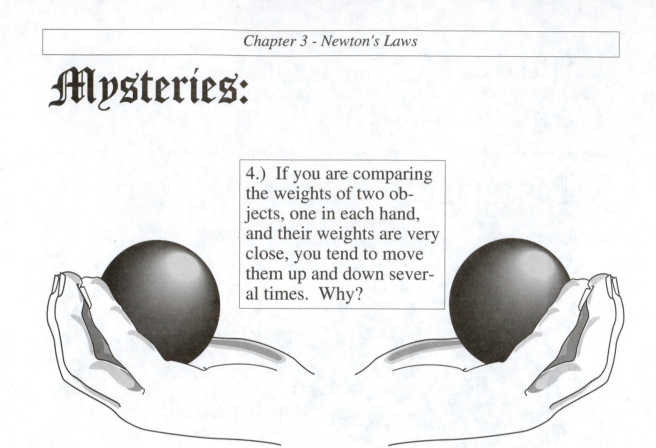

4.) If you are comparing the weights of two objects, one in each hand, and their weights are very close, you tend to move them up and down several times. Why?

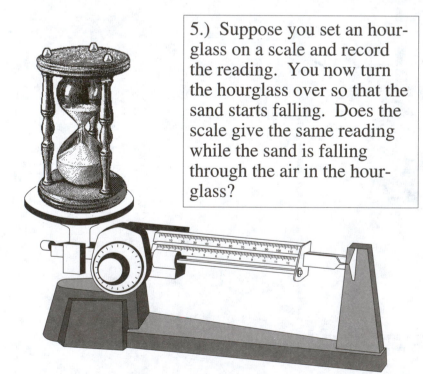

5.) Suppose you set an hourglass on a scale and record the reading. You now turn the hourglass over so that the sand starts falling. Does the scale give the same reading while the sand is falling through the air in the hourglass?

Magic:

Pull hard, fast and horizontally.

The Slippery Tablecloth

Cover a table with a tablecloth with <u>no hemmed edge</u>. Place some dishes on the tablecloth. You may want to use plastic dishes until you become an expert. Now, gather the tablecloth up at one edge and pull, as nearly horizontally as possible and as fast as possible. The tablecloth will leave the table, but the dishes will remain behind!

The Heavy Feather?

Drop, at the same time, a book and a feather. The book will hit the ground first, while the feather slowly wafts to the floor. Now, place the feather on top of the book and drop the book again. The feather drops just as fast as the book.

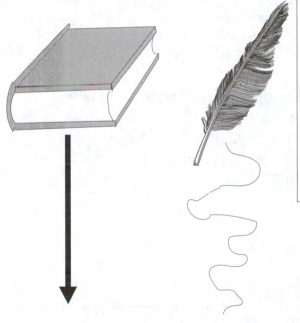

Have sheet held so that the cross section looks like this

Throw egg in here

Throwing an Egg with Confidence

Have two assistants hold a sheet between them so that it is stretched tightly across the top and the rest of the sheet hangs in a shape like the letter "J" (the diagram to the right shows a side view of the sheet.). Now, throw a raw egg into the vertical part of the sheet. It will not break!

Egg falls in here

Myth:

1.) Consider the horse and wagon - if the horse pulls forward on the wagon, then Newton's Third Law says that the wagon pulls back equally hard on the horse. Thus, the forces cancel and nothing happens. Right?!?

2.) If you are lucky enough to visit Disneyland or Disney World, be sure to ride "Star Tours". It is advertised as a wild and turbulent "ride" through space, but in reality, the "vehicle" that you enter never leaves its initial position!

3.) Rockets don't work if you turn them on in outer space because there is nothing for them to push against.

Concepts of Newton's Laws

Newton's Laws are the fundamental concepts in *dynamics*, which is the study of the <u>causes</u> of motion. These three laws describe the nature of *forces* and their relationships to the resulting motion of objects.

Newton's First Law: *An object in a state of motion remains in that state of motion unless acted on by an external force.*

First Law - if you don't do anything, the situation remains the same!

This is commonly referred to as the law of *inertia*. The state of motion referred to could be rest (zero velocity), so that an object tends to remain at rest if no forces act on it. The state of motion could also be an existing velocity. Unless a force acts, the velocity remains constant. This is the reason for <u>seat belts</u> in a car. A driver has an initial velocity, that of the car. If the car suddenly stops, there must be a force on the driver to change that velocity. Without a seat belt, the only force available is friction with the seat and the force of the steering wheel on the driver's hands. This generally is not sufficient and the driver will continue in the original motion (with the velocity slightly reduced by the small forces mentioned) and hit the windshield. The seat belt provides the large force necessary to change the driver's velocity and avoid this disaster.

Newton's Second Law: *The net force on an object is proportional to the mass of the object and its resulting acceleration.*

Second Law - if you do something, here's how the situation changes!

As alluded to in the First Law, a change in velocity requires a force. The Second Law gives more mathematical details. It says that the acceleration and force are proportional (for various forces applied to a <u>given</u> mass). If we apply twice the force (to the same mass), we obtain twice the acceleration. Similarly, force is proportional to mass (for causing various masses to have the <u>same</u> acceleration). It takes twice as much force to accelerate 2 kg than it does to give 1 kg the same acceleration.

This discussion of Newton's Second Law is summarized in mathematical form as follows*:

$$\mathbf{F} = m\mathbf{a}$$

This equation allows us to define the unit of force, which is called the *newton*, in terms of the basic units: 1 newton = $(1 \text{ kg})(1 \text{ m·s}^{-2})$ = 1 kg·m·s^{-2}.

Newton's Third Law: *If object A exerts a force on object B, then B exerts an equal and opposite force on A.***

This law is commonly stated, "Every action has an equal and opposite reaction". The Third Law demonstrates the fact that forces are always *interactions* between objects and thus always occur in pairs.

Third Law - If I push you, then you also push me!

---

*This is the common statement of Newton's Second Law. But see Myth #2, Chapter 8.
**This is the common statement of Newton's Third Law. But see Myth #4, Chapter 22.

Discussions; Chapter 3 - Newton's Laws

Mysteries:

1.) Why do helicopters always have *two* sets of blades?

The answer to this mystery lies in Newton's Third Law. Imagine that a helicopter only has one set of blades. As the engine on the helicopter starts, it applies a force to the blades to cause them to rotate. By the Third Law, the blades will apply a force back on the helicopter body. As a result, the body will be spinning in the opposite direction to the blades. This is undesirable - for the comfort of the passengers, we want the body of the helicopter to remain stationary! Thus, we must provide a second set of blades to apply a force opposite to that of the first set so that there is no tendency for the helicopter body to spin. On many helicopters, this is done by means of a small set of blades, on the back end of the body, which provides a torque on the body opposite to that applied by the primary blades. On others, it is done by providing a second set of primary blades rotating in the opposite direction to the first set.

2.) If you were in an elevator and the cable broke, what would you experience?

Let us assume that the elevator is not equipped with any sort of safety braking mechanism. If the cable were to break, then you would be in free fall. You and the elevator would be falling together. This is the same situation as that for the astronauts in the space shuttle in orbit around the Earth. You would be able to float freely in the elevator, pushing off the walls and drifting to the other side. Of course, your experience would be short-lived, as the elevator will eventually hit the ground! In reality, elevators are equipped with safety mechanisms that will not allow free fall in this situation.

Now, we can ask, why do both you and the elevator, or, for that matter, any two objects, fall with the same acceleration (ignoring air resistance)? This stems from the dual role that mass plays. If we consider the mathematical form of Newton's Second Law, $\mathbf{F} = m\mathbf{a}$, we see that the mass of a falling object plays a distinct role in this equation as *resistance to changes in motion*. But the mass also determines the force on the left side of the equation, since it is the gravitational force between the Earth and the object, which is proportional to the mass of the object. Thus, more massive objects are pulled down to the Earth with more force, but they also exhibit a proportionally larger resistance to being accelerated. These effects cancel, so that all objects fall with the same acceleration.

3.) How can you tell a raw egg from a boiled egg?

A simple way to tell if an egg is raw or boiled is to spin it on a horizontal surface. There are two possible investigations:

a.) The boiled egg will spin very easily and quickly. The raw egg will be difficult to spin and will only do so slowly. The reasons lie in the First Law - inertia. The boiled egg is solid all the way through, resulting in strong intermolecular forces within the egg.

When it is spun, forces from the surface (which you apply with your hand to initiate spinning) are easily transmitted to the interior of the egg, so that it all spins together. For the raw egg, there is liquid inside. The transmission of forces between layers of liquid from the outside of the egg to the inside is weak. Thus, when an attempt is made to spin the egg, the outside is spinning rapidly, but the inside is still stationary or spinning slowly. Thus, when it is released after the spinning attempt, the various layers of liquid inside the egg exert friction forces that rapidly transform much of the rotational kinetic energy into internal energy (see Chapter 7 for discussions of energy).

b.) Once an egg is spinning, grasp it quickly and then let it go. The boiled egg will no longer spin. The raw egg will begin to spin again. When the boiled egg is grasped and released, the quick transmission of forces stops the entire egg. If the raw egg is successfully spun and then grasped and released, the inside portions continue to spin, due to inertia and the weak transmission of forces from the outside. Upon release, the spinning inside portion exerts friction forces on the outside portion that cause the egg to start spinning again!

4.) If you are comparing the weights of two objects, one in each hand, and their weights are very close, you tend to move them up and down several times. Why?

If you hold objects stationary in your hand to weigh them, you are actually measuring the force necessary for you to hold the objects up, that is, to cancel out their *weight*. If the weights are close so that you can't tell which is heavier, you tend to move them up and down. This is a use of Newton's Second Law. By moving them up and down, you are giving the objects an acceleration. You then "measure" the force with which the objects resist your attempts to move them. The object that exerts more resistance to these attempts is the more massive. Thus, you have supplemented your weight measurement with a dynamic measure of the *mass* of the objects.

5.) Suppose you set an hourglass on a scale and record the reading. You now turn the hourglass over so that the sand starts falling. Does the scale give the same reading while the sand is falling through the air in the hourglass?

The scale will give the same reading. The hourglass does not exert a normal force on the sand that is falling through the air, so you may tend to think that the reading will be less. This "loss of weight" is balanced, however, by the force of the sand hitting the bottom of the hourglass. By Newton's Second Law, the bottom of the hourglass must exert an upward force, larger than the weight, to give this sand an upward acceleration, thereby stopping it from falling. This exactly balances the weight of the sand in the air, so that the reading is the same. There will be some fluctuations at the beginning and the end. For example, at the beginning, some sand is in the air, but none has hit the bottom yet. Thus, during this brief period, the reading will be lower.

𝔐𝔞𝔤𝔦𝔠:

The Slippery Tablecloth

This activity is often presented as a demonstration of Newton's First Law, with an explanation that the dishes on the table are in a state of rest and will remain at rest unless a force acts on them. In reality, there <u>is</u> a force on the dishes - kinetic friction between the dishes and the tablecloth as the tablecloth is pulled horizontally. If the tablecloth is pulled very quickly, however, then the time during which this force acts is very short, so that the dishes hardly move and any horizontal velocity of the dishes is rapidly removed by friction between the dishes and the table surface. If we imagine pulling a very long tablecloth, even the relatively small friction force will have enough time to accelerate the dishes to an appreciable velocity. Thus, although one may argue that this demonstrates Newton's First Law, it more correctly demonstrates the <u>impulse-momentum theorem</u>. Both the force and the time during which it acts are small, resulting in a small change in momentum (Chapter 8) of the dishes.

A simpler, but less dramatic, method to show this effect is to place a bottle on a strip of paper. The paper can easily be pulled quickly from under the bottle with no motion of the bottle.

The Heavy Feather?

In the absence of air friction, all objects fall in a gravitational field with the same acceleration. For a book and a feather, however, the relationships between air friction and the weight are very different - the book is hardly affected by air friction, while the feather is affected greatly. Thus, if they are dropped next to each other, the feather arrives at the floor long after the book. If the feather is placed on top of the book, however, then the book "leads the way" for the feather through the air. The air cannot reach the underside of the feather and the two drop together.

Throwing an Egg with Confidence

This is a demonstration of Newton's Second Law. When the egg hits the sheet, the sheet moves backward some distance. Thus, the egg is brought to rest over a distance of several centimeters. The resulting acceleration is small, small enough that the force on the shell is not large enough to break it. Compare this to throwing an egg into a wall. This egg is brought to rest over a very short distance, resulting in a large acceleration and, therefore, a large force, easily enough to break the shell of the egg.

𝔐𝔶𝔱𝔥:

1.) Consider the horse and wagon - if the horse pulls forward on the wagon, then Newton's Third Law says that the wagon pulls back equally hard on the horse. Thus, the forces cancel and nothing happens. Right?!?

This is a common brainteaser that is based on an incomplete understanding of Newton's Third Law. If the depth of understanding of the Law is of the form, "Every action has an equal and opposite reaction", then this myth is quite confusing. But consider the form given in the Concepts section, "If object A exerts a force on object B, then B exerts an equal and opposite force on A". This form indicates clearly that the two forces involved in the Third Law act on <u>different</u> objects. <u>Forces acting on different objects cannot cancel</u> since, if you are determining the motion of one object, you add the forces on that object alone. The force which accelerates the horse and wagon is the force of the Earth on the horse's feet, which is the Third Law reaction force to the force that the horse exerts backward on the Earth.

2.) If you are lucky enough to visit Disneyland or Disney World, be sure to ride "Star Tours". It is advertised as a wild and turbulent "ride" through space, but in reality, the "vehicle" that you enter never leaves its initial position!

The Star Tours ride is a flight simulator. It depends on an extension of dynamics that is called the *Principle of Equivalence*. This principle states that it is impossible to differentiate between a gravitational force and a perceived force caused by an acceleration. You have experienced this when you are in a braking car. You feel pushed forward into the seat belt. Is there a force pushing you? No; it is the effect of the car slowing down and the inertia of your body that makes you feel pushed forward - it is the First Law! The same is true of astronauts sitting on their backs in a vertical rocket on the surface of the Earth. They feel pushed into the backs of their seats. When on the surface of the Earth, this is due to gravity. But if they were in space and moving forward with an acceleration of 9.8 m·s⁻², they would feel exactly the same thing.

Now there are usually other clues that allow one to differentiate between the two situations. In the accelerating car, the sound of the engine, bumps in the road and the sight of the external world zipping by tell you that you are accelerating. But, if the car were tilted up on its back end, the sound of an engine piped in, a movie of the landscape going by played for you and a few bumps provided, you would not be able to tell that you were not accelerating! This is what is done in Star Tours. Accelerations are replaced with tilts in a gravitational field so that you feel pushed into your seat or thrown forward. In addition, visual and audio clues are added. A movie plays in the foreground which supposedly shows you a view out of the front window and sounds and bumps are added for realism. If you are willing to wait in line again to experiment, try riding Star Tours with your eyes closed so that you remove the visual clues. Can you sense that you are just being tilted rather than accelerating?

The effect is not perfect, since, even when you are tilted, you will still have a component of the gravitational force pulling you into the seat bottom. This would not be present for an actual spacecraft in space, far from any planets. The only apparent forces would be forward or backward, relative to the seat <u>back</u>, due to the acceleration of the rocket (Although an "upward" (relative to the person in the seat) acceleration due to a circular motion (with the center of the circle above the head of the person) would give some sense of being pushed into the seat bottom.). Since most of us have not been in space, we do not notice the force pushing us into the seat bottom as a discrepancy with the situation in space.

3.) Rockets don't work if you turn them on in outer space because there is nothing for them to push against.

This is another common misconception - that rockets fire up from the Earth because the exhausted gases push against the Earth. In reality, rockets work by Newton's Third Law. The rocket pushes the gases out the back and, in turn, the gases push the rocket forward. This works regardless of where the rocket is. In fact, rockets work <u>better</u> in the vacuum of outer space than in an atmosphere, where the exhaust gases must do work to push the air out of the way.

Chapter 4
Friction

𝔐ysteries:

1.) Why do drag racing tires have no tread - isn't tread supposed to increase traction?

2.) Why shouldn't you lock your brakes when sliding on ice?

3.) How do ice skaters slide so well over ice that is so dry?

𝕸𝖆𝖌𝖎𝖈:

The Styrofoam Parachutist

Drop a piece of styrofoam packing material. The best kind is in the form of a section of a sphere, so that it has the shape of a shallow cup, as shown to the right. For this type, drop it with the opening upward. For other types, drop them in any orientation. Watch the motion. Do all objects fall with the same *acceleration*?

𝕸𝖞𝖙𝖍:

1.) Microwave ovens cook food from the inside out.

𝕸𝖞𝖙𝖍:

2.) Friction always opposes the motion.

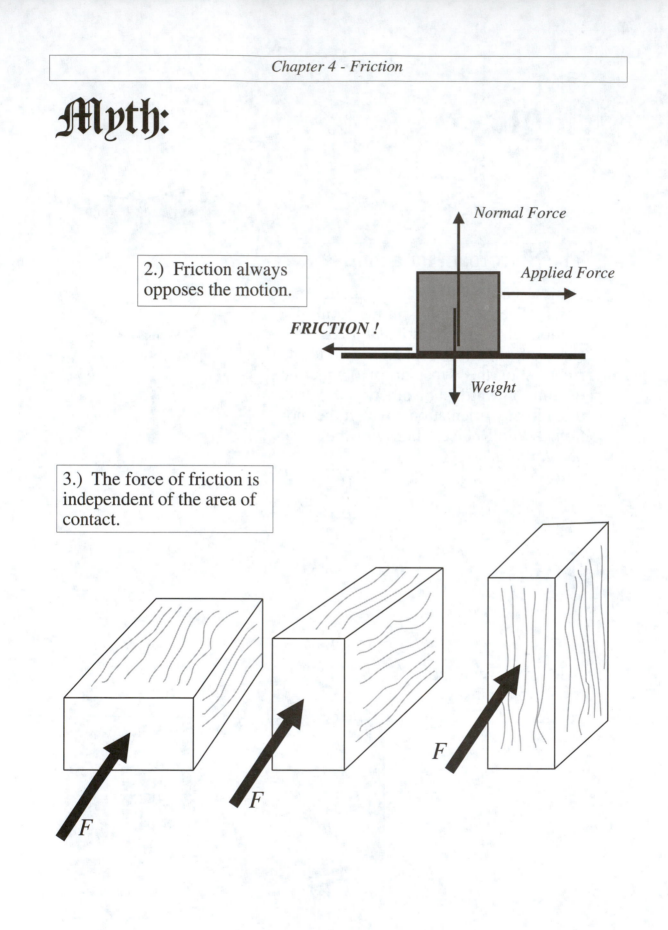

3.) The force of friction is independent of the area of contact.

Concepts of Friction

Friction is a force that is exerted when materials in "contact" move (or try to move) with respect to each other. Dry friction refers to the interaction of un-lubricated surfaces of solids. If there is no relative motion, a <u>static</u> friction force exists, while <u>kinetic</u> friction is in effect if the surfaces are sliding with respect to each other. The kinetic friction force depends on the strength of the normal force through a <u>kinetic coefficient of friction</u>, μ_k.

> *Dry Friction: The maximum Static Friction force is generally larger than the Kinetic Friction force.*

We can describe this dependence mathematically with the relation,

$$F_f = \mu_k N$$

where F_f is the friction force and N is the normal force.

The maximum possible static friction force (when the surfaces are just ready to break free - "impending motion") also depends on the strength of the normal force, through a <u>static coefficient of friction</u>, μ_s. This is described mathematically with the relation,

$$F_f = \mu_s N$$

The static coefficient is generally larger than the kinetic coefficient. In each case (kinetic and static), the coefficient of friction is the ratio of the appropriate friction force to the normal force. If, in the static case, however, we do <u>not</u> have impending motion, then the friction force is unrelated to the normal force. In this case, the friction force is simply equal and opposite to the component of the applied force parallel to the surface, so as to establish static equilibrium.

Viscous forces are frictional forces exerted in situations involving movement through fluids. These tend to be more complicated than dry friction. One complication is that the friction force depends on velocity, often in complex ways. One result of fluid friction is the existence of a *terminal velocity*. As an object moves faster through a fluid, the viscous force increases. Eventually, the viscous force is equal to the applied force and the acceleration goes to zero. The velocity then remains constant at the terminal velocity. This is the principle behind the <u>parachute</u>, which provides a large viscous force as it moves through the air.

> *Viscous forces depend on velocity.*

Discussions; Chapter 4 - Friction

Mysteries:

1.) Why do drag racing tires have no tread - isn't tread supposed to increase traction?

Tread is indeed supposed to increase traction but only in one situation - wet roads. The tread design on tires is aimed at directing water out efficiently from under the tires of a standard automobile so that good contact is made with the roadway in rainy weather. Drag racers do not race in the rain. Since the drag racer will only be operated on dry roadways, tread is not necessary and the treadless tires put more "rubber on the road" for increased friction.

2.) Why shouldn't you lock your brakes when sliding on ice?

This is standard driving advice and has an explanation based in physics. When a tire is rolling on the roadway, it is governed by <u>static</u> friction, since, during the moment that a particular piece of the tire is at the bottom of its rotation, in contact with the pavement, there is no relative motion (sliding) between the road and that piece of rubber. If the brakes are locked, however, then the rubber is <u>sliding</u> on the roadway and we are in the realm of <u>kinetic</u> friction. Since the coefficient of kinetic friction is smaller than the co-efficient of static friction, one does not have the control in the sliding case that one has in the rolling case. Thus, the advice is given to pump the brakes so that the wheels keep rolling and the driver can use the larger static friction to control the car.

The explanation is similar for the advice to *steer into a skid*. If you are skidding, you are experiencing kinetic friction. By steering away from the skid, the tires are placed at a large angle from the direction of motion so that they cannot possibly roll and kinetic friction is the <u>only</u> possibility. By steering into the skid, the tires are oriented so that rolling is possible, static friction can take over and the larger frictional force can be used to gently slow and/or change the direction of the car.

3.) How do ice skaters slide so well over ice that is so dry?

It is certainly possible to slide over ice when wearing shoes, but not as freely and with such low friction as with ice skates. How do ice skates reduce the friction? After all, it is said that the dry friction force is independent of area (see Myth #3), so just reducing the area must not be the answer.

A popular explanation of this effect is this - when high pressure is exerted on ice, the ice <u>melts</u>, a process called *regelation*. Thus, the effect of the thin blades is to produce a film of water which provides much less friction than the ice and allows the ice skater to glide very freely across the ice. This explanation, however, might have been placed in the Myth section - it cannot account for the details of the phenomenon. Edmiston

("Does Skating Melt Ice?", *The Physics Teacher*, **27**, 327 (1989)), calculates that, for an 80 kg skater, the melting point of ice is changed by only 0.6 °C by the pressure of the skate blade.

A more relevant effect for ice skating is that of *surface melting*. Faraday proposed long ago that solids have a thin layer of liquid on the surface but it was only in the 1980's that much attention began to be paid to this idea. Data show that ice has a surface layer of water varying in thickness from about 40 nm at 0°C to about 0.5 nm at - 35°C. It is primarily this layer of liquid on the surface that accounts for the low friction of ice. For more information and references, see J. D. White, "The Role of Surface Melting in Ice Skating", *The Physics Teacher*, **30**, 495 (1992).

Magic:

The Styrofoam Parachutist

Upon observation, you will notice that the styrofoam does not appear to accelerate, but falls at a constant velocity. This is a straightforward demonstration of *terminal velocity*. The styrofoam is so light and the air resistance so relatively large that terminal velocity is reached almost instantaneously.

Myth:

1.) Microwave ovens cook food from the inside out.

This is not true. Microwave ovens cook food from the outside in, just like normal ovens. But the mechanism for transferring the energy into the food is different. In conventional ovens, the energy is transferred to the food by means of air convection and conduction (See Chapter 15). In a microwave oven, energy is transmitted by microwave radiation. In response to the radiation, collections of water molecules in the food start to flip back and forth. A simple explanation of the heating effect that is often provided is that this motion of the water molecules results in kinetic friction within the food. The energy of vibration is converted to internal energy, resulting in a rise in the temperature of the food. While this may be qualitatively valuable in explaining microwave heating in a conceptual way, it is a severe simplification. See Mystery #13, Chapter 25 for a more detailed description of the heating process.

Regardless of the details of the process, the microwaves can only penetrate the food for a short distance, so the outside becomes hot first and the inside becomes hot by normal conduction of heat through the food material. Thus, the food cooks from the outside in.

2.) Friction always opposes the motion.

This statement is often heard or read and, indeed, friction does oppose the motion in many, many situations. But consider the following situation, with a block sitting on top of another block and the system moving to the right under the action of an applied force:

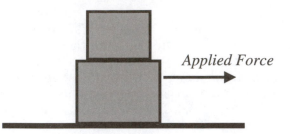

If we think about the forces on the system, we indeed find that friction between the lower block and the table is opposing the motion. But what about the upper block? What is causing it to move? In this case, the upper block is moving <u>because</u> of friction between it and the lower block. Friction is in the <u>same</u> direction as the motion! Imagine the behavior if we <u>lubricated</u> the surfaces between the blocks - the lower block might then slide out from under the upper one. In this case, then, we have <u>reduced</u> the motion of the upper block by <u>reducing</u> the friction!

In fact, we defy this myth everyday. The simple act of <u>walking</u> is a demonstration of friction acting in the direction of motion. We push back on the Earth and the friction force between our feet and the Earth moves us forward, with the help of Newton's Third Law.

3.) The force of friction is independent of the area of contact.

This is true only for the ideal, but non-existent case of perfectly <u>hard</u> surfaces. It would also require that all surfaces of the object be <u>identically</u> smooth and that the smoothing process not introduce any directionality in the surface. Real surfaces do not meet these requirements and are generally deformable to varying degrees. <u>Deformable</u> surfaces can conform to the shape of the other surface. In this case, then, the friction force increases with area. Rubber tires, for example, represent a deformable surface. Thus, the more tire area, the more friction with the roadway. If this were not true, then you would see drag racers with tires like bicycles, since they would be less massive and easier to accelerate!

Chapter 5
Equilibrium; Center of Mass

𝕸𝖞𝖘𝖙𝖊𝖗𝖎𝖊𝖘:

1.) Why can't you touch your toes if you are standing with your heels against a wall?

2.) If you are standing on the edge of a cliff, facing the drop, and someone bumps you toward the drop, you bend forward, in the <u>same</u> direction as the drop. Why would you bend in the same direction that you are trying to avoid? The same is true for balance beam gymnasts, for example. If they start to fall to the left, they bend their bodies to the left. Why?

Magic:

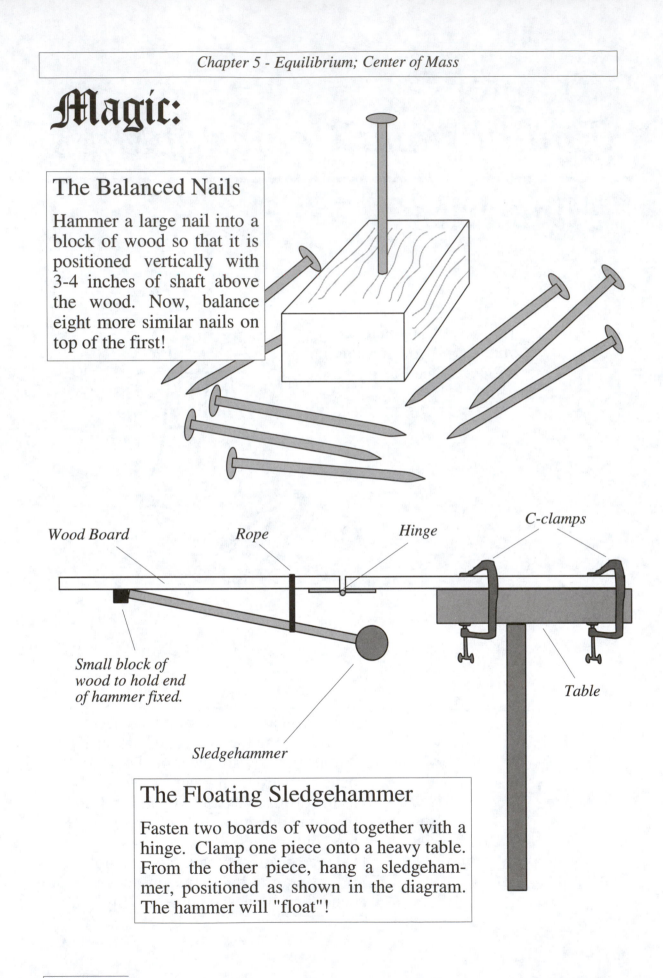

The Balanced Nails

Hammer a large nail into a block of wood so that it is positioned vertically with 3-4 inches of shaft above the wood. Now, balance eight more similar nails on top of the first!

Wood Board   Rope   Hinge   C-clamps

Small block of wood to hold end of hammer fixed.

Sledgehammer

Table

The Floating Sledgehammer

Fasten two boards of wood together with a hinge. Clamp one piece onto a heavy table. From the other piece, hang a sledgehammer, positioned as shown in the diagram. The hammer will "float"!

Myth:

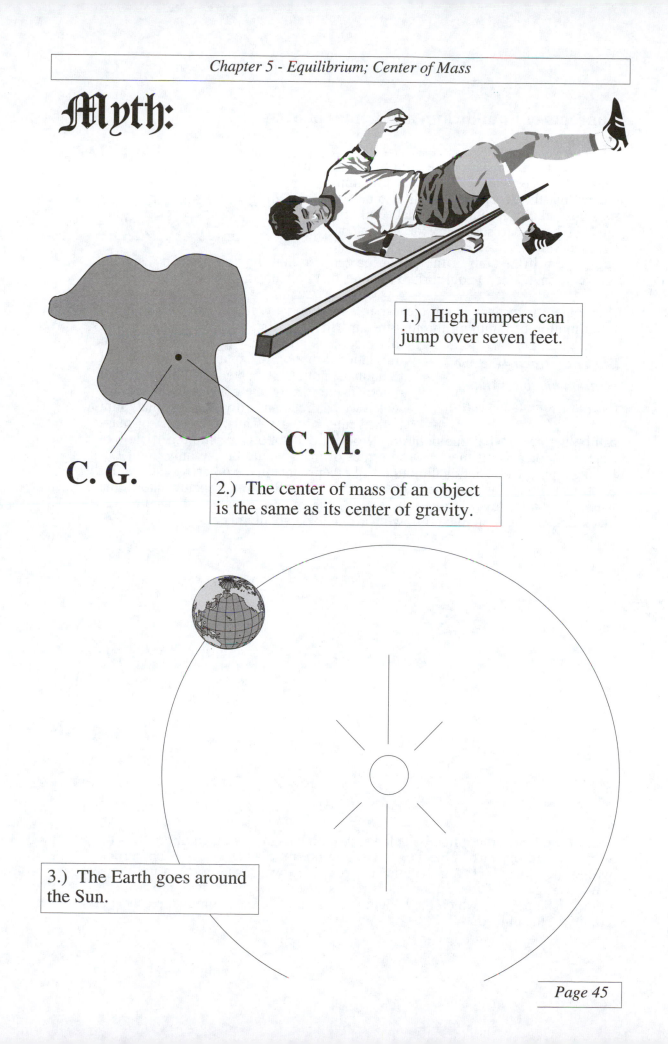

C. M.

C. G.

1.) High jumpers can jump over seven feet.

2.) The center of mass of an object is the same as its center of gravity.

3.) The Earth goes around the Sun.

Concepts of Equilibrium and Center of Mass

An object is in <u>translational equilibrium</u> if the sum of the forces on the object is zero. From Newton's Second Law, this means that the acceleration will also be zero. Thus, an object in translational equilibrium can be either stationary or moving at constant velocity. An object is in <u>rotational equilibrium</u> if the sum of the torques is zero, so that the object has no angular acceleration. General equilibrium requires that both conditions be met.

Equilibrium: Total Force = Zero
Total Torque = Zero

If an object is thrown as a projectile in a random way, it will exhibit a complicated motion of tumbling through the air. It is possible to show, however, that this motion is equivalent to a combination of translational motion (Chapter 2) of a point and rotational motion (Chapter 10) around that point. There is only one point in the object for which we can make this division and that point is the <u>center of mass</u>. Thus, the center of mass can be defined as that one point in the object that follows a simple particle-like trajectory for general motion. The center of mass would be the only point following a purely translational motion. All other points in the object would exhibit rotational motion around the center of mass. The illustration below shows the letter M being thrown through the air, and the center of mass, indicated by the black dot, can be seen to be following a simple parabolic trajectory.

Center of Mass: The one point on an object that follows a simple trajectory.

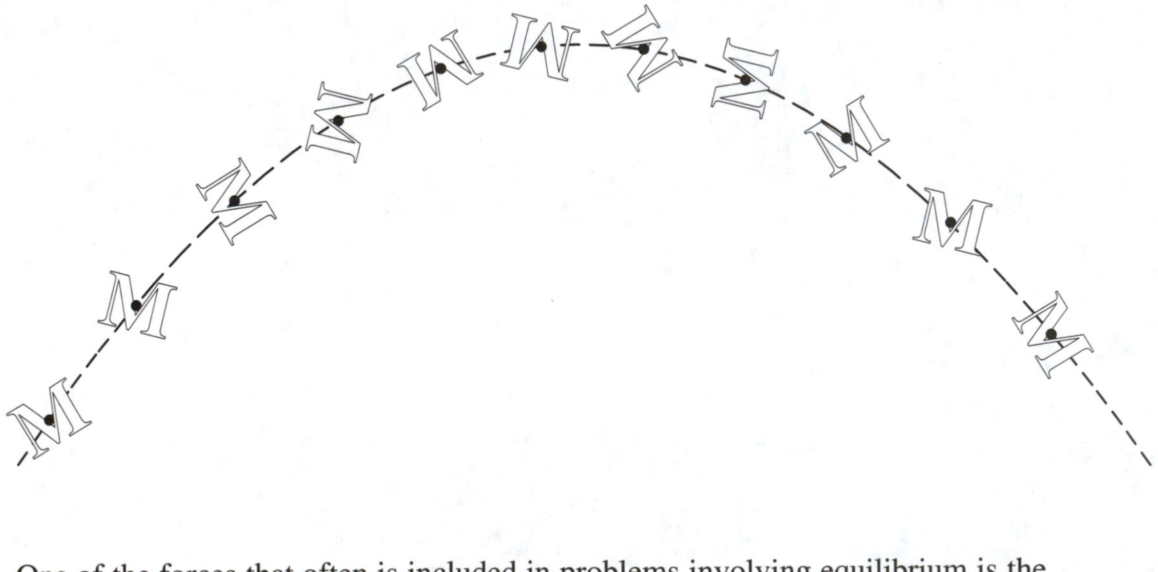

One of the forces that often is included in problems involving equilibrium is the weight of an object or objects. But where does the weight act? Each particle within the object feels a gravitational force at its own location. Can we somehow average all of these forces and represent the weight as acting at a single point? The point at which the weight is considered to be acting is the <u>center of gravity</u> of the object.

Discussions; Chapter 5 - Equilibrium and Center of Mass

𝔐𝔶𝔰𝔱𝔢𝔯𝔦𝔢𝔰:

1.) Why can't you touch your toes if you are standing with your heels against a wall?

In order to maintain your balance while you lean over and touch your toes, you must make sure that your center of gravity stays above the points of support, which are your feet. Normally, you do this by moving your back end backward while your top half moves forward and leans over your feet. If you are standing with your back against the wall, however, your back end cannot move backward. Thus, when your top half leans forward, your center of gravity is moved away from its location above your feet and you fall over.

2.) If you are standing on the edge of a cliff, facing the drop, and someone bumps you toward the drop, you bend forward, in the <u>same</u> direction as the drop. Why would you bend in the same direction that you are trying to avoid? The same is true for balance beam gymnasts, for example. If they start to fall to the left, they bend their bodies to the left. Why?

When one thinks about this, it might seem counterintuitive to bend toward the danger-ous direction. However, we are lucky that we <u>don't think</u> when this situation occurs! The desire is to move the body away from the undesired direction. In order to do this, a force must be applied in this direction. The only force available is the force of friction between your feet and the ground. By bending forward, the tendency is for your back end to move backward and your head and feet to move forward. If you performed this motion while floating in space, your center of mass would stay fixed and your head and feet would move one way while your back end would move the other way, as you "fold" your body at the waist. Doing this while standing on the Earth, however, your feet cannot move, since they are "fixed" by friction on the ground. Since you are trying to move your feet forward by "folding" your body, your feet apply a frictional force on the ground in the direction toward the cliff. By Newton's Third Law, then, the ground applies a force on your feet and, therefore, on your body, <u>away</u> from the cliff, which is the desired result.

𝔐𝔞𝔤𝔦𝔠:

The Balanced Nails

This may seem impossible and it is, unless you know the trick! Lay one nail from left to right in front of you on the table. Now, alternate six more nails on top of this nail as

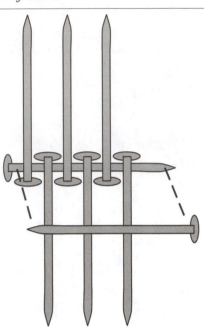

shown in the diagram at the right, which shows a top view of the arrangement. Finally, lay the last nail with its head to the right on top of the first nail and between the six heads of the perpendicular nails, as indicated by the dotted lines in the diagram. Carefully grasp the ends of the first and last nails and raise the combination upwards. The six nails will droop, but their heads will be "captured" by the cross nails. Now, lay the whole arrangement on top of the nail in the wood. The center of gravity of the drooping nails is below the support point, so the whole system is in equilibrium!

The Floating Sledgehammer

The center of gravity of the combination of the sledgehammer, the rope, the left-hand board and the small block of wood is right below the hinge when the hammer hangs in equilibrium. For a system of this size, the center of mass coincides with the center of gravity (see Myth #2). Thus, we can imagine the weight of the system described as acting at the center of mass of the system. The force of gravity has pulled this point to its lowest position, which is where it remains in equilibrium. Any motion of the system away from this position causes the center of mass to rise, so that it will swing back like a pendulum.

𝔐𝔶𝔱𝔥:

1.) High jumpers can jump over seven feet.

The bar may be seven or eight feet off the ground, but the height jumped is far less. First of all, the center of mass of the high jumper is already about three feet off the ground to begin with, since the center of mass is somewhere in the hip-stomach area. Secondly, the center of mass of the high jumper does not pass over the bar! If you watch a high jumper carefully, his or her body is never <u>entirely</u> higher than the bar. The body is contorted into a "roll" or a "flop" position, so that only part of the body is higher than the bar at a single point in time. The rest of the body is lower than the bar. As a result, the center of mass of the body remains lower than the bar at all times. Thus, a seven foot jump corresponds to an actual raise in the center of mass of something close to only four feet.

We can see that tall jumpers have an advantage, since their center of mass is already higher off the ground than that of a short jumper. Perhaps high jump competitions should be <u>normalized</u> to the original height of the center of mass of the jumper, to give short jumpers a chance!

2.) The center of mass of an object is the same as its center of gravity.

For most situations, this is so close to being true that it is not an item of concern. But it is only strictly true if there is no variation of g, the gravitational field, across the object. The center of mass of an object is purely a <u>geometrical</u> point, while the center of gravity is an "average" point for the <u>gravitational forces</u> on the object. If the gravitational field is not constant, then the two points will differ. For example, the center of mass of a mountain will be slightly different than the center of gravity, due to the variation of gravity with height. For an object floating in space, very far from any planets or stars, the center of mass is defined, since it is a geometrical average, but the center of gravity has no meaning, since there is no gravitational field.

3.) The Earth goes around the Sun.

In reality, both astronomical objects rotate around their common center of mass. Since the Sun is so much more massive than the Earth, however, the center of mass is very close to the center of the Sun. In fact, the center of mass is only 450 km from the Sun's center, which is much less than the Sun's radius of 700,000 km. Thus, the Sun moves in an orbit of radius 450 km, making a full rotation in one year (we are ignoring the effects of all the other planets in saying this.). The motion of the Sun is small, so it appears that the Earth goes around the Sun.

Chapter 6
Gravity

𝕸ysteries:

1.) What is a black hole?

2.) Why are there tides?

3.) Why does the moon always show us the same face?

Mysteries:

4.) Why does Halley's Comet "return" every 76 years?

5.) Why is there a meteor shower every August (as well as at other times)?

6.) If satellites are orbiting the Earth, how can a satellite receiving dish remain stationary? Wouldn't it have to "track" the satellite?

Magic:

The Anti-Gravity Brazil Nuts

Shake a can of mixed nuts for about a minute and then take off the top. Notice how many Brazil nuts are on top! Try pushing them down into the other nuts and then repeat the experience - the Brazil nuts will again rise to the top!

Gravity Assist

NASA regularly uses *Gravity Assist*, whereby a space probe passes by a planet and ends up going faster after it leaves the planet than it was going before it encountered the planet. How can this be?

Magic:

The Non-Falling Keys

This is originally a magician's trick, published in a magician's journal in 1926. At one end of a long piece of string, tie a set of keys. On the other end, fasten a light object such as a matchbook. Drape the end of the string with the keys over a pencil and let it hang. Hold the end of the string with the matchbook horizontally and release it. Does gravity pull the keys to the floor?

Myth:

1.) On huge planets like Saturn and Jupiter, the gravity would be so large as to crush a human who landed on the surface.

2.) Astronauts in an orbiting space shuttle are *weightless*.

𝔐𝔶𝔱𝔥:

3.) The following dialogue takes place in a radio commercial which is advertising a video camera that will record for a much longer time than other cameras. It is the narration of an owner of one of the "other" cameras for a showing of his videos.

"Now, here is a video of Mt. St. Helens. It's just ready to .... (White Noise). Oh... I ran out of tape."

"Now, here is a video of Old Faithful. It's just ready to... (White Noise). Oh...I ran out of tape."

"Now, here is a video of Halley's Comet. It's just ready to.... (White Noise). Oh...I ran out of tape."

4.) In describing the tides, some text-books explain the bulge of water on the side of the Earth away from the Moon by claiming that the Moon pulls the Earth away from the water.

Direction to the Moon

Water (Bulges are highly exaggerated)

Concepts of Gravity

Gravity is the attractive force between particles based on their possession of mass. It has perplexed scientists for centuries and continues to be an area of intense theoretical research.

Newton studied gravity and made a great leap of knowledge in realizing that the <u>same</u> force that causes an apple to fall to the ground is the force acting to hold the Moon in its orbit around the Earth. He used this understanding to develop the Law of Universal Gravitation. This Law states that the gravitational force between two point

Newton's Law of Universal Gravitation:

$$F = G \frac{m_1 m_2}{r^2}$$

(or spherical) masses m_1 and m_2 is proportional to their masses and inversely proportional to the square of the distance between them, r. The parameter G in the equation is a universal constant that converts the combination of masses and distance into the correct size of the newton that we have defined (Concepts, Chapter 3).

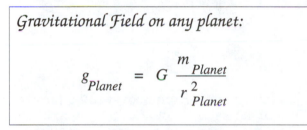

Gravitational Field on any planet:

$$g_{Planet} = G \frac{m_{Planet}}{r_{Planet}^2}$$

From the Law of Gravitation, we can determine an expression for the gravitational field on a planet in terms of its mass and radius. This is shown in the box at the left. A freely falling mass on the planet would fall with an acceleration given by this expression.

The concept of *potential energy* (see Chapter 7) is useful in gravitational situations. For movement of an object of mass m on the surface of the Earth through a small height difference h (relative to the radius of the Earth), the change in potential energy is adequately described by the simple equation,

$$\Delta PE_G = mgh$$

The more general expression for gravitational potential energy between two masses m_1 and m_2 is derived from the Law of Universal Gravitation, using calculus. The result is

$$PE_G = -G \frac{m_1 m_2}{r}$$

where we have identified the zero of potential energy as that situation in which the masses are infinitely far apart. Since this is the highest potential energy possible, all other potential energies, for smaller separations, are negative - hence the negative sign in the equation. It can be shown that this equation does indeed reduce to the simpler equation above for masses on the Earth and small variations in the separation distance.

Discussions; Chapter 6 - Gravity

Mysteries:

1.) What is a black hole?

A black hole is a possible final state of a very massive star. A star is formed by the gravitational attraction of a large number of dust and gas particles in space. As the particles come together to form a star, their gravitational potential energy is converted to internal energy within the star. Eventually, after nuclear fusion has begun, the force of the electromagnetic radiation emanating outward from the center of the star balances the inward force of gravity and the star achieves a stable equilibrium. Once the fuel has been exhausted, however, there is no longer an outward radiation force and the gravitational force causes the star to collapse once more. If the star is massive enough, the material in the star can be squeezed so tightly by gravity that electrons and protons combine to form neutrons - a *neutron star*. For a certain range of masses, an equilibrium can be reached between gravity and quantum mechanical repulsive forces. If the star is massive enough, however, gravity will continue to cause the star to collapse until all the material in the star ends up in an infinitesimally small region at the center. The gravitational field near this incredibly dense object is so high that even light cannot escape. Hence, the name *black hole*, since it will look completely black.

2.) Why are there tides?

The study of tides is very complicated and only a simple summary will be presented here. The tides are caused by the gravitational attraction of the moon (and, to a lesser extent, the Sun) on the Earth. Because the gravitational force depends on distance, those parts of the Earth near an imaginary line connecting the Earth and the moon are closest to the moon and will feel the strongest attraction. The Earth itself is not too pliable and, therefore, exhibits only a small distortion to this variation in forces. The water in the oceans is very pliable, however, and responds easily to this variation. As a result, the water "piles up" in this region of strongest attraction, forming a "bulge" of water facing the Moon as shown in the diagram below, in which the effect is <u>highly exaggerated</u>. Because of frictional effects, the bulges are actually displaced from the line connecting the Earth and the moon. This effect is ignored in this discussion for the sake of

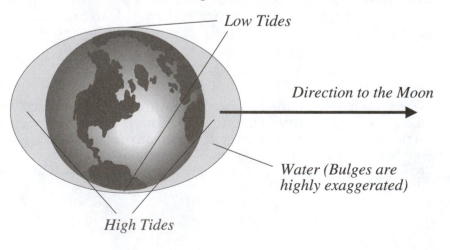

Low Tides

Direction to the Moon

Water (Bulges are highly exaggerated)

High Tides

simplicity. Another bulge appears on the other side of the Earth, due to inertia of the water and the centripetal acceleration of the Earth toward the Moon as they both revolve around their common center of mass (see Myth #4). These bulges stay fixed relative to the line between the Earth and the Moon. The Earth, however, rotates with respect to this line. Thus, as areas of the Earth pass through the bulge, they experience high tides. About six hours later, they pass through the regions between the bulges and they experience a low tide. Thus, there are two high tides and two low tides daily (although this is only approximately true, since the moon moves in its orbit during the period of a day).

3.) Why does the moon always show us the same face?

The standard answer to this question is that *the revolution and rotation periods are the same*. But is this just a cosmic coincidence? What are the chances that these two periods would be exactly the same? There must be more to it.

We continue with our discussion of the tides. In the discussion of Mystery #2, it was mentioned that the Earth exhibits only a small distortion due to the gravitational attraction of the moon. It is small, but measurable. In addition to the large water bulge on the Earth, there is a smaller land bulge. As the Earth rotates, this bulge stays fixed relative to the Moon. Thus, if an observer on the Earth were able to see this bulge, he or she would see it moving across the surface of the Earth, just as he or she sees the water bulge move across the Earth in terms of the ocean tides. This continuous distortion of the earth's surface represents a great deal of energy transformation from rotational kinetic energy of the Earth to internal energy, a phenomenon that is called *tidal friction*. As a result, the Earth's rotation is slowing down, so that each day is slightly longer than the previous. This is a very small effect and not one which will require us to continuously have our clocks repaired. Eventually, however, the Earth will stop rotating relative to the moon. When there is very little rotational energy left, the moon will grab onto some slowly moving hunk of land like the Himalayas and that point on the Earth will be captured so as to be always facing the moon.

Now, what does this have to do with the question? Well, this process has already occurred for the moon. Due to tidal friction on the moon, it has stopped rotating (relative to an observer on Earth) and a point on the surface has been gravitationally locked toward the Earth. Thus, we always see the same face. But, see Myth #2, Chapter 16.

4.) Why does Halley's Comet "return" every 76 years?

The general shape of an object in orbit around a star is an ellipse, with the star at one focus of the ellipse. The orbits of the planets are very gentle ellipses, so that they are almost circles. But some objects have very elongated elliptical orbits. Halley's Comet is one of these objects. For most of its elliptical orbit, it is very far from the Sun, reaching the orbital positions of the outer planets. According to Kepler's Laws, it is also moving very slowly when it is far from the Sun. Thus, most of its life is spent far away from the Sun, where it is invisible from Earth. During the time in its elliptical orbit that it is close to the Sun, it is moving quickly and is being vaporized (a comet is largely composed of ice) by energy from the Sun, giving rise to the comet "tail". Its orbital period is 76 years. Thus, every 76 years, it is close to the Sun in its orbit and spends a very brief time as a visible comet due to the vaporization of its surface.

5.) Why is there a meteor shower every August (as well as at other times)?

The answer to this is the same as that to the previous Mystery. In other elliptical orbits around the Sun are a large number of small meteoroids. The orbits of a group of these meteoroids and the orbit of the Earth intersect at one point in space. This point happens to be the location of the Earth when it is August. Thus, every August, the earth passes through this ring of meteoroids around the Sun and many of them fall into the atmosphere as the Perseid meteor shower. Orbits of other groups of meteoroids intersect the Earth's orbit at other points in space, resulting in additional meteor showers at other times of the year.

6.) If satellites are orbiting the Earth, how can a satellite receiving dish remain stationary? Wouldn't it have to "track" the satellite?

The period of a satellite around the Earth depends on its distance from the Earth. Thus, the trick to communications satellites is to put them at just the right place so that their orbital period is 24 hours. Thus, they will orbit the Earth once just as the Earth rotates once and are, therefore, described as being in a "geosynchronous orbit". To an observer on the surface of the Earth, then, they will appear to be stationary above a certain point on the surface at the equator. Based on the gravitational parameters of the Earth, the "correct" orbital radius for these satellites is about 22,000 miles above the surface.

For more information on this topic, see E. Zebrowski, "Aiming a Satellite Dish", *The Physics Teacher*, **26**, 153 (1988).

𝔐𝔞𝔤𝔦𝔠:

The Anti-Gravity Brazil Nuts

This actually has more to do with statistics than with gravity. As the can is shaken, openings between nuts momentarily open and close. It is far more likely for a small opening to occur than a large opening. Thus, during the shaking, small nuts migrate toward the bottom by falling into these small holes. The chances of a Brazil nut-size opening occurring is very small, so that the Brazil nuts migrate toward the top as the smaller nuts fall downward.

This same effect occurs in any mixture of different size objects. Have you noticed how the "dust" in a cereal box is at the bottom? For more information, see Magical Beans? in Chapter 13. In addition, there is more information in A. Rosato, et al, "Why the Brazil Nuts are on Top; Size Segregation of Particulate Matter by Shaking", *Physical Review Letters*, **58**, 1038 (1987).

The rising of rocks to the surface of the ground in the spring in areas with cold climates is another example of this phenomenon. For information on this aspect, see D. M. Raybin, "The Stones of Spring and Summer", *The Physics Teacher*, **28**, 500 (1990).

Gravity Assist

Unfortunately, this is not a demonstration that can be performed in the classroom (except by computer simulation), but it does appear fairly magical.   The effect depends on two factors - the mass of the planet must be large compared to the space probe, which is naturally true. Secondly, the space probe must approach the planet from an area inside the planet's orbit and exit the interaction outside its orbit and generally in the direction in which the planet is moving. In the reference plane of the planet, an observer would see the space probe approach and recede with no net transfer of energy. But an observer on the Earth will see the space probe attracted to the planet and leave the other side with a net increase in kinetic energy. The space probe has "stolen" a little energy from the orbital kinetic energy of the planet. The planet, due to its huge mass, is affected only <u>very</u> slightly. The satellite, due to its small mass, is greatly affected, moving much faster after the interaction (relative to the Earth observer) than before.

As an extreme Earth-bound example of this, imagine a truck approaching a VERY fast jogger from behind. The truck is moving 35 mph and the jogger at 30 mph. Thus, the relative velocity of the truck and jogger is 5 mph. A strong man on the truck reaches out as the jogger passes and throws the jogger forward so that the jogger is now moving 5 mph <u>faster</u> than the truck. According to an observer on the truck, the situation was symmetric. The jogger was moving at 5 mph toward the truck beforehand and 5 mph away afterward. But *relative to the ground, the jogger is now moving at 40 mph*. The jogger stole a little energy from the truck which did not affect the truck's motion significantly but changed the jogger's motion quite a bit.

The Non-Falling Keys

This trick was originally described in S. James, *Linking Rings*, 1926, and has been discussed by S. Morris (*OMNI*, p. 98, June 1990) and M. Gardner (*The Physics Teacher*, **28**, 390 (1990)). As the matchbook starts to fall and therefore rotate around the pencil, the keys also fall, shortening the string between the matchbook and the pencil and reducing the radius of rotation of the matchbook. As a result, the rotational speed of the matchbook increases rapidly, resulting in the string with the matchbook winding around the pencil and stopping the fall of the keys. A mathematical description in terms of Lagrangian mechanics is available in A. R. Marlow, "A Surprising Mechanics Demonstration", *American Journal of Physics*, **59**, 951 (1991).

𝔐𝔶𝔱𝔥:

1.) On huge planets like Saturn and Jupiter, the gravity would be so large as to crush a human who landed on the surface.

At first thought, this may seem reasonable, due to the very large mass of these planets. But a look at the equation for the surface gravity of a planet (as in the Concepts section) also shows a dependence on the <u>size</u> of the planet. Jupiter and Saturn are massive, but they are also very big. If we do the calculation, using known data (mass and equatorial radius) for the planets, we find that the gravitational fields are as follows:

Jupiter: $\quad g \ = \ 24.8 \text{ N·kg}^{-1}$ (due only to mass and radius)

Saturn: $\quad g \ = \ 10.5 \text{ N·kg}^{-1}$ (due only to mass and radius)

If we were to measure the weights of objects of known mass on these planets in order to calculate the gravitational field, we would <u>not</u> obtain these values. This is due to the fact that these planets are rotating relatively rapidly. Points on the surface of the planet have a centripetal acceleration (Chapter 9) and, according to the Principle of Equivalence (Myth #2, Chapter 3), the result is an apparent contribution to gravity away from the center of the planet. Some would be tempted to call this a result of *centrifugal force*, but see Myth #1, Chapter 9. If we include the effect of rotation, the effective gravitational fields on the surfaces of these planets are:

Jupiter: $\quad g_{eff} \ = \ 22.9 \text{ N·kg}^{-1}$ (due to mass, radius and rotation)

Saturn: $\quad g_{eff} \ = \ 9.05 \text{ N·kg}^{-1}$ (due to mass, radius and rotation)

The effective gravitational field on Jupiter is large (a little over twice that of the Earth), but not crushing, and that on Saturn is actually less than that on Earth where $g = 9.8 \text{ N·kg}^{-1}$.

2.) Astronauts in an orbiting space shuttle are *weightless*.

The astronauts in the orbiting space shuttle are just like the person in the falling elevator in Chapter 3. The astronauts and the space shuttle are all falling toward the Earth in free fall, so that they float around freely within the shuttle. But <u>they are not weightless</u>. There is still a gravitational force on them, only slightly reduced from that on the surface, due to their being another hundred or so miles farther away from the Earth's center.

For more information on weightlessness and the equally silly *microgravity*, see D. Chandler, "Weightlessness and Microgravity", *The Physics Teacher*, **29**, 312 (1991).

3.) The following dialogue takes place in a radio commercial which is advertising a video camera that will record for a much longer time than other cameras. It is the narration of an owner of one of the "other" cameras for a showing of his videos.

"Now, here is a video of Mt. St. Helens. It's just ready to .... (White Noise). Oh... I ran out of tape."

"Now, here is a video of Old Faithful. It's just ready to... (White Noise). Oh...I ran out of tape."

"Now, here is a video of Halley's Comet. It's just ready to.... (White Noise). Oh...I ran out of tape."

The question that arises here is this - what did the creators of this commercial expect Halley's Comet to *do*? The first two examples refer to phenomena that exhibit sudden and violent <u>events</u>. But a comet just appears as a fuzzy light in the sky - there is no sudden or violent event. Now, some people expect comets to go flying at high speed across the sky, so perhaps this was the misconception that led the creators of the commercial to choose a comet as an example. It is a sad statement about the science literacy in the general public that <u>no one</u> in the line of people who must have approved this commercial caught the error.

4.) In describing the tides, some textbooks explain the bulge of water on the side of the Earth away from the Moon by claiming that the Moon pulls the Earth away from the water.

This is not a very satisfying explanation and seems to be another example of trying to make the physics simple and, in the process, making it incorrect. The bulge of water on the side facing the Moon seems easy to understand, since the gravitational force on it is larger, due to its relative closeness to the Moon, compared to the center of the Earth. If this were the only contribution, it would seem that this larger force would pull a bulge of water toward the Moon, and water would move from the side facing away from the Moon. This would result in a <u>depression</u> on the side away from the Moon, not a bulge. But we have another factor to consider - in order for the Moon to stay at its distance from the Earth, it must be in <u>orbit</u> around the Earth. As we saw in Myth #3, Chapter 5, however, this is not quite the correct way to say it. In reality, the Earth and Moon both rotate around their common center of mass. Thus, the Earth is moving in a (relatively small) circular orbit. The water on the side of the Earth away from the Moon is in a smaller gravitational field than is the center of the Earth and is on the outside of a circular motion. Thus, by Newton's First Law, it tends to be "flung out" from the Earth observer's point of view (But this is not "centrifugal force" - see Myth #1 in Chapter 9.). Thus, we see a bulge on the side of the Earth away from the Moon.

Another approach to explaining the origin of two bulges is as follows. If you imagine falling feet first into a black hole (Mystery #1), then as you become very close to the black hole, you encounter a significant <u>gradient</u> in the gravitational field between your feet and your head. Since the field is stronger at your feet, they will accelerate faster than your head. As a result, your body is <u>stretched</u> along its length. We generalize this situation as follow - an object accelerating toward the source of a non-uniform gravitational field is stretched along the direction of the acceleration. Now, as mentioned above, the Earth is in a small orbit around the Earth-Moon center of mass. Thus, it is accelerating toward the Moon. It is also in a non-uniform gravitational field due to the moon. Thus, it stretches along the direction of the acceleration. Since the water is a fluid, it can respond easily and "stretches", resulting in an oval shape - hence, two bulges.

Chapter 7
Energy

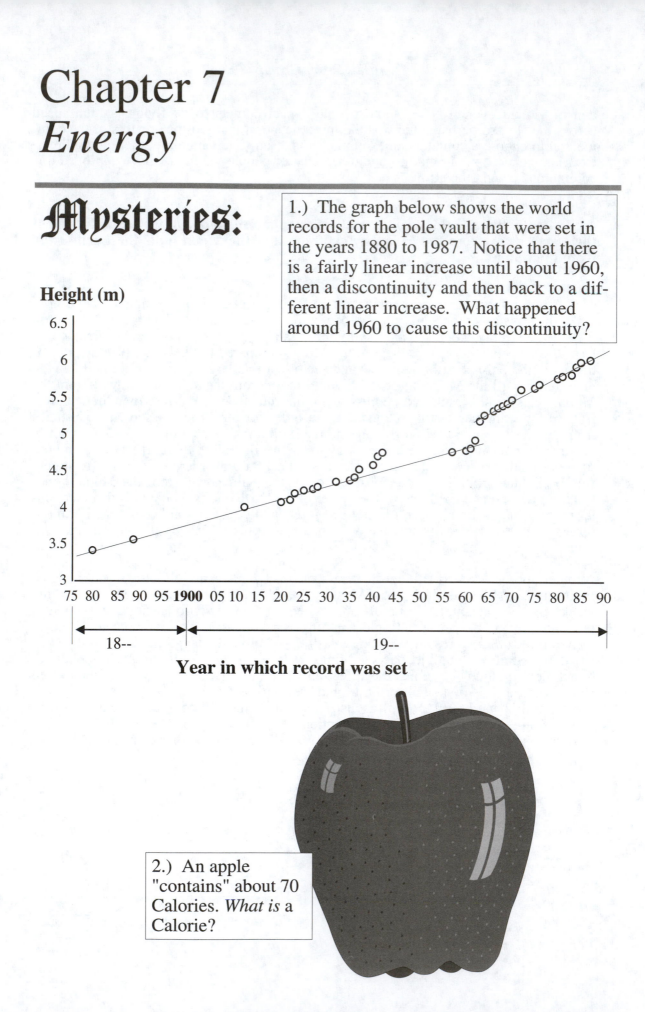

Mysteries:

1.) The graph below shows the world records for the pole vault that were set in the years 1880 to 1987. Notice that there is a fairly linear increase until about 1960, then a discontinuity and then back to a different linear increase. What happened around 1960 to cause this discontinuity?

Height (m)

Year in which record was set

2.) An apple "contains" about 70 Calories. *What is* a Calorie?

Mysteries:

3.) Why is it easier to bicycle a mile than it is to run a mile?

4.) Cloth burns very easily and quickly. So why is it that a candle wick, which is made of cotton, burns so slowly and lasts for so long?

Magic:

Use heavy rope

Fasten very securely into hole

The Pendulum of Death

Fasten a bowling ball (<u>securely!</u>) to a rope and hang it from the ceiling. Pull the bowling ball to the side and rest it against the chin of a willing volunteer. Let it go so that it swings as a pendulum. As long as the volunteer does not move, the bowling ball will come back and stop just short of his/her chin.

Myth:

E   F

1.) We have an energy crisis in today's world - we are running out of energy. But wait a minute - energy is <u>conserved</u>. How can we be *running out* of energy?

2.) A lot of work is done in a tug-of-war.

𝕸𝖞𝖙𝖍:

3.) Solar energy is a recently developed resource.

4.) The work-energy theorem is a fundamental principle in physics.

$$W = \Delta KE$$

5.) Power is the rate at which work is done.

Concepts of Energy

Energy is a very difficult concept to define. It is like trying to define <u>money</u>. We all know what to do with money, but we simply cannot define it without referring to what we do with it. The same is true of energy, although we have not generally developed the familiarity with the concept of energy as we have with money in our everyday lives. But we can think of energy as *the currency of nature*. Just as transfers and storage of money are necessary for processes to occur in our financial world, transfers and storage of energy are necessary for processes to occur in our natural world.

> *Energy = The Currency of Nature*

> *Energy is either being stored or it is being transferred and/or transformed.*

We can divide our discussion of energy into two sections: forms of *storage* of energy and forms of *transfer and/or transformation* of energy.

Storage of Energy

Just as money can be stored in checking accounts, savings accounts, cash, certificates of deposit, etc., energy can be stored in various ways:

Kinetic Energy: energy stored in the motion of a mass.

Potential Energy: energy stored because of the relative positions and/or orientations of particles or objects and associated with the fundamental forces. Examples include gravitational potential energy (associated with the gravitational force), nuclear potential energy (the nuclear strong force) and electrical and magnetic potential energies (the electromagnetic force). Another category is spring (or elastic) potential energy which is fundamentally related to the electromagnetic force between the atoms making up the material of the spring.

Internal Energy: energy stored in the kinetic and potential energies of molecules and often related to the temperature of the collection of molecules.

Other types of energy that may be encountered in everyday life are examples of some member of the above list. For example, the chemical potential energy in gasoline is actually electrical potential energy stored in the chemical bonds of the molecules of the gasoline.

Transfer/Transformation of Energy

Energy can be in transit or transformation, just as in the case of money when we write checks, make deposits or withdrawals, etc. The forms of energy transfer/transformation are as follows:

Work: the transfer of energy from one object to another, due to the application of a force by the first object and the resulting displacement of the second object. Transformation can also take place, as in the case of a falling object, where gravitational potential energy is being transformed to kinetic energy due to the work of the gravitational force of the Earth on the object.

Heat: the transfer of energy due to collisions between molecules; thus, this can be considered to be related to <u>work</u> on a microscopic level. This restricted view of heat is different from the treatment presented in most textbooks and by many physics teachers. See Chapter 15 for further exploration of this concept.

Mass Transfer: the transfer of energy along with moving mass. An example is a bowling ball transferring energy along the lane to the pins - the energy is transferred along with the mass of the ball. Another example is the energy transfer associated with carrying a tank full of gas around in an automobile.

Sound: the transfer of energy by means of mechanical disturbances in materials.

Electromagnetic Radiation: the transfer of energy by means of electromagnetic waves, such as light.

Electrical Transmission: the transfer of energy by means of electrical current, such as the energy transmission from the electrical power plant to your home.

Chemical Reactions: the transformation of electrical potential energy in chemicals to other types of energy. Often, the transformation is to internal energy, as in the case of burning fuel, or exploding weapons. Other possibilities include transformation to the transfer method of electrical transmission, as in the case of a battery, or to electromagnetic radiation, as in the case of chemiluminescent light sticks.

Nuclear Reactions: the transformation of nuclear potential energy in the nuclei of atoms to other types of energy. One common transformation is to internal energy and transfer by electromagnetic radiation, as in the case of nuclear explosions or nuclear power generation. Radioactive decay also represents the transformation of nuclear potential energy into kinetic energy of particles (alpha, beta decay) and/or electromagnetic radiation (gamma decay).

It should be noted that this taxonomy of energy transfers refers specifically to the <u>movement</u> of the energy through space, without regard for the <u>pickup</u> or <u>delivery</u> processes at the beginning and end of the transfer. For example, <u>convection</u> (see Chapter 15) is categorized as mass transfer rather than heat, since the energy is carried with the mass of a fluid during the transfer process. The pickup of the energy <u>before</u> the transfer likely involves conduction (heat) of energy from, for example, an automobile engine block into the coolant. Similarly, conduction is involved again in the <u>delivery</u> of the energy to the air surrounding the radiator. Focusing on only the movement through space, however, results in convection being assigned to the mass transfer category.

Discussions; Chapter 7 - Energy

𝔐𝔶𝔰𝔱𝔢𝔯𝔦𝔢𝔰:

1.) The graph below shows the world records for the pole vault that were set in the years 1880 to 1987. Notice that there is a fairly linear increase until about 1960, then a discontinuity and then back to a different linear increase. What happened around 1960 to cause this discontinuity?

The discontinuity in the graph is due to the introduction of a *new material for the poles* in the pole vaulting event. In the early part of the century, pole vaulters used metal or bamboo poles. When the metal pole was used, the collision between the pole and the pit when the pole was planted for takeoff was quite inelastic, with much of the kinetic energy stored in the run transformed into internal energy of the pole and metal of the pit. Bamboo poles were more flexible than the metal poles and some of the kinetic energy stored in the run was transformed into elastic potential energy in the bending of the pole. As the vaulter approached the top of the jump, this elastic energy was transformed to gravitational potential energy, resulting in a higher jump. Overall, the collision was more elastic than one with a metal pole. Bamboo, however, had a high probability of snapping during the vault and possibly injuring the vaulter. The early 1960's saw the introduction of the <u>extremely flexible</u> *fiberglass* pole. When the fiberglass poles were introduced, a dramatic jump in world records occurred. This is due to the large amount of energy that can be stored in the bending of the pole, which is returned as potential energy at the top of the vault, with minimal danger of breaking.

The graph on the next page shows a different sort of record. It is the best vault of a given year plotted against the year. It is remarkable that the best vault of a given year increases <u>linearly</u> until 1960, despite several perturbations, as indicated by the light line drawn through the points. The slope of this line is about 1.8 cm/year. In the early years of pole vaulting, the sport was not well regulated. As a result, there is quite a bit of scatter in the data points at the lower end of the graph until about 1880. The data points settle down nicely to follow the linear increase until 1889. Before this year, vaulters were allowed to use a <u>climbing</u> technique, in which the jumper could "climb" the pole on the way up with a hand-over-hand technique. This was banned in 1889. As a result, the year's best performances fell and remained below the line until about 1910. It is interesting to note, however, that the records returned to the <u>same</u> linear increase by this date as was defined by the climbing technique, <u>even though the allowed technique had changed</u>.

Shortly after the records returned to the line, World War I depleted the ranks of vaulters and the year's best records dropped. By the mid-1920's, the records were once again following the linear increase. The line was followed fairly faithfully until the early 40's, when there was a sharp upturn. This perturbation is primarily the work of one man, Cornelius (Dutch) Warmerdam. He was a graduate student at Stanford University who wrote a master's thesis on the approach and takeoff in pole vaulting. He held the pole at a grip point much higher than his contemporaries. By means of this higher grip and his academic understanding of the techniques, Warmerdam set seven outdoor and five indoor pole vaulting world records between 1939 and 1943. By the time he stopped setting records in 1943, the ranks of pole vaulters had once again been depleted by war, this time World War II, and the year's best records fell below the line. By 1960, the

Height of Best Pole Vault of the Year

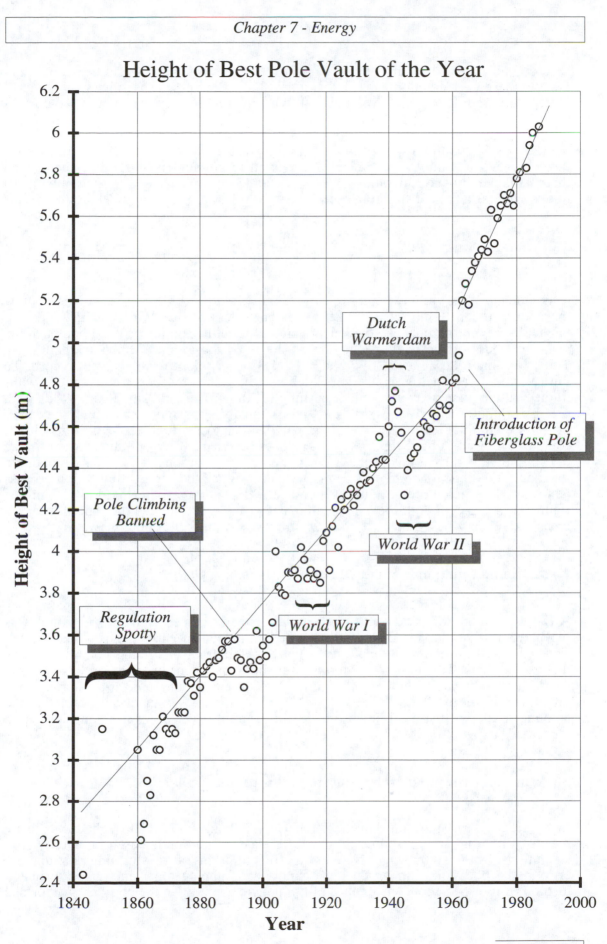

Dutch
Warmerdam

Introduction of
Fiberglass Pole

Pole Climbing
Banned

World War II

Regulation
Spotty

World War I

Height of Best Vault (m)

Year

records had returned to the line again, but the advent of the fiberglass pole quickly caused a large perturbation and, finally, a new line of increase. The slope of this recent increase in records is 3.5 cm/year, almost twice the rate of increase as before the fiberglass pole.

We can ask if this trend will continue. As a gross estimate, let us imagine that the kinetic energy of running is completely converted to gravitational potential energy and that this is the only consideration. We estimate the maximum speed of the runner at 10 m·s⁻¹, which is a typical sprinting speed. Using conservation of energy, then,

$$\frac{1}{2}mv^2 = mgh \quad => \quad h = \frac{v^2}{2g} = \frac{(10 \text{ m·s}^{-1})^2}{2(9.8 \text{ m·s}^{-2})} = 5.1 \text{ m}$$

This indicates that the maximum possible height would be only 5.1 m, which is below the current record! But, let us make our estimate less gross. We have imagined the vaulter as a point particle. To the height calculated, let us add the original height of the center of mass of the vaulter when he or she is on the ground, which is about 1 m. Thus, we obtain an adjusted figure of 6.1 m, which is close to the current record. To this, we should add additional height which may be gained by energy conversions within the vaulter, such as work done in pushing off the ground and any efforts made with the arms to pull the body upward on the pole. Finally, we should realize that the center of mass of the vaulter passes under the bar (similar to the high jumpers in Myth #1, Chapter 5), whereas the height recorded is that of the bar. If we optimistically estimate the combined contributions of these last factors as an additional 1 m, then we obtain a maximum theoretical height of 7.1 m (23.3 feet). Since the energy conversion process is not totally efficient, the eventual highest record should be somewhat lower than this figure. If we accept both the 7.1 m value as the highest value and the 3.5 cm/year increase noted previously, then we should see the record peak out about the year 2017.

2.) An apple "contains" about 70 Calories. *What is* a Calorie?

A Calorie is a unit of energy, just like a joule. Its name comes from *caloric*, which was the name for the supposed fluid that bodies contained in early theories about heat. Thus, the 70 Calories in an apple represent the storage of this much potential energy that can be transformed into other forms in the body by means of digestion. It is important to notice the capital letter on the word. A Calorie is not the same as a calorie. They differ by a factor of 1000: 1 Cal = 1000 cal = 1 kcal. The calorie is defined as the amount of energy that must be absorbed by 1 gram of water to result in an increase in the water's temperature of 1°C.

3.) Why is it easier to bicycle a mile than it is to run a mile?

In both cases, running and bicycling, the body is being moved through one mile by means of internal energy sources. The difference in exertion is based on energy considerations. Riding a bicycle represents an efficient means of transforming potential energy in the body into kinetic energy of motion. Running represents a less efficient method of transportation, since much of the potential energy in the body is transformed into internal energy of the ground and the feet. In riding a bicycle, the body is primarily moving only the legs and the motion is fairly uniform as the feet move in continuous circles. Compare this to running. When running, the legs are moving also, but are continuously starting and stopping during each stride. Work that was done to accelerate the leg is transformed into internal energy of the ground and the feet as the foot hits the ground and stops. In addition, the arms are pumping when running, representing more energy from the muscles transformed into internal energy. And, finally, when running, the body is moving up and down. Work done by the legs transforms potential energy in the body into gravitational potential energy as the body lifts, which is transformed into internal energy as the body falls and hits the ground again.

For additional information about the physics of the bicycle, see R. G. Hunt, "Bicycles in the Physics Lab", *The Physics Teacher*, **27**, 160 (1989).

4.) Cloth burns very easily and quickly. So why is it that a candle wick, which is made of cotton, burns so slowly and lasts for so long?

The wick in the candle is not the fuel and is not the source of the energy. It is only a fuel delivery system. Wax is the fuel, but only gaseous wax can burn - you cannot ignite the liquid or the solid form. Energy transferred from the candle flame melts wax at the top of the candle which is then drawn up the wick by capillary action. As long as the wick remains soaked with liquid wax, it cannot become hot enough to burn (This is similar to Boiling Water in a Paper Cup, Chapter 15.). The wax near the top of the wick vaporizes and participates in the combustion. Only the very tip of the wick burns, since it is the only part depleted of liquid fuel. Thus, the rate of burning of the wick is the same as the rate of descent of the wax level at the top of the candle.

Magic:

The Pendulum of Death

This demonstration vividly verifies the law of conservation of energy for the willing volunteer. A certain amount of energy has been stored in the Earth-pendulum system by the work done in lifting the ball to the chin. There is no way that the ball can return to a higher point without some type of extra input of energy. Therefore, be sure that the bowling ball is not *pushed* as it is let go, as this will store more energy in the system!

Myth:

1.) We have an energy crisis in today's world - we are running out of energy. But wait a minute - energy is conserved. How can we be *running out* of energy?

The amount of energy in the universe is fixed and constant - we can't possibly have an energy crisis. We do have a crisis, however. The crisis arises due to the conversion of energy from various types of storage into internal energy. According to the Second

Law of Thermodynamics, it is impossible to completely convert energy stored as internal energy to other forms. It is easy to go the other way - rub your hands together and the potential energy transferred from your body to move your hands has been <u>completely</u> converted to internal energy - your hands are warmer, as are the walls of the room (slightly!) due to the absorption of energy transferred from your hands as sound. Now place your warm hands together and see if the internal energy will cause your hands to start moving! We are continuously transforming our energy from "useful" forms of storage, such as the chemical potential energy of gasoline, into the "non-useful" form of internal energy. This is equivalent to saying that we are taking ordered systems such as gasoline molecules and converting them into disordered systems, such as exhaust gases and increased molecular motion. Thus, we have an *entropy* crisis, not an *energy* crisis. Entropy is a measure of <u>disorder</u> and appears again in Myth #2 in Chapter 15.

2.) A lot of work is done in a tug-of-war.

The truth of this statement depends on your definition of the work done. In a tug-of-war in which the system (the rope and the tuggers) is in equilibrium and at rest, then large forces may be in action, but there is no motion. Thus, there is no net work done on the system from external sources. From this point of view, the statement is a Myth. This is only the case, however, if we are simply looking at the *physical* system from the outside. The answer is different if we consider the *biological* systems that are participating in the tug-of-war. In order to exert the forces on the rope, the bodies of the participants must be undergoing internal processes. Within the bodies of the tuggers, there are hearts that are applying forces to move blood through the arterial system. Thus, work is done on the blood. Similarly, the body must apply forces to move the chest outward so as to expand the lungs to allow for inspiration of air. Work is done in this process. Other processes also occur, so that a great deal of work is done within the body even though no work appears to be done on the system of tuggers.

3.) Solar energy is a recently developed resource.

Even though we are only now finding ways to efficiently convert the incoming radiation from the sun into "useful" forms of energy, we have been "solar-powered" from the beginning. Almost all "sources" of energy can be traced back to the sun and then extended back to the Big Bang. As an example, consider the motion of your automobile. The chart on the next page shows the storage forms and transfer and transformation methods associated with the car's energy, tracing it back to the sun and beyond.

The splitting of the tracing near the bottom deserves some special attention. Two forms of energy are necessary for a star to evolve into a life-giving agent. First, we must have a source of nuclear potential energy, which is represented by the hydrogen and helium atoms in the primordial stellar material. This energy is in the nuclei by virtue of the Big Bang, by means of a process that is at present not clear. This is the right-hand branch in the tracing. Secondly, in order to release this energy, we must have a high enough temperature for nuclear fusion to occur. The required internal energy is provided by the conversion of gravitational potential energy into internal energy as the material of the newly forming star falls to the star core, as indicated in the left-hand branch. Once nuclear fusion occurs, we have a transformation of the nuclear potential energy into even

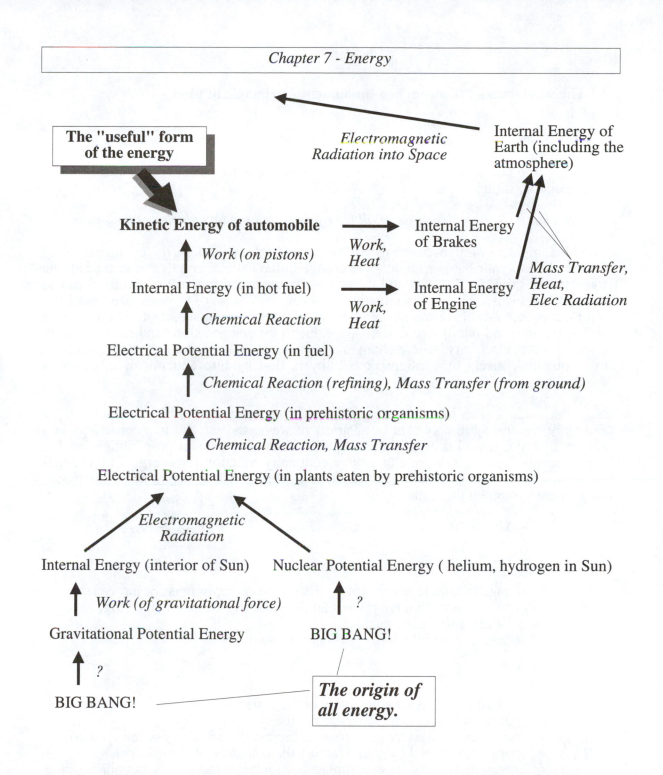

more internal energy and into the extremely high amount of electromagnetic radiation that emanates from the star.

This kind of tracing could be done for any energy process on our planet with one variation. If a nuclear process for heavy elements is involved, such as fission, then the energy came from the supernova explosion of some other star than our Sun, since the formation of heavy elements requires <u>enormous</u> energy transfer, available only in a supernova. While a tracing of this process would bypass our Sun, it still originates in the Big Bang, which was the original source of energy for the star undergoing the supernova explosion.

4.) The work-energy theorem is a fundamental principle in physics.

The work-energy theorem as described in most physics textbooks is actually a special case of a more general idea, which is often expressed by means of a <u>continuity equation</u>. The idea of a continuity equation, which we will apply to energy, but which is much more general, is this :

total change in system = sum of all transfers across the boundaries of the system

The idea might be brought home to you very clearly if we think of the continuity equation applied to your bank account - the change in the balance equals the sum of all the transfers, positive for deposits and interest payments and negative for checks, fees and other withdrawals. This kind of situation exists in many areas of physics: studies of liquid flow, where liquid is flowing into and out of a region of space; acoustics, where air is flowing into and out of a region of space due to the compressions and rarefactions of a sound wave; electricity, where charge is flowing into and out of a region of space; and even quantum mechanics, where probability is flowing into and out of a region of space.

Now, let us apply this idea to energy flow. It is complicated only by the fact that we can store the energy in a system in a variety of ways, and we can transfer energy in a variety of ways. If we list all of the changes in storage and transfer possibilities listed in the Concepts section, then we can write a continuity equation for energy with changes in the <u>amount</u> of energy stored in the system on the left and <u>transfers</u> across the boundary of the system on the right:

$$\Delta KE + \Delta PE + \Delta U = W + Q + E_{MT} + E_S + E_{ER} + E_{ET}$$

where

ΔKE = change in kinetic energy (of or around the center of mass of the system)
ΔPE = change in potential energy within the system (gravitational, spring, electromagnetic, nuclear)
ΔU = change in internal energy within the system

and

W = work (done across the boundary of the system)
Q = heat (passed across the boundary of the system)
E_{MT} = energy transferred by mass transfer (across the boundary of the system)
E_S = energy transferred by sound (across the boundary of the system)
E_{ER} = energy transferred by electromagnetic radiation (across the boundary of the system)
E_{ET} = energy transferred by electrical transmission (across the boundary of the system)

We are not including the *transformation* processes of chemical and/or nuclear reactions, since these generally represent <u>transformation</u> of energy <u>within</u> the system rather than <u>transfers</u> across the boundaries of the system. The continuity equation is a <u>transport</u> equation which incorporates energy <u>transfer,</u> not internal transformation.

Now, let us reduce this equation to a simpler form by looking at appropriate special cases. Suppose we apply a force on an object (the system), causing it to move through a distance, resulting in a change in kinetic energy. Then we ignore all terms in the continuity equation except the <u>work</u> and <u>kinetic energy</u>, since these are the only variables appropriate in this situation. We represent this below by showing the work and kinetic energy in regular print and the rest of the variables in a lighter tint:

$$\Delta KE \ + \ \Delta PE \ + \Delta U \ = \ W \ + \ Q \ + \ E_{MT} \ + \ E_S \ + \ E_{ER} \ + \ E_{ET}$$

and we arrive at

$$W \ = \ \Delta KE,$$

the work-energy theorem! Thus, the work-energy theorem is simply a special case of the more general idea of the continuity equation. *But, wait, there's more!* Suppose we look at a gas system, where <u>heat</u> is added (e.g., by conduction from a candle flame) and we do <u>work</u> on the system (e.g., by pushing in a piston, compressing the gas), with the result that the <u>temperature</u> (which we will choose to relate to the internal energy of the system) of the gas changes. Then we eliminate those energy members of the equation that show no change on the left, and transfers that do not occur on the right. We arrive at,

$$\Delta KE \ + \ \Delta PE \ + \Delta U \ = \ W \ + \ Q \ + \ E_{MT} \ + \ E_S \ + \ E_{ER} \ + \ E_{ET}$$

and we obtain,

$$\Delta U \ = \ Q \ + \ W$$

which is the First Law of Thermodynamics! (It is often written as $\Delta U \ = \ Q \ - \ W$, where the minus sign arises because the work, in this case, is considered to be done <u>by</u> the gas rather than <u>on</u> the gas.)

Now, suppose we consider a particle (so there is no meaning to internal energy U) as an isolated system, so that there are <u>no transfers</u>. We arrive at,

$$\Delta KE \ + \ \Delta PE \ + \Delta U \ = \ W \ + \ Q \ + \ E_{MT} \ + \ E_S \ + \ E_{ER} \ + \ E_{ET}$$

which gives us

$$\Delta KE \ + \ \Delta PE \ = \ 0$$

which is the equation for the conservation of mechanical energy!

As a final example, consider the very special case of <u>heating</u> a solid, which results in a <u>temperature</u> change. We assume a negligible volume change, so that we ignore any work done on the atmosphere. Writing out our energy continuity equation again and eliminating the inappropriate terms, we have,

$$\Delta KE \ + \ \Delta PE \ + \Delta U \ = \ W \ + \ Q \ + \ E_{MT} \ + \ E_S \ + \ E_{ER} \ + \ E_{ET}$$

which gives us

$$Q \ = \ \Delta U$$

Now, if we express the internal energy change in terms of mass, heat capacity and temperature change, ($\Delta U = mc\Delta T$), we have

$$Q = mc\Delta T$$

another equation that appears as a "separate" idea in many physics textbooks.

Now, let us return to the work-energy theorem. There is a subtlety over which we have glossed. The work-energy theorem as applied in many physics textbooks is expressed for a <u>particle</u> or an <u>object</u>. The work-energy theorem that we generated from the continuity equation is expressed for a <u>system</u>. If the system consists of a single object, then the two approaches are equivalent. If an object on a frictionless table is pushed by a horizontal force, for example, then the work done by the force equals the change in kinetic energy of the object as it slides. From the continuity equation point of view, the object is the system and the work done crosses the boundary of the system (although in both cases, the system implicitly includes the Earth, so that we have a reference point against which to measure the velocity!).

Where the approaches will differ somewhat is that case is which the work done does not appear as a change in the kinetic energy. For example, let us consider the simple act of lifting an object to a height h. From the traditional work-energy theorem approach, the change in kinetic energy of the object is zero, and the work represents that work done by <u>all forces on the object</u>. There are two forces acting - the applied force of the hand lifting the object, and gravity. Thus, the traditional work-energy theorem approach would be as follows:

$$W = \Delta KE \quad => \quad F_{applied} \, h \; - \; (mg) \, h \; = \; 0$$

From the continuity equation point of view, we consider the Earth and the object as a <u>system</u>. The applied force (applied by the human) is then doing work on the system from the outside, and the energy change is <u>gravitational potential</u>, so we have (eliminating all the inappropriate terms from the general continuity equation),

$$\Delta KE + \Delta PE + \Delta U \; = \; W + Q + E_{MT} + E_S + E_{ER} + E_{ET}$$

and we arrive at

$$W = \Delta PE,$$

or,

$$F_{applied} \, h \; = \; mgh$$

which is mathematically equivalent to the equation generated from the traditional work-energy theorem, but generated from a conceptually different point of view.

So finally, after this long-winded explanation, we see that the work-energy theorem, as well as several other equations in a physics textbook, stems from a more fundamental idea, the continuity equation.

The continuity equation is a statement of a conservation principle. Since the total energy of a system is conserved, the only way we can change it is to transfer energy across the boundary of the system - hence, the continuity equation for energy that we have been discussing. We also have a <u>momentum</u> conservation principle (Chapter 8). Is there a corresponding continuity equation? The answer is yes. The only way to change the momentum of a system is to apply an external force. The continuity equation describing this is the *Impulse-Momentum Theorem*:

$$\overline{F}\,\Delta t \;=\; \Delta p$$

where impulse (average force multiplied by the time interval) can be imagined to cross the system boundary, resulting in a change in momentum. As a final example, conservation of charge can be described with the following continuity equation:

$$\overline{I}\,\Delta t \;=\; \Delta q$$

where average current $\overline{I}$ crosses the system boundary and results in a change in the charge in the system. The long energy equation that we discussed above can be abbreviated to look just like these last two continuity equations if we cast it in the form,

$$\overline{P}\,\Delta t \;=\; \Delta E$$

where average power $\overline{P}$ is the power crossing the system boundary by means of all transfer mechanisms and E is the sum of all types of energy storage in the system.

For more information and other points of view on problems with the work-energy theorem, see A. J. Mallinckrodt and H. S. Leff, "All About Work", *American Journal of Physics*, **60**, 356 (1992), R. P. Bauman, "Physics the Textbook Writers Usually Get Wrong - I; Work", *The Physics Teacher*, **30**, 264 (1992) and A. B. Arons, "Developing the Energy Concepts in Introductory Physics", *The Physics Teacher*, **27**, 506 (1989).

5.) Power is the rate at which work is done.

This is another statement that often appears in physics textbooks. It is possible that this is done because power is introduced in a chapter on work and energy (in most textbooks, somewhere around Chapter 6 - 8), and work is the only transfer/transformation mechanism discussed so far. Unfortunately, however, this is another case where trying to state something in a simple way makes it incorrect. It is certainly true that the rate at which work is done is power. But the reverse is too restrictive. The general concept of power refers to the rate at which any transfer or transformation of energy is made. Thus, the rate of *heat flow* can be expressed as power. The rate of energy movement due to *mass transfer* can be expressed as power (this is how the horsepower was first defined). The rate of *electrical transmission* can be described as power, as the power company will tell you. Thus, the concept of power can be applied as the rate of any of our transfers or transformations on the right side of the continuity equation in the discussion of Myth #4, not just work.

Chapter 8
Momentum

Mysteries:

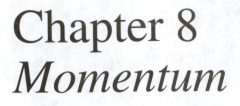

1.) Why is it safer to drive a large vehicle than a small one?

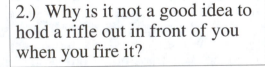

2.) Why is it not a good idea to hold a rifle out in front of you when you fire it?

3.) What if the rifle were firing blanks? Would it be as dangerous to hold the rifle out in front of you?

Magic:

Higher than a Super Ball

Drill a hole halfway into the center of a rubber ball of diameter about 2-3 inches. Into this hole, insert a straight piece of metal rod. Now, drill holes all the way through a smaller rubber ball and a ping pong ball. "Stack" these two balls on top of the larger ball, with the ping pong ball on top, by sliding them onto the metal rod. Now, drop the combination from a given height, with the rod pointing upward. The ping pong ball will bounce much higher than the height from which it was dropped.

Physics for Executives

A popular "executive office toy" consists of five balls hanging from a frame by strings and forming a straight line. When one ball is pulled out and released, it hits the others and one ball bounces out the other side. When this is repeated with two balls, two balls bounce out. With three, three bounce out, etc. Why is it always the same number of balls? If we drop one ball, why don't two balls come out, each with half the velocity? After all, this would conserve momentum.

Myth:

1.) You are driving along the freeway and encounter an unavoidable situation. You have a choice of hitting, head-on, another <u>identical</u> car coming toward you with the <u>same</u> speed or hitting a large, solid tree. Think fast - which do you choose?

2.) **F** = m**a** is Newton's Second Law

3.) An inelastic collision is one in which the kinetic energy after the collision is less than that before the collision.

Concepts of Momentum

The concept of momentum comes to us from Newton, who called it *quantity of motion*. Momentum is defined as the product of mass and velocity:

> *Momentum = Mass x Velocity*

$$\mathbf{p} = m\mathbf{v}$$

The concept of momentum allows us to express Newton's Second Law in a more general form:

$$\overline{\mathbf{F}}_{net} = \frac{\Delta\mathbf{p}}{\Delta t}$$

> *Force: How fast is the momentum of an object changing?*

Thus, the average net force can be interpreted as a measure of how fast the momentum of an object is changing due to the force. If we push harder on an object, its momentum changes more quickly.

This is actually a more general form than that given in Chapter 3, since the above form can be used to describe situations in which the mass changes, such as a rocket, which is constantly ejecting fuel, or a snowball rolling down a hill and picking up more snow. The form in Chapter 3 ($\mathbf{F} = m\mathbf{a}$) can only be used in situations in which the mass is constant.

The concept of momentum also gives us a new conservation principle, the conservation of linear momentum:

> *For an isolated system of particles (no external forces), the momentum of the system is conserved.*

This principle is particularly useful in *collisions*, where two or more particles interact by means of forces. Since the forces only act between the particles and are therefore internal to the system of particles, the momentum of the system is conserved. An <u>elastic</u> collision is one in which kinetic energy is also conserved, while <u>inelastic</u> collisions result in a different amount of kinetic energy than that before the collision.

> *The momentum of a system of particles is conserved if the system is isolated from external forces.*

A <u>totally inelastic</u> collision represents the largest possible transformation of kinetic energy into other forms for the given initial conditions and occurs when the initial objects "stick together" and leave the collision as a single object.

Discussions; Chapter 8 - Momentum

𝔐𝔶𝔰𝔱𝔢𝔯𝔦𝔢𝔰:

1.) Why is it safer to drive a large vehicle than a small one?

There are many contributions to the additional safety of a larger vehicle, but one of the most important has to do with conservation of momentum. In general, because of the conservation principle, in a collision between two vehicles, the larger the mass of the vehicle you are driving, the less will be your change in velocity. We can imagine the extreme example of a collision between a truck and a ping pong ball, initially traveling in opposite directions. The truck driver is completely unaware of the collision, while the ping pong ball has suffered a complete reversal in direction of motion. If the ping pong ball had a driver, this would result in severe forces on the driver! If we now imagine the more realistic example of a collision between a heavy car and a light car, the heavier car will suffer a smaller change in velocity. The driver of the light car will suffer a larger change in velocity and, more likely, more injury.

Studies have shown that driver fatalities decrease with increasing mass for one-car crashes as well. For details on these studies and references to additional literature, see L. Evans, "The Science of Traffic Safety", *The Physics Teacher*, **26**, 426 (1988).

2.) Why is it not a good idea to hold a rifle out in front of you when you fire it?

Again, conservation of momentum is the answer. As the bullet is fired from the gun, conservation of momentum requires that the rifle exhibit a recoil in the opposite direction, which is toward you. If you are holding the rifle gingerly in front of you, it will recoil with a respectable velocity into your body. It is safer to hold the rifle securely against your shoulder when you shoot. In this case, the mass of the object that is recoiling is the rifle mass plus your mass, and the resulting recoil velocity is correspondingly lower.

3.) What if the rifle were firing blanks? Would it be as dangerous to hold the rifle out in front of you?

A blank is a rifle shell with an explosive charge, but no bullet. Thus, if the rifle is firing blanks, then there is not a bullet mass being fired forward, as is the case in Mystery #2. Thus, there is very little forward momentum. As a result, there is much less recoil of the rifle and the danger is much lower.

𝔐𝔞𝔤𝔦𝔠:

Higher than a Super Ball

This apparent violation of conservation of energy can be explained with conservation of momentum. It is fairly easy to understand if we imagine that the balls separate slightly as they fall. Thus, the bottom ball hits the table first and reverses its motion. When the second ball collides with the bottom ball, the bottom ball is moving <u>upward</u>. Thus, the change in momentum of the second ball is much larger than it would be if it collided with the first ball while the first ball was moving in the same direction. Now, when the third ball hits the second ball, the very light ping pong ball bounces off the second ball (moving in the upward direction) to a very large height. Energy conservation is certainly not violated. The extra gravitational potential energy of the ping pong ball is accounted for by the fact that the other two balls achieve heights slightly lower than they would in the absence of the ping pong ball.

This demonstration can be performed with a combination of only two balls, but the results are not as dramatic.

For further discussion of this demonstration, see R. A. Egler, "Supernova Core Bounce: A Demonstration", *The Physics Teacher*, **28**, 558 (1990) and J. S. Huebner and T. L. Smith, "Multi-Ball Collisions", *The Physics Teacher*, **30**, 46 (1992).

Physics for Executives

Let us address the question raised at the end of the description of the demonstration: *If we drop one ball, why don't two balls come out, each with half the velocity? After all, this would conserve momentum.* This would indeed conserve momentum. But the collisions in this device are very close to elastic, so we can make the approximation that kinetic energy is conserved also. Let us imagine that we drop one ball and two come out with half the velocity. Momentum is conserved as follows:

Initial: $p_i = mv$
Final: $p_f = (2m)(\frac{1}{2}v) = mv \quad \Rightarrow \quad p_i = p_f$

Now, in this same hypothetical situation, let us test for kinetic energy conservation:

Initial: $KE_i = \frac{1}{2}mv^2$
Final: $KE_f = \frac{1}{2}(2m)(\frac{1}{2}v)^2 = \frac{1}{4}mv^2 \quad \Rightarrow \quad KE_i \neq KE_f$

Thus, kinetic energy is <u>not</u> conserved in this situation. We can show that the <u>only</u> way that we can conserve <u>both</u> momentum and kinetic energy in this device is for the same number of balls to leave the collision as were sent into the collision. A useful followup to this demonstration is to tape or glue the last two balls together and then send one ball into the collision. In this way, we <u>force</u> two balls to leave the collision. What happens?

𝔐𝔶𝔱𝔥:

1.) You are driving along the freeway and encounter an unavoidable situation. You have a choice of hitting, head-on, another <u>identical</u> car coming toward you with the <u>same</u> speed or hitting a large, solid tree. Think fast - which do you choose?

The myth here is that you have any choice at all. The result on you will be the same.

Imagine the effect on your momentum if you hit the tree - the front of your car becomes immobile and the rest of the car crushes toward the front. As a result, you are brought to rest in a short period of time. Now, think about hitting the other identical car traveling at the same speed. Since the cars have the same mass and the same speed (in opposite directions), the net momentum of the system of the cars is <u>zero</u>. Furthermore, the net momentum of the system of any similar <u>portions</u> of the cars is also zero. If we imagine the portion in question to be the front grill of each car, when the grills collide, they <u>stop</u>, so that the final momentum of this system matches the initial momentum of zero. Thus, the front of your car becomes immobile and the rest of the car crushes toward the front. Again, you are brought to rest in a short period of time, in fact the same period of time as if you hit the tree. Thus, your change in momentum and, therefore, the forces on your body are the same in both cases.

2.) F = *m*a is Newton's Second Law

This is the common form of Newton's Second Law as known by freshman physics students all over the world. But it is <u>not</u> the Second Law as stated by Newton. The mathematical statement of the Law as formulated by Newton is,

$$\bar{\mathbf{F}} = \frac{\Delta \mathbf{p}}{\Delta t}$$

as we saw in the Concepts section. If the mass of the system we are studying is constant, then the mass can be factored out and the "usual" form of the law results, as follows (in terms of *average* force and acceleration):

$$\bar{\mathbf{F}} = \frac{\Delta \mathbf{p}}{\Delta t} = \frac{\Delta(m\mathbf{v})}{\Delta t} = \frac{m\,\Delta \mathbf{v}}{\Delta t} = m\left(\frac{\Delta \mathbf{v}}{\Delta t}\right) = m\mathbf{a}$$

3.) An inelastic collision is one in which the kinetic energy after the collision is less than that before the collision.

This is the case in most collisions studied with conservation of momentum. But look back at the definition of an inelastic collision: "Inelastic collisions result in a different amount of kinetic energy than that before the collision." This says that the kinetic energy afterwards is *different*, not necessarily *less*. It may be hard to imagine an increase in kinetic energy, but this can be the case if there is potential energy stored in the particles of the system. Suppose for example, that, in a collision between two small carts, one of the carts is equipped with a spring-loaded plunger that is compressed before the collision. When the two carts collide, the plunger is released and the potential energy is transformed into additional kinetic energy of the carts. Another example would be a collision between projectiles that explode in midair. The total kinetic energy of the particles afterwards will be larger than that before the collision, the additional kinetic energy coming from the potential energy stored in the explosives.

These types of collisions, in which the kinetic energy increases, are sometimes referred to as *hyperelastic*, as a special subcategory of inelastic collisions.

Chapter 9
Circular Motion

𝔐ysteries:

1.) Why are starting positions on a racetrack staggered?

2.) Why do tall chimneys and/or smokestacks break near the middle as they topple over? What if the same chimney were to fall over on Mars? Would it break at the same point?

3.) How do you stay in amusement park rides that go upside-down?

Mysteries:

Oh, Daddy, it's so cold in the air! Please put me back in the water again!

Oh, Daddy, the water's so cold! Please don't put me back in the water again!

4.) Why is the water at Southern California beaches colder than at northeastern states' beaches even though it is located further south?

5.) Why are curved roadways *banked*?

6.) Why are there so many pieces of tire tread lying on the freeway?

7.) Why are rockets launched toward the East? And why are they launched from Florida? After all, California has better weather.

E

𝔐agic:

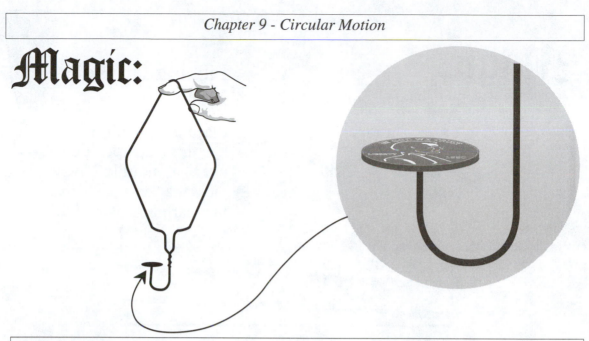

The Rotating Penny

Find a standard wire coat hanger. Pull on the middle of the long straight portion so that the hanger forms a diamond shape. Now, hang the hanger from your finger with your finger located at the bend that you just made. The hook of the hanger should now be pointing upward at the bottom of your device. Balance a penny on the end of the hook and then swing the whole combination in a vertical circle. The penny will stay on the hanger!

This requires <u>much</u> practice, especially in the stopping stage. You may have to file the end of the hook in order to obtain a flat surface on which the penny sits.

The Rotating Water

Cut a piece of thin wood (about 1/8 inch) about four inches square. Drill holes in the corners and tie a long string in each hole. Gather the strings together and tie them at the other end so that when you pick up the tied end, the piece of wood hangs horizontally and level. Now, place a plastic glass filled with water on the wood and swing it in a vertical circle.

𝕸𝖆𝖌𝖎𝖈:

Picking Up the Marble

Place a jar with a wide opening upside down over a marble. Now pick the jar up so that the marble rises with it!

More Rotating Water

Place a clear cup or glass of water at the center of a rotating turntable, so that the center of the glass and the center of the turntable coincide. Gently drop some food coloring in the water at a point away from the center of rotation. Watch the subsequent behavior of the color.

Myth:

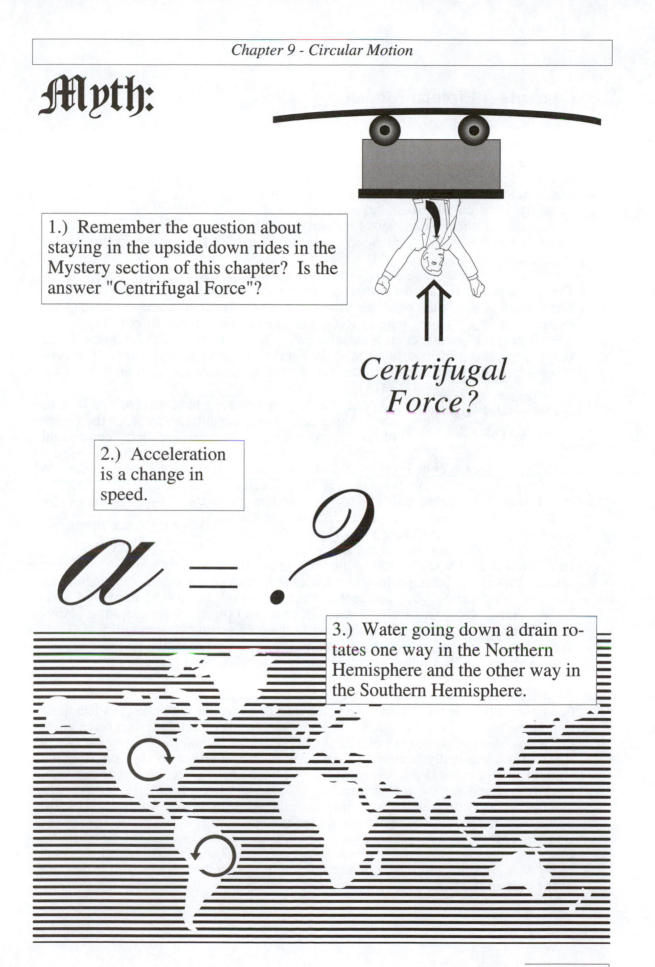

1.) Remember the question about staying in the upside down rides in the Mystery section of this chapter? Is the answer "Centrifugal Force"?

Centrifugal Force?

2.) Acceleration is a change in speed.

$$a = \text{?}$$

3.) Water going down a drain rotates one way in the Northern Hemisphere and the other way in the Southern Hemisphere.

Concepts of Circular Motion

The most important concept in circular motion is that, even though an object is traveling in circular motion at a constant speed, it is still accelerating. Acceleration is a change in <u>velocity</u>. For a particle traveling in a circular path at constant speed, the magnitude of the instantaneous velocity vector is constant but the <u>direction</u> changes. This *is* a change in the vector, so there *is* an acceleration.

> *An object moving in a circle at constant speed is accelerating.*

The acceleration is called *centripetal* and points toward the center of the circle. The force that keeps the particle in the circular path is often called the *centripetal force*. <u>This is not a new type of force</u>, but one of our familiar forces playing the role of keeping an object in a circular path. For example, the centripetal force keeping the Earth in its orbit around the Sun is the gravitational force. The centripetal force keeping an object sitting on a rotating turntable is friction.

> *Centripetal - Yes!*
> *Centrifugal - No!*

Anyone standing on a rotating platform feels as if he or she is being forced in a direction away from the center of the circle. This is popularly called "centrifugal force". It is actually the result of inertia and is called a "fictitious" force. Your body would like to travel in a straight line according to Newton's First Law. As the platform floor accelerates toward the center and you try to continue on your straight line motion, it *feels* as if a force pushes you to the outside. But there is no force. The same thing occurs when you punch the accelerator of your car. You *feel* as if you are pushed into the seat of the

> *An observer in an accelerating reference frame feels "fictitious" forces.*

car, but there is no force pushing you backward. The back of the seat is pushing on you and accelerating you; you feel pushed backward due to the inertia of your body. You are in your original state of motion and the car is accelerating. Anytime there is an acceleration, an observer in the accelerating system will feel "fictitious" forces. The <u>effects</u> of these situations are, of course, real - just ask any astronaut about the "fictitious" effects of the high acceleration upon takeoff!

A strange aspect of circular motion is the *Coriolis force*. This is another "fictitious" force that results when objects moving in a circular path change their position along the radius of the circle. The apparent result is that there is a force causing a deviation of a moving object perpendicular to its motion. For example, a rocket fired vertically appears to an Earth-bound observer to deviate from its straight line path, as if a sideways force had been applied. An observer in space, however, would see the rocket traveling in a straight line. Similarly, if two individuals on opposite sides of a rotating circular platform attempt to throw a ball to each other, the would-be catcher would see the ball appear to swerve to the side as it approaches, as if a force acted on the ball, lateral to its direction of motion. An observer hanging above the platform, however, would see the ball travel in a perfectly straight line, while the circular motion of the catcher moves him or her out of the path of the ball.

Discussions; Chapter 9 - Circular Motion

𝔐𝔶𝔰𝔱𝔢𝔯𝔦𝔢𝔰:

1.) Why are starting positions on a racetrack staggered?

Staggered starting positions are only used in races where the runners must remain in their lanes throughout the entire race and the race is long enough to occur on at least a part of the curved portion of the track. Since the race track has some semicircular portions, there will be some lanes on the "outside" of the track and others on the "inside". Since the circumference of a circular arc depends on the radius, the distance around the track in the outside lane will be longer than the distance around the track in the inside lane. To compensate for this, the runners in the outer lanes start from positions further along the track. For races in which the runners do not need to stay in their lanes, the starting positions are not staggered, since the runners will attempt to move to the inside lane before the first turn.

2.) Why do tall chimneys and/or smokestacks break near the middle as they topple over? What if the same chimney were to fall over on Mars? Would it break at the same point?

As a smokestack falls, gravity is the force causing it to accelerate. But the falling smokestack is undergoing a <u>rotational</u> motion (Chapter 10). The linear acceleration of any point on the smokestack is equal to the angular acceleration multiplied by the distance to the rotation axis, which we will assume to be at the base of the smokestack. Thus, points high on the smokestack will exhibit large linear accelerations. In fact, the vertical acceleration components of the upper portions of the smokestack can be larger than that of gravity. It is impossible for the force of gravity to cause a vertical acceleration larger than g, so some other force must be causing these larger accelerations. The force that does this is the force between the bricks of the smokestack. The lower portion of the stack is trying to pull the upper portion down faster than it would naturally fall. This creates great stresses in the structure of the smokestack which can cause a fracture as it falls. For a detailed study of falling chimneys as well as some photographs, see F. P. Bundy, "Stresses in Freely Falling Chimneys and Columns", *Journal of Applied Physics*, **11**, 112 (1940).

Although gravity is the force that causes the chimney to fall, the breaking point is determined solely by the geometry and structural factors within the chimney. Thus, even though the Martian gravity would be different than Earth's, the chimney falling on Mars would break at the same point (if it breaks at all).

3.) How do you stay in amusement park rides that go upside-down?

The trick to staying in your seat on an upside-down ride is to move across the top of the circular track with enough velocity that your centripetal acceleration is larger than the acceleration due to gravity. Then there must be a centripetal force larger than your weight, with the difference provided by the normal force from your seat. Another way to look at it is in terms of kinematics. At the top of the circle, if you were simply projected horizontally, you would follow a parabolic path to the ground. As long as the circular path of the ride curves downward faster than this parabola, you will stay in the car. See Myth #1 for more information. For information on a number of amusement park rides, see R. S. Speers, "Physics and Roller Coasters - The Blue Streak at Cedar Point", *American Journal of Physics*, **59**, 528 (1991), C. Escobar, "Amusement Park Physics", *The Physics Teacher*, **28**, 446 (1990) and J. McGehee, "Physics Students' Day at Six Flags/Magic Mountain", *The Physics Teacher*, **26**, 13 (1988).

4.) Why is the water at Southern California beaches colder than at northeastern states' beaches even though it is located further south?

This is a result of the Coriolis force. Because of water movements in the ocean, the Coriolis force causes a generally clockwise rotation of water in the North Atlantic and Pacific Oceans, as shown in the diagrams below. In the Atlantic, this results in the Gulf Stream off the coast of Florida and results in warm tropical water moving upward along the East Coast. In The Pacific, the clockwise rotation causes cold Arctic water to move downward along the West Coast. Thus, despite the latitude differences, water at California beaches is colder than at northeastern states' beaches.

Pacific Ocean: *Atlantic Ocean:*

   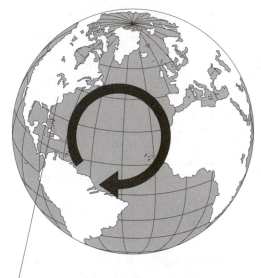

Rotation brings cold water from Arctic down the Pacific Coast.

Rotation brings warm water from tropics up the Atlantic Coast.

In reality, the flow of ocean currents is much more complicated than this simple discussion implies. The tendency of water to flow due to the Coriolis force is highly affected by interactions between the water and the continents. For example, in the North Atlantic Ocean, the strong counterclockwise Labrador Current causes Arctic water to flow southward along the coast of Canada and the extreme northern coast of the United

States. As a result, beaches south of Cape Cod, Massachusetts, tend to be warm due to the Gulf Stream, but there is a rapid decrease in temperature of the water as one moves northward from Cape Cod.

For details on this and other movements of water in the oceans, see A. B. C. Whipple, *Restless Oceans*, Time-Life Books, Alexandria, Virginia, 1983.

5.) Why are curved roadways *banked*?

If a car is traveling on a curved roadway, it must have an acceleration oriented toward the center of the circle that represents the curved roadway. There must be a force to provide this acceleration. If the roadway is level, the only force that this can be is the static friction between the tires and the road. If the road is wet or icy, however, the friction force may not be sufficient to provide the required acceleration and the car will move in a straight line tangent to the circle, that is, it will slide off the roadway. To provide an additional force in the centripetal direction, the roadway is banked. In this way, a component of the normal force is in the centripetal direction to provide, in cooperation with the friction force, the centripetal acceleration of the car. The diagram below demonstrates this concept for a car moving into the page.

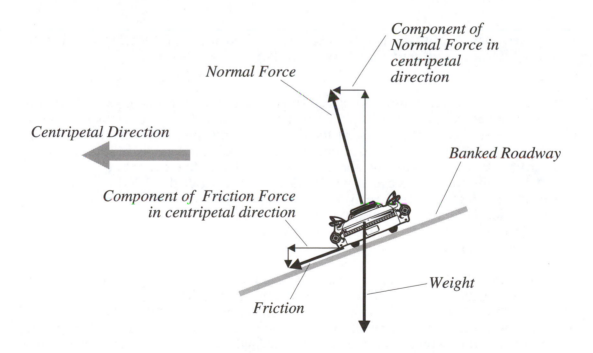

The design angle of banking is determined by demanding that it be possible to negotiate the curve at the design speed <u>without dependence on friction</u> - only the component of the normal force would provide the centripetal acceleration. Thus, in drawing a diagram such as that above to determine the angle, the friction force would not be included.

6.) Why are there so many pieces of tire tread lying on the freeway?

Many tires on trucks are <u>retreads,</u> in which a new tread is bonded onto an old bald tire. Let us do a calculation of the centripetal acceleration of the retread. We will estimate the radius of a truck tire as 0.5 m and we will assume a speed of 50 mph, which we can easily convert to about 22 m·s^{-1}. Thus, the centripetal acceleration of the tread is approximately,

$$a_c = \frac{v^2}{r} = \frac{(22 \text{ m·s}^{-1})^2}{0.5 \text{ m}} = 968 \text{ m·s}^{-2}$$

which is almost 100 times the acceleration of gravity! If we assign a guess to the total mass of the tread, say, 10 kg, this translates to a centripetal force of about 9680 N (over 2000 lbs, or one ton, if you don't have a good feel for a Newton!) that the bonding must provide. It often occurs that the bonding releases under this large force and the tread separates from the tire and is deposited on the roadway. It is interesting to note what happens if the truck speeds up to 70 mph. The centripetal force increases as the square of the velocity. The higher speed of 70 mph is just about equal to the square root of 2 times the 50 mph speed. Thus, the centripetal force on the tread at 70 mph is about twice that at 50 mph, over two tons of force on the entire tread. It is not surprising that treads readily separate from the tire!

6.) Why are rockets launched toward the East? And why are they launched from Florida? After all, California has better weather.

Very high velocities are required for space shuttles and other rockets to reach Earth orbit. A typical velocity is 17,000 mph. Any assistance to achieve this velocity is most welcome. The surface of the Earth (at the equator) is traveling at about 1000 mph *toward the East* due to the Earth's rotation. Thus, if a rocket is launched from the equator toward the East, it already has about 6% of the velocity that it needs to achieve orbit. If it were fired to the West, it would have to cancel out the 1000 mph and also achieve the additional 17,000 mph velocity, requiring more fuel. Since the United States is not located on the equator, we cannot take full advantage of this velocity, but we maximize our advantage by launching toward the East from Florida, which is the southernmost state in the continental US.

But, wait a minute. California is almost as far south as Florida. Why don't we take advantage of the better weather in California and launch from there? The reason is <u>safety</u> - launching to the East from Florida takes the rocket over the Atlantic Ocean, so that any mishap which occurs will result in debris falling into the water. Launching to the East from California would take the rocket over land, where an accident could cause danger from falling debris for population centers.

Rockets are launched from Vandenburg Air Force Base in California, but only for <u>polar</u> orbits. In these cases, launching the rocket to the North or the South results in a Coriolis deflection (see Mystery #4 and Myth #3) to the <u>West</u>, taking the rocket safely over the Pacific Ocean.

𝕸𝖆𝖌𝖎𝖈:

The Rotating Penny
The Rotating Water

Both of these demonstrations involve the same concepts as the upside-down amusement park ride in Mystery #3. If the systems are rotated quickly enough, the centripetal acceleration at the top of the circular path is larger than the acceleration of gravity and the water or penny do not fall. In the case of the penny, the normal force from the tip of the hanger hook and the force of gravity provide the centripetal force. In the case of the water, the force of gravity is assisted by the normal force from the bottom of the cup in providing the centripetal acceleration.

Picking Up the Marble

To perform this demonstration, be sure that the jar has a lip just under the opening (i.e., the part of the jar on which the top is screwed must be smaller in radius than the rest of

the jar). The secret to picking up the marble is to "swirl" the jar (grasp it and rotate the center of mass of the jar around a vertical axis displaced from the center of mass) so that the marble rolls around the inner edge. If the jar is given a small jerk during this rapid rotation, the marble will rise up over the lip of the jar as shown in the diagram to the right. Once this has occurred, the jar can be picked up while still being rotated and the marble will remain in its rolling motion around the lip of the jar.

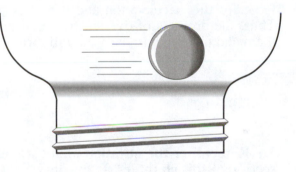

A similar demonstration with an olive and a wine glass is described in M. Gardner, "Physics Trick of the Month - Transporting an Olive", *The Physics Teacher*, **29**, 51 (1991).

More Rotating Water

The dye will spread downward in the water due to gravity, but will be compressed into a thin circular sheet in the radial direction, as indicated in the diagram accompanying this *M*. This phenomenon was described by Taylor in the early 1920's (G. I. Taylor, "Experiments with Rotating Fluids", *Proceedings of the Royal Society* (London), **A100**, 114 (1921)). It is also discussed in D. J. Baker, "Demonstrations of Fluid Flow in a Rotating System", *American Journal of Physics*, **34**, 647 (1966), although the discussion is not satisfying, as it incorporates the "fictitious" centrifugal force (see Myth #1) in force diagrams so that a water droplet in the rotating system is in equilibrium. Of course, a particle in a rotating system is not in equilibrium, it is accelerating toward the center of the circle.

We first must ask about what is supplying the centripetal force on the water as the water rotates. If we mentally compare the stationary glass to the rotating glass, we can see that the wall of the rotating glass must provide an additional normal force (over that of the stationary glass) on the layer of water in contact with the wall in order to provide the centripetal acceleration of this water. In turn, each layer of water must apply an additional force on the layer radially inward from it to provide the centripetal force on that layer. These forces will become smaller as we move toward layers closer to the center, since the centripetal force depends on the radius. Thus, we will generate a <u>pressure gradient</u> in the rotating fluid, with the pressure increasing as we go from the center to the wall of the glass. At any given point, a water droplet satisfies Newton's Second Law - the force inward due to the pressure gradient is equal to the mass of the droplet times the centripetal acceleration.

Now, as the dye falls into the water, it makes room for itself by displacing some water toward the center of the circle and some toward the outside. The water that moved toward the center is now at a smaller radius than before and is thus moving with a tangential velocity larger than appropriate for its new radial position. It thus has a centripetal acceleration too large for its new position. The pressure gradient is not sufficient to provide this acceleration, so the droplet moves outward, by Newton's First Law, toward its original position. For the water that was displaced outward, the centripetal acceleration is smaller than appropriate for its new position. The pressure gradient is now too large for this acceleration and it pushes the droplet back toward its original position. Thus, the droplets of dye are squeezed into a thin sheet by the water on either side of them and the thin circular wall will form.

𝕸𝖞𝖙𝖍:

1.) Remember the question about staying in the upside down rides in the Mystery section of this chapter? Is the answer "Centrifugal Force"?

Centrifugal force is a common misconception among scientific laypersons. It is a "fictitious" force as discussed in the Concepts section. Some arguments are made that centrifugal force is the Newton's Third Law reaction force to centripetal force. This is not reasonable, since the centrifugal and centripetal forces then act on the same body, which is impossible for Third Law reaction force pairs, as we saw in Chapter 3.

Another argument against centrifugal force can be made on the basis of forces as <u>interactions</u> between entities. What is the second entity that is interacting with the object undergoing circular motion, resulting in the centrifugal force? No such second entity can be identified.

Any reference to centrifugal force as a force to be included in a free body diagram is erroneous. There are also recommendations (P. A. Smith, "Let's Get Rid of 'Centripetal Force'", *The Physics Teacher*, **30**, 316 (1992)) for postponing the term *centripetal force* until later in a student's education, since it often is interpreted as a separate force that must be included in a free body diagram.

2.) Acceleration is a change in speed.

The problem with this statement is in the last word. Acceleration is a change in *velocity*, not *speed*. If acceleration were just a change in speed, then a particle moving in a circular path at constant speed would not exhibit a centripetal acceleration.

3.) Water going down a drain rotates one way in the Northern Hemisphere and the other way in the Southern Hemisphere.

If (and these are big "ifs") the drain of a sink is pulled in such a way that <u>absolutely no rotation is imparted to the water during the pull,</u> and <u>if there is no residual rotation due to the filling process or anything else,</u> then the Coriolis force will cause the water to rotate in different directions in the two hemispheres. In reality, however, pulling the plug generally tends to impart some rotation to the water and then conservation of angular momentum causes the rotation to increase in angular velocity as the water moves toward the drain. The direction of this rotation depends simply on the original direction of rotation imparted to the water. The Coriolis force is far too weak to reverse this direction if it happens to be opposite to the Coriolis direction.

Even if the conditions described above were achieved, the rotation of the water due to the Coriolis force would hardly be noticed. H. R. Crane ("How Things Work - A Tornado in a Soda Bottle and Angular Momentum in the Wash Basin", *The Physics Teacher*, **25**, 516 (1987)) estimates a total rotation of less than 3° of the water due to the Coriolis force during the time that it takes to drain a typical sink.

Chapter 10
Rotational Kinematics

𝔐ysteries:

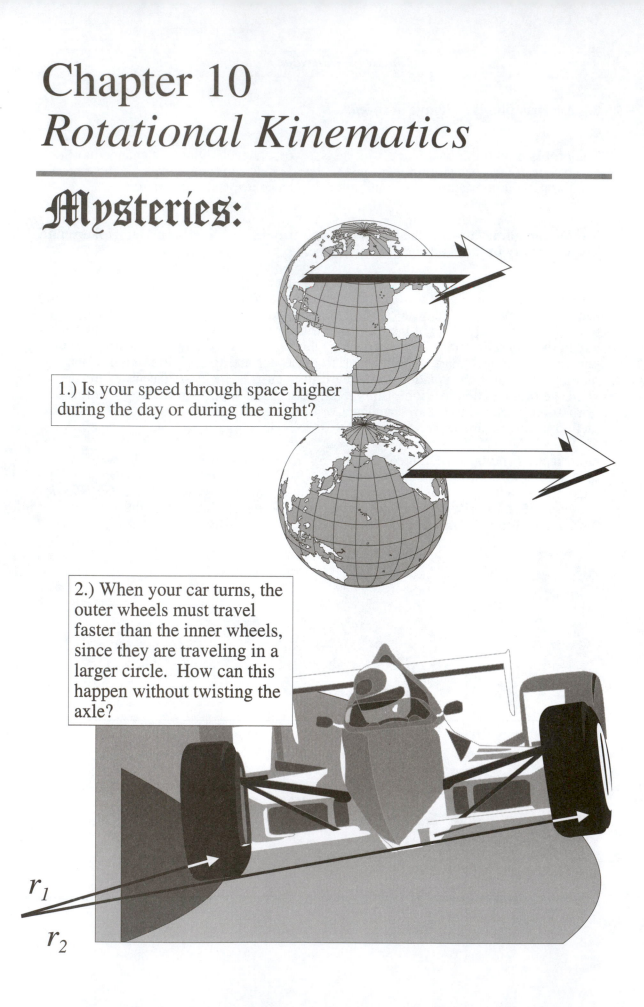

1.) Is your speed through space higher during the day or during the night?

2.) When your car turns, the outer wheels must travel faster than the inner wheels, since they are traveling in a larger circle. How can this happen without twisting the axle?

r_1

r_2

𝕸𝖆𝖌𝖎𝖈:

Adding up Rotations

Lay a book on the table in a position as if you were going to read it. We are going to turn the book around rotation axes with rotations which we expect to be representable by vectors. Rotate the book by 90° <u>three times</u> in the following order. The expected results of your rotations are shown in the diagram below.

    1.) Rotate around an axis perpendicular to your stomach as you face the book.

    2.) Rotate around an axis going from left to right.

    3.) Rotate around an axis going up and down

Record the final position of the book. Now, repeat the demonstration, but in a different order, such as:

    1.) Rotate around an axis going from left to right.

    2.) Rotate around an axis going up and down

    3.) Rotate around an axis perpendicular to your stomach as you face the book.

Now notice the final position of the book. Is it the same as before? It isn't. But wait a minute, isn't vector addition supposed to be commutative, that is, the order of addition doesn't matter? What's going on here?

First Trial:

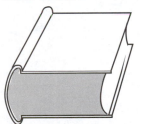 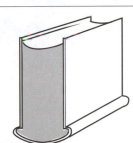 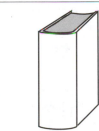 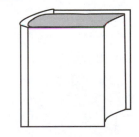

Starting Position   *After First Rotation*   *After Second Rotation*   *Final Position*

Second Trial:

 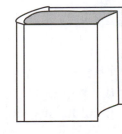 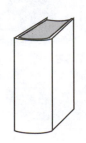 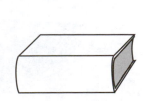

𝔐𝔞𝔤𝔦𝔠:

The Semi-Blurred Wheel

Take a photograph of a passing car, bicycle or other vehicle. Look carefully at the wheels of the vehicle. Do you notice anything different about the bottom of the wheels and the top?

𝔐𝔶𝔱𝔥:

1.) Have you noticed, when watching wagons move in Western movies, that sometimes their wheels turn the "wrong" way?

2.) A compact disk (CD) operates just like a record (LP), except that the sound is picked up optically instead of mechanically.

Concepts of Rotational Kinematics

Rotational kinematics is very similar to translational kinematics (Chapter 2), except that the object is rotating rather than translating. In fact, the equations describing rotational kinematics (and dynamics!) are of the same mathematical form as translational kinematic (dynamic) equations, with rotational variables replacing translational variables.

Rotational *position* is described by the angle variable, θ, which is the rotational analog to position, x. The angular position is measured in <u>radians</u>, where we define an angle for a sector of a circle, measured in radians, as

> *Angular Position - where is it? (θ)*

$$\theta = \frac{s}{r}$$

where s is the arc length of the sector and r is the radius of the circle. Note that the unit "radian" is actually dimensionless.

The rate of change of the angle is ω, the *angular velocity*. This is the rotational analog to velocity, v, and is expressed as follows:

> *Angular Velocity - how fast is the angular position changing? (ω)*

$$\omega = \frac{\Delta \theta}{\Delta t}$$

and has units of rad·s^{-1}, or simply s^{-1}. The angular velocity is also called the *angular frequency* and is related to the normal frequency as follows:

$$\omega = 2\pi f$$

The rate of change of angular velocity is α, the *angular acceleration* and the rotational analog to acceleration a, and is expressed as follows:

> *Angular Acceleration - how fast is the angular velocity changing? (α)*

$$\alpha = \frac{\Delta \omega}{\Delta t}$$

and has units of rad·s^{-2}, or simply s^{-2}.

Discussions; Chapter 10 - Rotational Kinematics

𝔐𝔶𝔰𝔱𝔢𝔯𝔦𝔢𝔰:

1.) Is your speed through space higher during the day or during the night?

We can imagine two <u>translational</u> velocity vectors that can be added to give a velocity relative to "space", or at least relative to a reference frame in which the Sun is at rest. These are the tangential velocity of the Earth in its orbit around the Sun and the tangential velocity of a point on the surface of the Earth due to the spin of the Earth on its axis. The <u>angular</u> velocity vectors of the orbital motion and the spin motion are in the <u>same</u> direction. Thus, at night, when you are on the side of the Earth away from the Sun, the two translational velocity vectors are in the same direction and they add, giving rise to a large relative velocity. During the day, the two translational vectors are in opposite directions and subtract, giving a smaller velocity. The difference in speed between night and day is about 3%, as can be calculated from the distance to the Sun, the radius of the Earth and the length of the year and the day.

2.) When your car turns, the outer wheels must travel faster than the inner wheels, since they are traveling in a larger circle. How can this happen without twisting the axle?

We focus our attention on a rear-wheel drive car. The front wheels on such a car are on separate axles, so there is no problem in the front. The rear wheels are also separated, each being mounted on a "half-axle". Unlike the front wheels, however, these half-axles must be <u>joined</u>, so that both rear wheels receive torque from the drive shaft. The two half-axles are joined by the apparatus known as the *differential*. The differential is a system of gears that allows the two half-axles to turn at the same rate when the car is going straight, but at different rates when the car is turning while still applying torque. For a nice explanation of the differential, see D. Macauley, *The Way Things Work*, Houghton Mifflin, Boston, 1988.

𝔐𝔞𝔤𝔦𝔠:

Adding Up Rotations

One of the requirements for a mathematical entity to qualify as a vector is that it must obey a <u>commutative property</u> under addition, that is, two vectors **A** and **B** must obey the relation,

$$\mathbf{A} + \mathbf{B} = \mathbf{B} + \mathbf{A}$$

or, more generally, the result of the addition of <u>any</u> number of vectors must not depend on the order in which they are added. This demonstration shows that <u>finite angular displacements do not act as mathematical vectors</u>, since they do not obey the commutative law. We performed three angular displacement operations in different orders and obtained different results. If the experiment is repeated with very small rotations, rather than 90° rotations, one can convince oneself that, as the individual rotations become smaller and smaller, the final orientations of the book become indistinguishable. Thus, <u>infinitesimal</u> rotations *do* act as mathematical vectors. Thus, even though θ cannot be expressed as a vector, the instantaneous angular velocity, ω, can, since it is a ratio of an infinitesimal rotation to an infinitesimal change in time.

The Semi-Blurred Wheel

The camera is at rest with respect to the ground. Thus, the velocities as observed with the camera will be those with respect to the ground. Any point on the wheel has a tangential velocity with respect to the center, to which must be added the translational velocity of the center of the wheel. Thus, at the top of the wheel, both velocity vectors point forward, so that the velocity relative to the ground is high. At the bottom, the velocity vectors point in opposite directions and cancel (if the motion is pure rolling, the two velocity vectors have the same magnitude). Thus, when a picture is taken with a short but finite time during which the shutter is open, the top of the wheel will move through some distance and be blurry and the bottom will be clearer.

𝕸𝖞𝖙𝖍:

1.) Have you noticed, when watching wagons move in Western movies, that sometimes their wheels turn the "wrong" way?

The wheels do not, of course, actually turn the wrong way. The appearance that they are turning backward is a <u>stroboscopic</u> effect due to the "quantized" nature of the motion picture. Movies are normally shot at 24 frames per second. Thus, a movie is equivalent to viewing a scene by means of a strobe light flashing at 24 flashes per second. If the rotation of the wheels is such that a wheel does not make a full rotation between frames, (or does not quite turn by an angle equal to an integer multiple of the angle determined by adjacent spokes), then the wheel will appear to have rotated the opposite way between the frames. Over a finite length of film, then, the wheel appears to turn backward.

This phenomenon is related to the humming activity that we performed in Chapter 1 (A Real Humdinger). You could make the lines on the computer screen go up or down by changing the frequency of your humming. In that case, the frequency of the observed phenomenon is constant (the sweep frequency of the monitor) and you change the stroboscope's frequency by humming. In the movie, the stroboscope frequency is constant (at the rate of 24 per second), but the wheels will look like they are going forward sometimes and backward at other times as the wagon velocity (and, therefore, the wheels' angular frequency) changes.

A particularly interesting example of this effect occurs in the movie, *Class Action*, (Twentieth Century Fox, 1991). In the later part of the movie (if you are searching on your VCR, it is at 1 hour and 22 minutes after the opening studio logo), the lead actress is driving on the highway and the scene begins with a close-up of the front left wheel of the car. The hubcap has a design near its edge with 12 repeating features around the circumference, while there is a pattern of 4 repeating features in a circle near the center. The result is that the edge of the hubcap appears to rotating in the correct direction, but much more slowly than the tire. In addition, the center region of the hubcap appears to be rotating *backward*, faster than the edge, but slower than the tire. It is an interesting experience to see two parts of the same hubcap rotating in opposite directions!

2.) A compact disk (CD) operates just like a record (LP), except that the sound is picked up optically instead of mechanically.

There are many similarities between a record and a compact disk. The major difference, of course, is the method of extracting the information from the disk. But there is another operational difference that makes the question fit in this chapter on rotational kinematics. A record is played at a constant angular velocity, often $33\frac{1}{3}$ or 45 revolutions per minute. Given this fact, think about the tangential velocity of a point on the record as the stylus moves from the outside to the inside. Since the radius of the circular motion is changing, but the angular velocity remains constant, the tangential speed is decreasing. Thus, a given wavelength of sound will represent a shorter distance near the center of the record than near the edge. This puts some limitations on high frequency sounds. For the part of the record at 8 cm from the center, for example, the record surface is traveling at about 0.28 m·s^{-1} when the record is played at $33\frac{1}{3}$ rpm. If a 20,000 Hz sound is recorded at this point, then one wavelength of the sound must be reproduced in a length of about <u>14 microns</u> (This can be calculated from the frequency of the sound and the angular speed of the record.). This requires very careful and accurate cutting and reproduction of the record surface. Also, since the high frequencies are represented by very tiny pieces of vinyl jutting into the groove, it is easy to break off these pieces, which occurs as the needle slams into them during play. As a result, records readily lose their high frequency reproduction as they are played.

Now, what about CD's? In a compact disk, the <u>tangential</u> velocity is held constant, at about 1.3 m·s^{-1}, while the angular velocity changes as the CD is read from inside to outside. Thus, all of the "pits" that represent the encoded musical information are the same distance apart along the track, but there are more of them per revolution near the outside of the disk than near the inside.

It is an interesting exercise to calculate the length of a track on a record and a CD. For a 40 minute program, the length of the record track is about one kilometer, while the CD track is about 3 km long.

Chapter 11
Rotational Dynamics

𝕸ysteries:

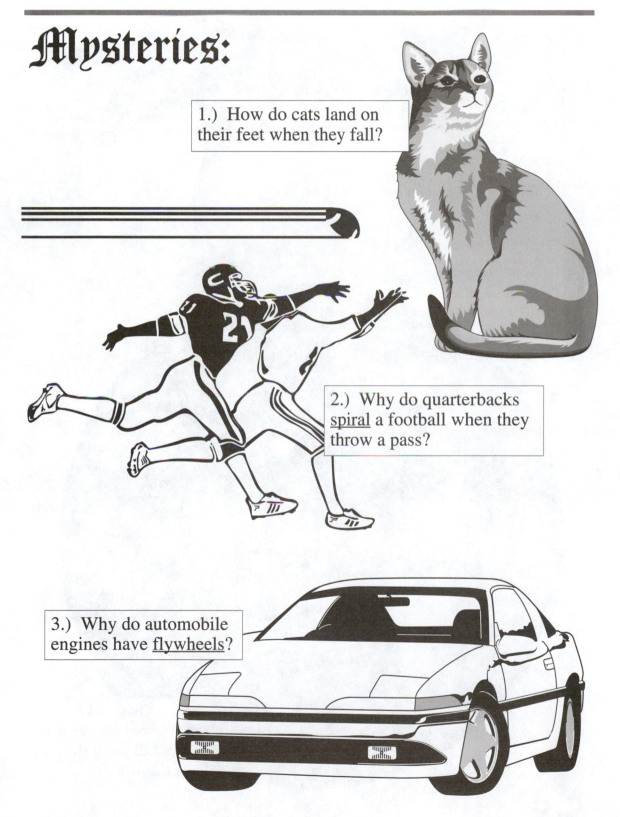

1.) How do cats land on their feet when they fall?

2.) Why do quarterbacks <u>spiral</u> a football when they throw a pass?

3.) Why do automobile engines have <u>flywheels</u>?

Mysteries:

4.) How are ice skaters able to spin so fast?

5.) What makes a roulette ball eventually fall from the rim of the wheel?

Mysteries:

6.) You are trying to loosen a stubborn bolt. As a person who works with your hands, it is almost intuitive to make your efforts more effective by using pliers with longer handles or a longer wrench or by putting a pipe over the wrench handle. Why does this help?

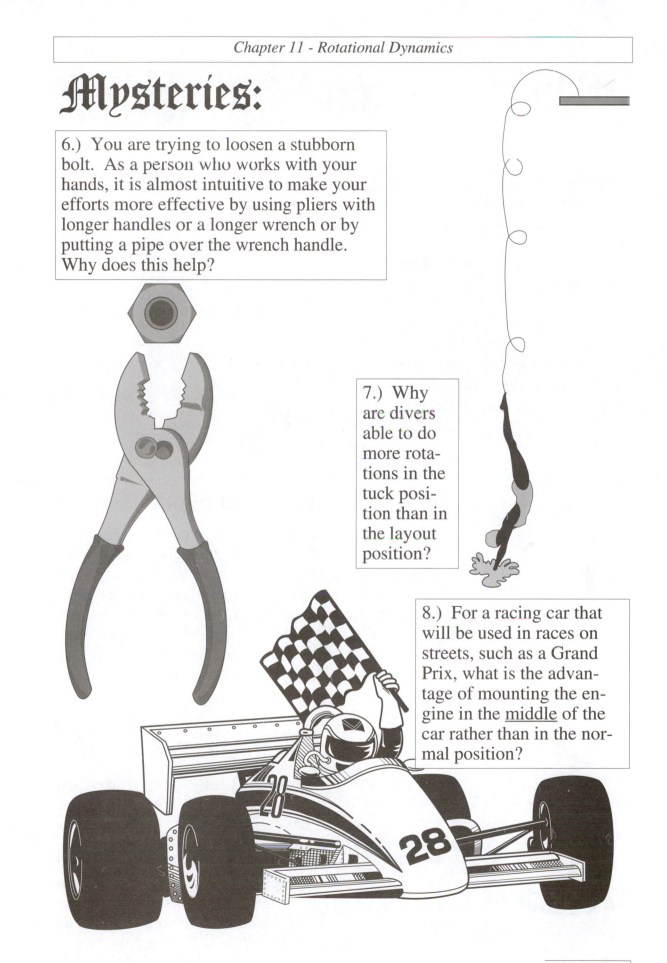

7.) Why are divers able to do more rotations in the tuck position than in the layout position?

8.) For a racing car that will be used in races on streets, such as a Grand Prix, what is the advantage of mounting the engine in the <u>middle</u> of the car rather than in the normal position?

𝔐𝔞𝔤𝔦𝔠:

The Soup Can Race

Place a long board on a table with one end raised. Choose cans of soup that are identical in mass and size, but different types, such as cream of chicken, chicken gumbo and chicken broth. Place two cans of soup on their sides at the high end of the board and let them roll down the hill. They will not roll at the same speed! Which type of soup is fastest?

The Sliding Meter Stick

Support a meter stick, yard stick, or other long smooth rod on one outstretched finger of each hand and move your hands inward slowly. The finger on which the meter stick slides alternates, first on one, then on the other! Try pulling both hands inward, and then one, holding the other stationary. The alternating effect is the same! Notice also that your fingers always meet at the exact middle of the rod! Why does this happen?

The Rolling Penny

Insert a penny into a deflated round balloon. Now, blow up the balloon with the penny inside. Hold the balloon by the end with the valve and twirl it in a small circle. The penny will stand up on its edge and roll around inside the balloon! What makes it stand up like this?

Magic:

Getting a Clamp on Moment of Inertia

Fasten two identical C-clamps to a meter stick, one at 40 cm and one at 60 cm. To another meter stick, fasten two more identical C-clamps, one at 10 cm and one at 90 cm. Grasp the meter sticks, one in each hand, at the 50 cm mark and wiggle them back and forth. Notice the difficulty with which you stop the system from rotating in one direction and start it up again in the reverse direction. Notice the difference in your effort between the two hands!

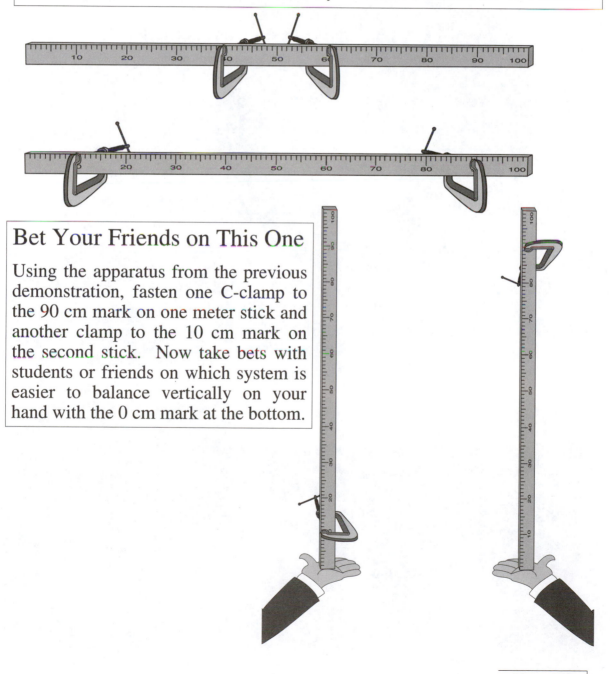

Bet Your Friends on This One

Using the apparatus from the previous demonstration, fasten one C-clamp to the 90 cm mark on one meter stick and another clamp to the 10 cm mark on the second stick. Now take bets with students or friends on which system is easier to balance vertically on your hand with the 0 cm mark at the bottom.

Myth:

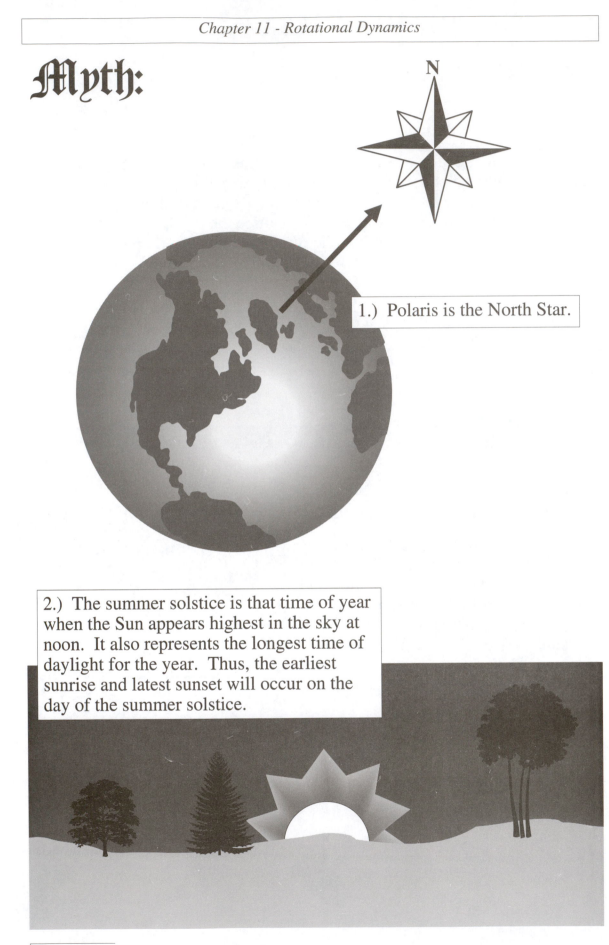

1.) Polaris is the North Star.

2.) The summer solstice is that time of year when the Sun appears highest in the sky at noon. It also represents the longest time of daylight for the year. Thus, the earliest sunrise and latest sunset will occur on the day of the summer solstice.

Concepts of Rotational Dynamics

In the previous chapter, we discussed the fact that concepts and variables involved in translational motion have analogs in rotational motion. We only looked there at the kinematics of rotational motion, however. Let us now consider the dynamic variables that are analogs to linear dynamic quantities:

Force ⇒ ***Torque***

Torque is often assigned the symbol, τ, and has units of N·m. The torque is the product of the force and the perpendicular distance from the reference axis to the line of action of the force. The diagram below indicates the force **F**, the position vector **r**, the line of action of the force and the

> *Torque = Cause of changes in Rotational Motion*

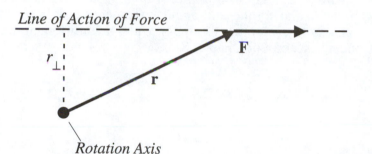

perpendicular distance $r_\perp$. The torque for such a situation is given by,

$$\tau = Fr_\perp$$

The introduction of the position vector **r** indicates the importance of where the force is applied in terms of the resulting rotational motion. For example, many individuals have had the experience of pushing on a door on the wrong side - near the hinges. The result is little, if any, rotation of the door. The same force applied farther from the hinges, however, can easily open the door.

Mass ⇒ ***Moment of Inertia***

> *Moment of Inertia = Resistance to changes in Rotational Motion*

The resistance to changes in translational motion is represented simply by the mass. Mass also plays a role in resistance to changes in rotational motion, but the situation is a bit more complicated - the resistance also depends on how the mass is distributed around the reference axis. The variable which describes this distribution and resulting resistance is the moment of inertia, indicated by the symbol, I, and having units of kg·m^2.

Momentum ⇒ ***Angular Momentum***

If an object is in translational motion, it has what Newton called a "quantity of motion" that we called linear momentum, as discussed in Chapter 8. If an object is in

Concepts of Rotational Dynamics (continued)

rotational motion, it will have a corresponding quantity of motion that we call angular momentum, with the symbol L, and with units $kg \cdot m^2 s^{-1}$. For rotation about a fixed axis, the angular momentum is the product of the moment of inertia of the object around the axis and its angular velocity. The direction of the angular momentum vector is perpendicular to the plane of rotation of the particles making up the object. Thus, it is parallel to the

> *Angular Momentum = "Amount" of Rotational Motion*

rotation axis and in the direction given by a right hand rule - by curling the fingers of your right hand in the direction of rotation of the object, your thumb points in the direction of the angular momentum vector.

Exploring these analogous relationships between translational and rotational variables, we can show that torque and angular momentum are related in the same mathematical way to each other as are force and linear momentum (as described in the Concepts section of Chapter 8):

Translational Motion:
$$\bar{\mathbf{F}}_{net} = \frac{\Delta \mathbf{p}}{\Delta t}$$

Rotational Motion:
$$\bar{\tau}_{net} = \frac{\Delta \mathbf{L}}{\Delta t}$$

In Chapter 8, we found that the linear momentum of a system of particles is conserved when there are no external forces on the system. Since torque and angular momentum are mathematically related to each other in the same way as are force and linear momentum, then we have another new conservation principle, conservation of angular momentum:

> *For an isolated system of particles, (no external torques), the angular momentum of the system is conserved.*

Conservation of angular momentum plays an important role when something in the system *changes*. Just as linear momentum depends on mass and velocity, angular momentum depends on moment of inertia and angular velocity. In a rotating system, we can redistribute the mass and change the moment of inertia. We have no such flexibility in linear systems, since the role of resistance to changes in motion is played by the mass, which cannot be altered. In a rotating system, if moment of inertia goes up, angular velocity goes down proportionally and vice versa. Thus, on a system where the moment of inertia changes for some reason, the angular velocity must change in the opposite direction in order to compensate.

Discussions; Chapter 11 - Rotational Dynamics

Mysteries:

1.) How do cats land on their feet when they fall?

The understanding of cat dynamics is still not complete. There are competing theories, based on high speed photography of falling cats. One popular theory claims that the rotation is done in two steps. The cat twists its body with its legs (hind legs, that is) extended and its arms (front legs) pulled in. Thus, the front half of the body has a smaller moment of inertia and rotates through a larger angle than the back half. The cat then pulls its legs in and extends its arms and twists the other way. This time, the front rotates through a smaller angle than before and the back half is brought around. The net result is a rotation of the body without there ever having been a net angular momentum.

Other theories involve rotational motion of the tail of the cat, which cancels out the angular momentum of the twisting body. For a series of photographs of falling cats with and without tails, see J. E. Fredrickson, "The Tail-less Cat in Free Fall", *The Physics Teacher*, **27**, 620 (1989). This article also lists 16 additional references to descriptions of falling cats, published from 1894 to 1986.

2.) Why do quarterbacks spiral a football when they throw a pass?

The spiraling football conserves angular momentum. Thus, the direction of the axis of rotation stays fixed, preventing the ball from tumbling and being more subject to air resistance and harder to control. Compare this to an arrow, which does not spin. The shaft of the arrow always remains tangent to the trajectory during the flight, so that the point implants in the ground upon landing. The football does not do this. The direction of the rotation axis stays fixed and does not remain tangent to the trajectory.

3.) Why do automobile engines have flywheels?

The flywheel in an engine uses the principle of inertia to "smooth out" the discontinuous impulses from the engine's individual cylinders. Without the flywheel, the crankshaft would jerk every time a spark plug fired. The same principle is used in a potter's wheel.

4.) How are ice skaters able to spin so fast?

Conservation of angular momentum is the answer again here. As the skater pulls his/her arms in after beginning to rotate, the moment of inertia decreases and, to conserve angular momentum, the angular velocity must increase. The skater will often hold his/her arms in a graceful arc over the head. This allows the mass of the arms to be brought even closer to the center of rotation, increasing the angular velocity even further than if the arms were just brought next to the body.

5.) What makes a roulette ball eventually fall from the rim of the wheel?

The diagram below gives a glimpse of the situation for the roulette ball as it rolls around the rim of the roulette wheel before it drops. The center of the wheel is to the right.

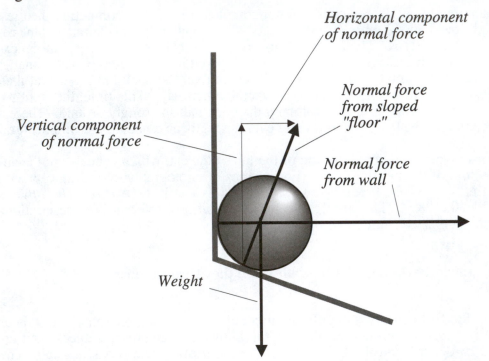

When the ball is moving rapidly, it has a large centripetal acceleration. This is provided by the normal force from the wall of the wheel and the component of the normal force from the sloped "floor" in the horizontal direction. In the vertical direction, the ball is in equilibrium, due to the balance between the weight and the vertical component of the normal force from the floor. As the ball slows its rotation due to rolling friction, the centripetal acceleration becomes smaller and the normal force from the wall will decrease. The normal force from the floor will also decrease, but not as rapidly, since only the horizontal component is related to the centripetal acceleration. Eventually, the centripetal acceleration will be small enough that it can be provided solely by the component of the normal force from the floor. At this point, the normal force from the wall will vanish and the ball will lose contact with the wall. As the ball slows further, it starts to roll down the sloped floor toward the numbered slots on the wheel.

6.) You are trying to loosen a stubborn bolt. As a person who works with your hands, it is almost intuitive to make your efforts more effective by using pliers with longer handles or a longer wrench or by putting a pipe over the wrench handle. Why does this help?

This is a clear example of increasing the <u>torque</u> on the rotating system by applying the same force at a larger perpendicular distance from the rotation axis.

7.) Why are divers able to do more rotations in the tuck position than in the layout position?

This is another example of conservation of angular momentum. As a diver leaves the platform or diving board, he or she has angular momentum from the torque exerted on the body by pushing off. By going into the tuck position, where the knees are brought close to the chest, the moment of inertia of the diver is significantly reduced. In order to conserve angular momentum, the angular velocity of the diver must increase. Thus, during the time determined by the height of the platform and the translational kinematics of falling to the water, the diver can do more rotations than in the layout position, where the moment of inertia is large and the angular velocity correspondingly small.

8.) For a racing car that will be used in races on streets, such as a Grand Prix, what is the advantage of mounting the engine in the <u>middle</u> of the car rather than in the normal position?

The moment of inertia of a system depends on the position of mass with respect to the axis of rotation. For an automobile with the engine in the normal position at the front of the car, the engine has significant moment of inertia around a vertical axis running through the center of the car, which we will choose as the rotation axis for rotations associated with turns. Thus, the engine would provide significant resistance to changes in its rotational motion around this axis. The engine mounted in the middle of the car is much closer to the rotation axis and thus has a much smaller moment of inertia around this axis. Thus, the mid-engine car will offer less resistance to changes in angular velocity, which are necessary for the tight and sudden turns when racing on the streets.

𝔐𝔞𝔤𝔦𝔠:

The Soup Can Race

You will find that the cream of chicken will roll faster than the chicken gumbo. The chicken broth is the fastest of all. Moment of inertia plays some role in the differing rolling speeds of the soups. Cream of chicken rolls essentially as a disk or a cylinder, due to the high viscosity of the soup. One might argue that the chicken gumbo begins to roll more like a hoop, since it takes a while for friction in the fluid to cause all of the inner layers of fluid to rotate. This hypothesis might lead one to a prediction that the chicken broth would roll slowly compared to the cream of chicken, since hoops roll more slowly than disks (more on this later). This is inconsistent, however, with the high speed of the chicken broth. The slowness of the chicken gumbo is not easily explained by means of a moment of inertia argument, since that is not the major factor. The major factor is the <u>internal friction</u> between the fluid and the <u>chunks</u> of food in the chicken gumbo. This internal friction transforms (by means of work) much of the original gravitational potential energy into internal energy rather than kinetic energy of rolling, so that the can rolls more slowly. The chicken broth does not have these chunks, so much less energy is converted by friction. Notice that you could actually <u>cook</u> the chicken gumbo just by rolling it down the hill a few (thousand) times!! A modification of this demonstration that will show the effect more clearly is to compare the rolling speed of two cans or jars, one containing a couple of handfuls of sand. The sand will result in internal friction, which will slow the rolling, just like the chunks of food in the chicken gumbo soup.

Now, what about our prediction that the broth will roll slowly because it rolls like a hoop? We need to be careful here. Physics teachers often roll disks and hoops of the same radius and same mass down inclined planes and we do indeed see the disk roll faster than the hoop, since the disk has a smaller moment of inertia and thus less resistance to changes in rotational motion. Now, in comparing the chicken broth to the cream of chicken, we see the same radius and the same mass printed on the can. But, be careful - remember that mass has a dual role - it determines the gravitational force <u>and</u> it represents resistance to changes in motion. Thus, the chicken broth has the same gravitational force causing it to accelerate as the cream of chicken. But, only the outer layers of the chicken broth are rotating initially. Thus, only <u>part</u> of the mass of the chicken broth is contributing to the moment of inertia and offering resistance to changes in rotational motion (while it offers the <u>same</u> resistance to changes in <u>translational</u> motion as the cream of chicken). With the same driving force and less rotational resistance, the chicken broth ends up rolling down the incline faster than the cream of chicken!

The Sliding Meter Stick

The basic concept is that torque depends on both force and distance. Since the meter stick is not rotating, the torques must balance. As one finger slides, the distance between it and the center of the meter stick decreases. In order for the torques to balance, if the distance decreases, the force must increase. Thus, the force on the sliding finger (it's a normal force!) increases until the frictional force is enough to stop the sliding. Then the other finger slides until the frictional force is enough to stop it from sliding and the alternation continues.

As far as meeting in the middle is concerned, the weight force, at the center of gravity, must stay between the fingers. If the weight acted outside of the region between your fingers, the meter stick would suddenly start rotating off one of your fingers. This obviously doesn't happen, so the center of gravity and, therefore, the center of the meter stick must stay between your fingers.

The Rolling Penny

Initially, the penny tends to bounce around inside of the balloon. Due to torques from these collisions, the penny will begin a rotational motion. If the angular momentum vector corresponding to this rotation (see the discussion of the right hand rule in the Concepts section) is not parallel to the side of the balloon, torques from further collisions with the walls of the balloon will oppose the motion. Eventually, a rotation will occur with an angular momentum vector parallel to the surface of the balloon, which will result in a rolling motion along the surface. Once this happens, the normal force from the wall passes through the center of the penny, parallel to a radial line on the penny. In this situation, the torque is zero and conservation of angular momentum will maintain the direction of the angular momentum vector. The rolling motion now has a period, corresponding to the time for one revolution of the balloon's interior. The operator senses this period and feeds energy in by twirling the balloon at the appropriate frequency. This further stabilizes the rolling by feeding energy in by resonance (See Chapter 16), which causes the penny to roll even faster.

Getting a Clamp on Moment of Inertia

With the C-clamps at 40 cm and 60 cm, the moment of inertia is relatively small and the meter stick can be wiggled rather easily. For the clamps at the 10 cm and 90 cm points, their distance from the rotation axis has increased by a factor of 4 compared to the first system. Since the moment of inertia depends on the square of the distance from the rotation axis, the moment of inertia has increased by a factor of 16! The resulting increase in difficulty in wiggling the meter stick is readily felt. Since both meter stick-clamp systems contain the same total mass, this is a vivid demonstration of the dependence of the resistance to rotational motion on the <u>positions</u> of the mass relative to the rotation axis.

Bet Your Friends on This One

Many people will believe that the meter stick with the clamp near the bottom will be easier to balance, since they are familiar with the idea of a "low center of mass" representing stability. In fact, however, <u>the stick with the high center of mass is easier to balance</u>. Because of the location of the C-clamp, it has approximately 9 times as much torque due to gravity as the stick with the C-clamp at 10 cm (This is an approximation, since we are assuming a massless meter stick; in reality the mass of the meter stick will alter these numbers slightly.). Thus, there is 9 times the "tendency" to fall over. But, torque is proportional only to distance, while moment of inertia is proportional to the <u>square</u> of the distance. Thus, the meter stick with the clamp at 90 cm has <u>81 times the resistance to rotational motion</u>. As a result, it is much easier to balance than the stick with the clamp at the bottom.

As a follow-up demonstration, place the two meter sticks, with clamps attached, with their lower end on the floor against a wall. Hold them at the same angle to the wall and let them fall. The stick with the clamp near the bottom will fall to the floor first.

Myth:

1.) Polaris is the North Star

The sun exerts a torque on the bulging Earth (is the bulge due to centrifugal force? See Chapter 9). This torque causes the axis of rotation of the Earth to <u>precess</u>. This is very much like a spinning top leaning over from the vertical, so that its weight results in a torque on the top around the pivot point at the bottom. As a consequence, the top precesses around a vertical axis. The torque of the Sun on the bulging Earth causes the

Earth to precess with a period of several thousand years. Thus, the axis of the Earth does not always point toward Polaris - it just happens to be close to that direction presently.

2.) The summer solstice is that time of year when the Sun appears highest in the sky at noon. It also represents the longest time of daylight for the year. Thus, the earliest sunrise and latest sunset will occur on the day of the summer solstice.

In reality, the days of the earliest sunrise and latest sunset are about a week apart, even after correcting for atmospheric refraction (Myth #5, Chapter 26). This difference is due to the tilt of the Earth's axis and the fact that the Earth's orbit is elliptical. By conservation of angular momentum, the Earth is moving faster in its orbit when it is closest to the Sun in January and slowest when it is farther away in July. According to our observations, then, the Sun is moving against the stars at successive noons faster during January than in July. The result of this and the tilt of the orbit is that clock time and Sun time do not match at all times of the year - they can differ by up to about 15 minutes. An "equation of time" is necessary to calculate a time interval for each day to be subtracted from the local Sun time to give the local "mean" time (Greenwich Mean Time is the time as indicated by a "mean" Sun, not the actual Sun). For information on the equation of time and its application to the sunrise and sunset, see D. Jacobs, "Why Don't the Earliest Sunrise and the Latest Sunset Occur on the Same Day?", *Physics Education*, **25**, 275 (1990).

Chapter 12
Pressure

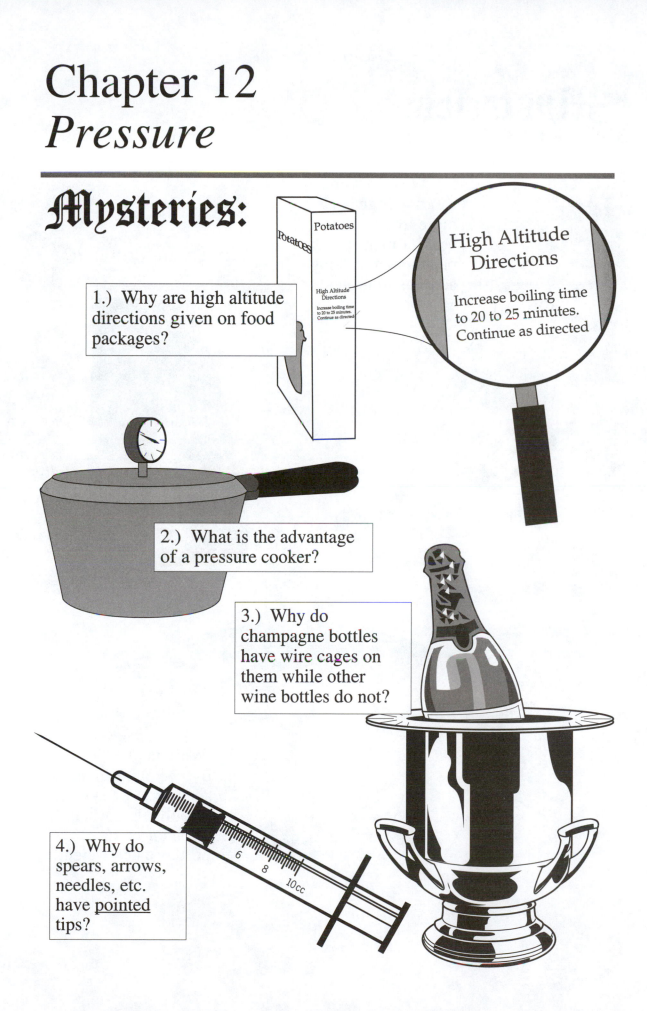

𝕸ysteries:

Potatoes

High Altitude Directions

Increase boiling time to 20 to 25 minutes. Continue as directed

1.) Why are high altitude directions given on food packages?

High Altitude Directions

Increase boiling time to 20 to 25 minutes. Continue as directed

2.) What is the advantage of a pressure cooker?

3.) Why do champagne bottles have wire cages on them while other wine bottles do not?

4.) Why do spears, arrows, needles, etc. have <u>pointed</u> tips?

Mysteries:

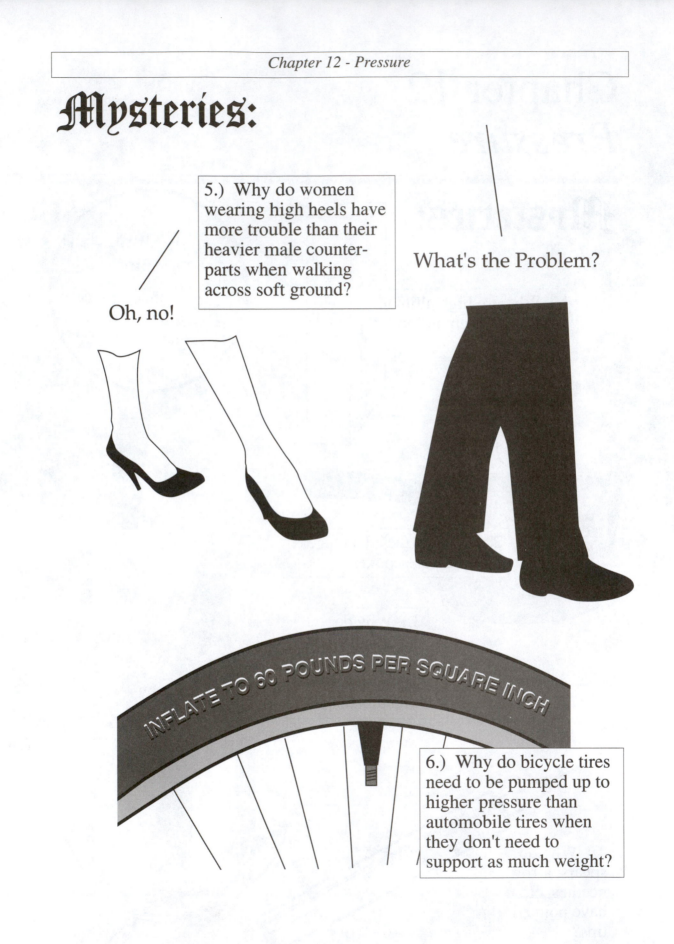

5.) Why do women wearing high heels have more trouble than their heavier male counterparts when walking across soft ground?

Oh, no!

What's the Problem?

INFLATE TO 60 POUNDS PER SQUARE INCH

6.) Why do bicycle tires need to be pumped up to higher pressure than automobile tires when they don't need to support as much weight?

Mysteries:

7.) Why do your ears "pop" when you travel in an airplane or drive in the mountains?

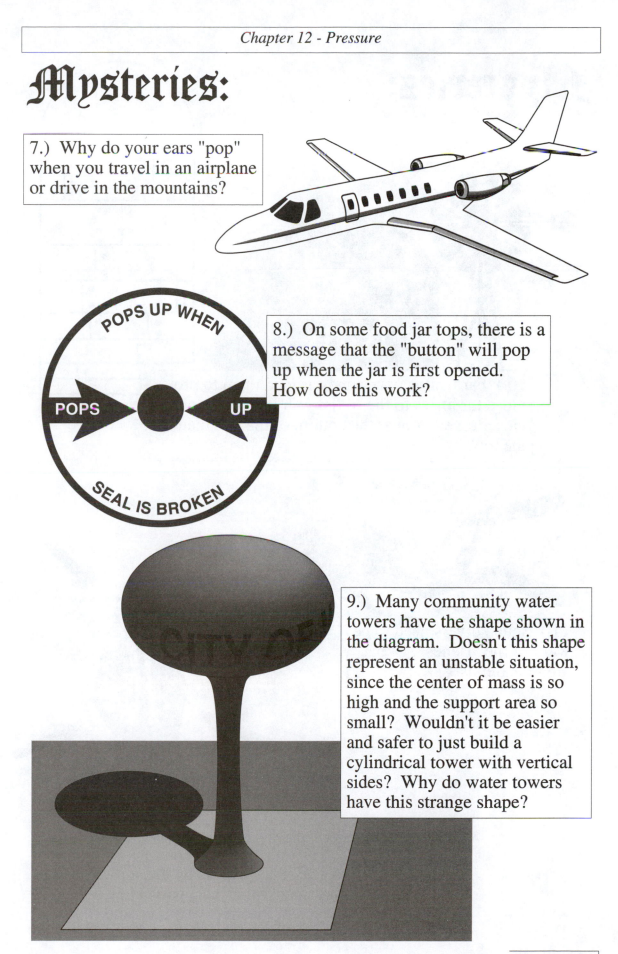

POPS UP WHEN

POPS   UP

SEAL IS BROKEN

8.) On some food jar tops, there is a message that the "button" will pop up when the jar is first opened. How does this work?

9.) Many community water towers have the shape shown in the diagram. Doesn't this shape represent an unstable situation, since the center of mass is so high and the support area so small? Wouldn't it be easier and safer to just build a cylindrical tower with vertical sides? Why do water towers have this strange shape?

Mysteries:

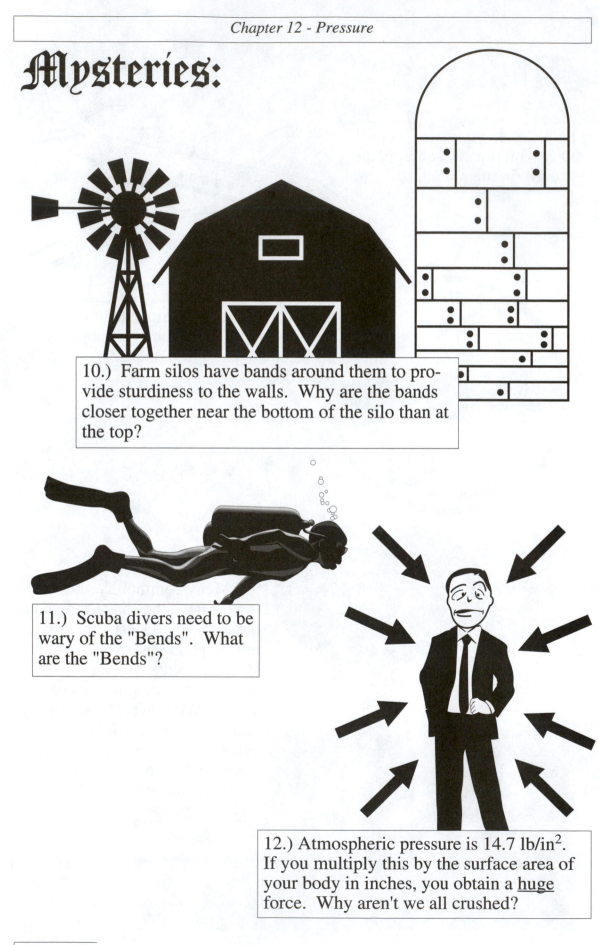

10.) Farm silos have bands around them to provide sturdiness to the walls. Why are the bands closer together near the bottom of the silo than at the top?

11.) Scuba divers need to be wary of the "Bends". What are the "Bends"?

12.) Atmospheric pressure is 14.7 lb/in^2. If you multiply this by the surface area of your body in inches, you obtain a <u>huge</u> force. Why aren't we all crushed?

𝕸ysteries:

13.) Take a look at a <u>non-retractable</u> ball point pen. Somewhere on the barrel you are very likely to find a small hole. What's this for?

There it is

14.) If you are caught in quicksand, what is the best way to escape?

Uh-oh!

𝕸agic:

The Indestructible Balloon

Dip a long, sharpened knitting needle into petroleum jelly. Insert the end of the needle into a blown up, tied off round balloon, <u>just next</u> to the valve. The balloon doesn't pop! Now, keep pushing the needle so that it goes through the end of the balloon opposite the valve, passing through the region where there is a small dark circle. If you are careful, you can successfully push the needle through this end of the balloon also without causing it to pop!

The Non-Inflatable Balloon

Insert a balloon into a bottle with a narrow neck and stretch the mouthpiece over the opening of the bottle. Now try to blow the balloon up.

Blow in here.

The Small but Powerful Balloon

Blow up two balloons, one fully and one only about half as much. Holding them so that the air does not escape, connect their mouthpieces together with a short piece of hollow tubing (You may need a third hand here!). Now, release your grip on the balloons. Wouldn't you expect the larger balloon to have more pressure and to push air into the smaller balloon until they are of equal size? Is that what happens?

Magic:

The Way to Larger Balloons

Place a blown up, tied off round balloon in a bell jar in a vacuum system and evacuate the bell jar. The balloon will grow larger as the air is pumped out. Now, allow the air back in and the balloon returns to its original condition.

The Way to Larger Marshmallows?

Place a marshmallow in a bell jar in a vacuum system and evacuate the bell jar. The marshmallow will grow larger as the air is pumped out. Now, allow the air back in and the marshmallow does not return to its original condition! It ends up much smaller!

The Way to More Shaving Cream??

Put some shaving cream on a paper or glass plate. Place the plate in a bell jar in a vacuum system and evacuate the bell jar. The mound of shaving cream will grow larger as the air is pumped out. Be careful that no shaving cream is pumped into the vacuum pump. Now, allow the air back in and the cream does not return to its original condition! It ends up as a gooey mess!

𝕸𝖞𝖙𝖍:

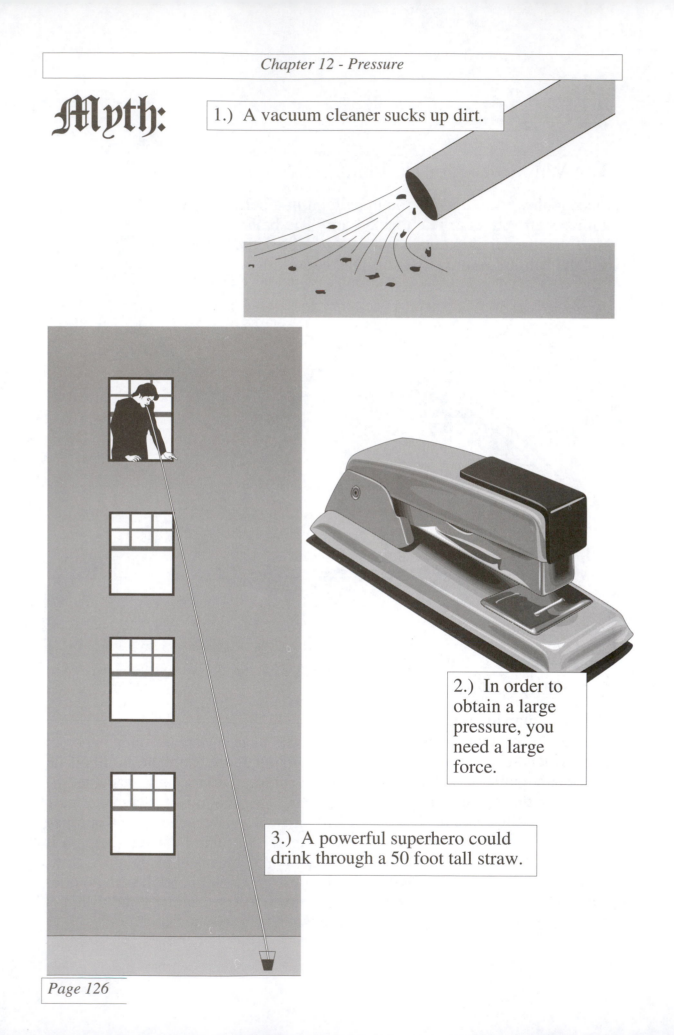

1.) A vacuum cleaner sucks up dirt.

2.) In order to obtain a large pressure, you need a large force.

3.) A powerful superhero could drink through a 50 foot tall straw.

𝕸𝖞𝖙𝖍:

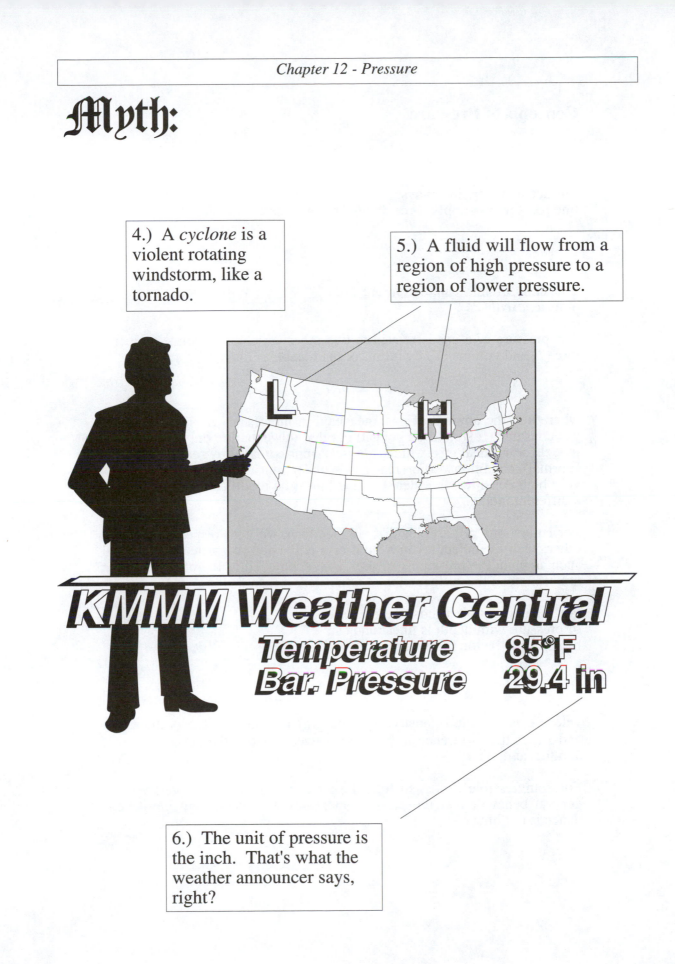

4.) A *cyclone* is a violent rotating windstorm, like a tornado.

5.) A fluid will flow from a region of high pressure to a region of lower pressure.

KMMM Weather Central
Temperature   85°F
Bar. Pressure   29.4 in

6.) The unit of pressure is the inch. That's what the weather announcer says, right?

Concepts of Pressure

Pressure is very easy to define, but is misunderstood by many people and can act in surprising ways. We de-

> *Pressure = Force Divided by Area*

fine pressure as simply force divided by the area over which the force acts:

$$P = \frac{F}{A}$$

The units of this calculation are N·m^{-2}. We name this combination of units a *pascal* (Pa).

The pressure of a gas is related to two other variables, absolute temperature (T) and volume (V) by means of the Ideal Gas Law:

$$PV = nRT$$

where n is the amount of gas, measured in moles, and R is the universal gas constant. It should be pointed out that this equation only represents a model for an ideal gas, which is closely approximated by real gases under "normal" conditions. For real gases under unusual conditions, such as very high densities, the Ideal Gas Law must be modified or replaced with a different model.

For collections of particles that are free to move with respect to one another, such as molecules in a fluid or a collection of particulate material such as sand, the pressure increases with depth. Thus, the pressure of the atmosphere at the surface of the Earth is larger than that high in the air.

For incompressible fluids, the increase of pressure with depth is linear and can be shown to be the following:

> *Pressure in an incompressible fluid increases linearly with depth.*

$$\Delta P = \rho g h$$

where ρ is the fluid density, g is the gravitational field, h is the depth and ΔP is the difference in pressure between that at the surface ($h = 0$) and that at depth h.

For compressible fluids, such as the air in the atmosphere, we have the general behavior that the pressure increases with depth, but the dependence is not linear.

Discussions; Chapter 12 - Pressure

𝕸𝖞𝖘𝖙𝖊𝖗𝖎𝖊𝖘:

1.) Why are high altitude directions given on food packages?

This is a result of the dependence of the boiling temperature of water on pressure. Boiling depends on the temperature of the water being hot enough to provide a vapor pressure higher than the surrounding atmospheric pressure, so that bubbling can be supported. As the pressure at the surface of the water drops, the water does not have to be raised to as high a temperature to provide the sufficient vapor pressure for bubbling, and the boiling temperature is correspondingly lower. At high altitudes, the atmospheric pressure on the surface of the water is lower than at sea level. Thus, the boiling point of water is reduced below the standard 100°C. Details on the dependence of boiling point on altitude are given in J. P. Negret, "Boiling Water and the Height of Mountains", *The Physics Teacher*, **24**, 290 (1986).

If food requiring boiling were cooked for the normal time, it would be cooked at a lower temperature and would not be finished in the normal time. Thus, high altitude directions for boiled items generally call for extended cooking times to compensate for the lower temperature.

For baked items, it becomes a little more complicated. Again because of the reduced pressure, more water will evaporate from the food, so the amount of water mixed in the beginning is increased. For expanding items like cakes, the reduced pressure will result in larger volumes, possibly surpassing the ability of the air cell walls to sustain the cells and the cake will fall. The recipe calls for extra flour to compensate for this effect. Finally, cakes will not brown as well under reduced pressure because of the lower boiling point of water discussed above. To compensate, some high altitude directions call for higher oven temperatures.

2.) What is the advantage of a pressure cooker?

This situation is in contrast to that in Mystery #1. In a pressure cooker, the top makes a tight seal with the bottom so that the pressure above the water can increase as the temperature goes up. Thus, by the time the normal boiling temperature is reached, the pressure is high enough that boiling will not occur. The temperature must be raised <u>beyond</u> the normal boiling point for the water to boil. As a result, the food is boiled at a higher temperature and the cooking time is therefore <u>reduced</u>.

The pressure cooker can also be used to cook foods by steam rather than boiling, by placing just a small amount of water in the cooker. This method was used before the invention of the microwave oven to reduce the cooking time for many meals.

3.) Why do champagne bottles have wire cages on them when other wine bottles do not?

Champagne is in a category called *sparkling wines*. The "sparkle" is achieved in these wines in a way similar to that for soft drinks: there is carbon dioxide in solution in the liquid. The carbon dioxide is under pressure, especially if the bottle were to become warm. Thus, there is a large force pushing outward on the stopper in the bottle. The wire cage provides an additional force holding the stopper in the bottle so that it does not shoot out.

4.) Why do spears, arrows, needles, etc. have <u>pointed</u> tips?

The intent of an arrow, spear or needle is to <u>pierce</u> the surface that it encounters. This can be performed much easier if the spear or arrow can exert a large pressure on the surface. By bringing the end to a sharp point, the force at the point is applied over a small area, resulting in the needed large pressure. The same idea exists for a straight pin - it is easy to pierce your skin with the pointed end, but try to poke the end with the pinhead into your skin!

5.) Why do women wearing high heels have more trouble than their heavier male counterparts when walking across soft ground?

This is similar to Mystery # 4. The woman's weight is applied over the very small area of the heel, especially as the foot just hits the ground while taking a forward step. Thus, the pressure is large and the soft ground is easily pierced by the heel. The man's heel is many times larger, so that the pressure is many times smaller and the soft ground is not pierced.

6.) Why do bicycle tires need to be pumped up to higher pressure than automobile tires when they don't need to support as much weight?

It is certainly true that bicycle tires do not need to support as much weight as automobile tires. The difference in weight may be perhaps a factor of 20 or 30. But, consider the difference in *area* of the contact region between the tire and the ground. Bicycle tires are much thinner than automobile tires. In addition, there are only two tires on a bicycle and four on an automobile. Thus, the total area of contact is much less for a bicycle than for a car, by a factor of 40 or 50. So, for the bicycle, even though the required force is less, the force is spread out over a smaller area, which more than compensates for the smaller force and the resulting pressure must actually be higher.

Bicycle tires will work at lower pressures, of course, but the flatter tire will have more area in contact with the ground, so that it still provides the same force. Similarly, we could pump an automobile tire to a higher pressure. The result would be that the edges of the tire would lift off the road, resulting in a reduced area.

7.) Why do your ears "pop" when you travel in an airplane or drive in the mountains?

Your middle ear is separated from the outside world by the tympanic membrane, the eardrum. If you fly upward in an airplane, the outside pressure drops and the eardrum will be bowed out by the higher inner pressure in your middle ear. This results in the uncomfortable feeling that you experience. In order to compensate, you can yawn or swallow. These actions temporarily open the *Eustachian tube*, which connects the middle ear to the oral cavity. Thus, with the Eustachian tube open for a moment, the pressures can equalize and the discomfort disappears.

8.) On some food jar tops, there is a message that the "button" will pop up when the jar is first opened. How does this work?

This type of top exists on jars on which the top is placed while the food inside is still warm, such as baby foods and apple sauce. As the food cools, the pressure in the air above the food drops. The difference between the outside atmospheric pressure and the lower inside pressure will overcome the "springiness" of the jar top and push the "button" in. When the jar is opened, air enters the jar and raises the inside pressure back to that of the atmosphere and the springiness of the top pushes the button back out.

9.) Many community water towers have the shape shown in the diagram. Doesn't this shape represent an unstable situation, since the center of mass is so high and the support area so small? Wouldn't it be easier and safer to just build a cylindrical tower with vertical sides? Why do water towers have this strange shape?

The consideration in building the strange shape is the water pressure that is required for the community. As mentioned in the Concepts section, pressure varies linearly with depth for an incompressible fluid such as water. Imagine that the water tower has the cylindrical shape suggested in the Mystery. Let us also imagine that it is a time of low water supply and/or peak water usage, so that the tower is half full. In this case, the depth of the water is half the full depth and the water pressure will be only half that of the full tower. This will be quite evident at the faucet. Now, imagine the same situation for the bulbous water tower. If it is half full, the water surface is still up in the round part of the tower. Thus, the depth of the water is still large and the water pressure is still high. This is a nice practical example of the fact that water pressure depends only on the depth and not on the cross sectional area.

10.) Farm silos have bands around them to provide sturdiness to the walls. Why are the bands closer together near the bottom of the silo than at the top?

This is a direct result of the fact that the pressure in the grain in the silo is larger at the bottom of the column of grain than at the top, just as for liquids. Thus, the silo must be stronger at the bottom to withstand the larger forces pushing outward.

11.) Scuba divers need to be wary of the "Bends". What are the "Bends"?

This condition occurs if the pressure outside the body is rapidly reduced. In this situation, nitrogen, which is in the blood and other body fluids, will form bubbles in body tissues and the blood stream. It is similar to the bubbles of carbon dioxide which form when a champagne bottle is opened. In the body, the nitrogen bubbles can cause tissue damage and/or circulation problems and can be fatal. Treatment is performed by increasing the outside pressure and driving the nitrogen back into solution. The pressure is then reduced slowly, allowing the nitrogen to escape slowly.

12.) Atmospheric pressure is 14.7 lb/in². If you multiply this by the surface area of your body in inches, you obtain a <u>huge</u> force. Why aren't we all crushed?

This calculation will give the force on the <u>outside</u> of the body. But there is pressure from the air on the inside, also, since humans are not sealed systems. Thus, the force from the inside pressure will balance the force from the outside.

13.) Take a look at a <u>non-retractable</u> ball point pen. Somewhere on the barrel you are very likely to find a small hole. What's this for?

The action of the pen depends on atmospheric pressure pushing on the top of the column of ink in the supply. If the pen were a sealed system, as ink was used up, the pressure inside the pen would drop and it would cease to write. The hole in the side of the pen connects the inside of the pen to the outside world and allows the pressure inside to remain at atmospheric pressure at all times.

14.) If you are caught in quicksand, what is the best way to escape?

The best way is to <u>lie down on your back</u>. This reduces the pressure on the quicksand by spreading your weight out over a large area. Your are then less likely to sink into the quicksand, in the spirit of Mystery #5. While lying on your back, you should slowly pull your feet out. If you pull quickly, the viscosity of the quicksand increases and holds you even tighter. Once your legs are free, you should roll to solid ground, in order to keep your weight spread out over as large an area as possible.

𝔐𝔞𝔤𝔦𝔠:

The Indestructible Balloon

The region just next to the valve and the dark region at the other end are areas of very low tension in the balloon material. This is why they are <u>dark</u>. The high tension regions are stretched and are thus thinner and more transparent. By pushing the needle through the dark regions, a small hole is produced and sealed by the petroleum jelly. The material does not tend to be pulled away from the hole, because of the low tension. If the needle is inserted in any other area, the high tension in the balloon material will pull the material away from the rupture and result in a violent tearing of the material.

The Non-Inflatable Balloon

As the balloon starts to blow up, it displaces air in the bottle. The pressure of the air in the bottle increases in response and this pressure on the outside of the balloon is enough to keep the balloon from inflating.

The Small but Powerful Balloon

The major contributing factor to the pressure in a balloon is not the amount of air, but the elastic forces applied by the rubber material of the balloon itself. The diagram below shows portions of the balloon material for two balloons blown up to different radii. When the balloon is small, the curvature is large but the radius of curvature of the surface is small. Thus, the components of the tension forces on a given area element toward

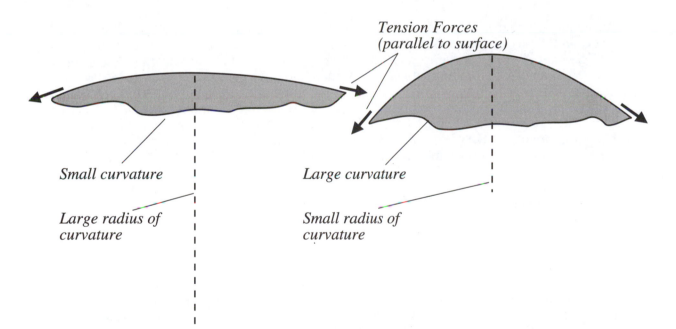

Tension Forces (parallel to surface)

Small curvature

Large radius of curvature

Large curvature

Small radius of curvature

the center of the balloon are large and the pressure in the balloon is large. This is why it is sometimes difficult to start blowing up a balloon, but it is easier to blow air into it once it is partially inflated. When the balloon is bigger, as on the left in the diagram, the radius of curvature is larger, and the components of the forces toward the center of the balloon are smaller, leading to a reduced pressure.

Thus, when the two balloons are attached, the higher pressure in the smaller balloon causes air to be pushed from the smaller balloon into the larger one.

The Way to Larger Balloons
The Way to Larger Marshmallows?
The Way to More Shaving Cream??

All three of these demonstrations exploit the difference between external and internal pressure. By use of the vacuum pump and bell jar, we have removed much of the air from the outside of the balloon, marshmallow or shaving cream, thus reducing the pressure. The pressure inside the balloon can then expand the balloon against the reduced outside pressure. In the marshmallow, there are small air bubbles (the word "puffed" is sometimes used to describe marshmallows!), which expand upon reducing the outside pressure. In the shaving cream, there are also air bubbles (to make it "foamy"!) which expand in the same way. Upon replacing the air, the balloon returns to its normal size, since the air inside can not diffuse through the rubber material very rapidly. The air in the marshmallow and shaving cream, however, does escape rapidly and the final state is different than the initial state, as indicated in the descriptions of these activities.

𝔐𝔶𝔱𝔥:

1.) A vacuum cleaner sucks up dirt.

A vacuum cleaner is one device that exhibits the common "sucking force" that is used freely in the physics layperson's vocabulary. Another such device is the soda straw. In reality, *there is no sucking force*. A vacuum or a reduced pressure area does <u>not</u> apply a force on an object that pulls it into the reduced pressure region. The correct conceptual treatment is that, because of the reduced pressure, there is less force on the piece of dirt on the side close to the vacuum cleaner. Thus, the force on the piece of dirt on the other side, due to the air farther from the vacuum cleaner, is no longer balanced. The net force, then, is toward and into the vacuum cleaner. Thus, the atmosphere actually <u>pushes</u> the dirt into the vacuum cleaner!

2.) Large force means large pressure.

Again, in everyday language, force and pressure are used quite freely and often interchangeably, so that a large force is interpreted to mean a large pressure. But we can easily think of counterexamples to this idea. As an example, consider the force and pressure of a phonograph needle on a record (For you younger readers, you may have heard of *records*, in which music was recorded in grooves on a black plastic disk!). A typical effective mass of the needle on the record is a few grams. If this is converted to an effective weight, which is then divided by the area of the tip of the needle, the result is several <u>hundred</u> atmospheres of pressure on the record. Thus, we have a tiny force resulting in a huge pressure. A reverse example to this is the use of snowshoes. Here, the large weight of a person is spread out over a large area, so that the pressure is reduced enough so that the walker does not break through the surface of the snow. This is similar to Mystery #14, in which the force of the body is spread out over a large area of quicksand.

The stapler whose picture accompanies this myth is another example of a large pressure associated with a small force. A large pressure is necessary to cause a piece of metal to pierce a pile of papers. This large pressure is supplied by a relatively small force in the stapler, due mainly to the sharpness of the staple points so that the force is concentrated over a very small area, similar to the situation in Mystery #4.

3.) A powerful superhero could drink through a 50 foot tall straw.

As mentioned in the explanation of the Myth #1 above, a straw is a device that is commonly thought of as exerting a "sucking force". But, once again, this does not exist. When using a straw, the drinker creates a low pressure area inside the straw. The surface of the liquid inside the straw then sees an unbalanced force - a large upward force from the pressure of the liquid due to the atmosphere pushing on the surface outside of the straw and a smaller downward force due to the reduced pressure inside the straw. Thus, there is a net force upward and the liquid rises in response. It is actually the <u>atmosphere</u> that pushes the liquid up the straw! Now, what about the superhero? Let us assume that he or she could create a perfect vacuum inside the straw (and we also assume that this is a thick-walled straw that would not collapse due to the difference in pressure between the inside and the outside!). Then, the atmosphere will push the liquid up the straw. But, the force due to the atmosphere on the liquid is limited by the value of atmospheric pressure. As it turns out, atmospheric pressure is sufficient to raise water up an evacuated tube to a height of about 34 feet. Under these conditions, the weight of the water column balances out the upward force from the atmospheric pressure. Nothing can be done to overcome this limit set by atmospheric pressure. The superhero can suck as hard as possible, but he or she cannot make the liquid rise any more than 34 feet. The superhero is not pulling on the liquid - the atmosphere is pushing!

4.) A *cyclone* is a violent rotating windstorm, like a tornado.

This is another term which is used very often and incorrectly in the layperson's vocabulary. The word *cyclone* refers to the normal, gentle rotation of air around a low-pressure area in the atmosphere. It does not refer to a violent storm of any sort, although falling barometric pressure often signals an upcoming rainstorm.

5.) A fluid will flow from a region of high pressure to a region of lower pressure.

In many cases, this is true, but fluid flow is affected by another force besides that due to the pressure difference - gravity! Imagine that we have a pipe of uniform cross section, carrying an incompressible fluid, and that at some section of the pipe, the level drops:

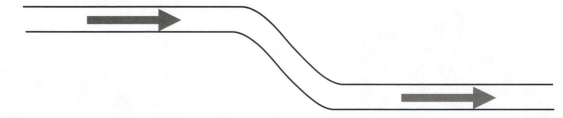

According to the equation of continuity for fluids (see Chapter 13), the cross sectional area has not changed, so that the velocities in the upper and lower portions of the pipe must be the same. Then, from Bernoulli's Principle (see Chapter 13 again), we see that, since the height of the fluid decreased, the pressure must have increased in going from the upper pipe to the lower pipe. Thus, this fluid is flowing from <u>low</u> pressure to <u>high</u> pressure!

6.) The unit of pressure is the inch. That's what the weather announcer says, right?

It is indeed true that weather reports often give the barometric pressure in inches, but that is not a valid unit for pressure. The more complete description would be "inches of mercury", referring to the height of a column of mercury in an evacuated barometer. Thus, normal atmospheric pressure would result in a column of mercury 29.9 inches or 760 mm high. This is similar to what the superhero was doing in Myth #3. He or she was creating a vacuum in the straw and atmospheric pressure raised the column of water to 34 feet. Mercury barometers are a little less cumbersome than water barometers, since the higher density of mercury results in a shorter barometer!

As mentioned in the Concepts section, the correct SI unit for pressure is the pascal (Pa).

Chapter 13
Fluids

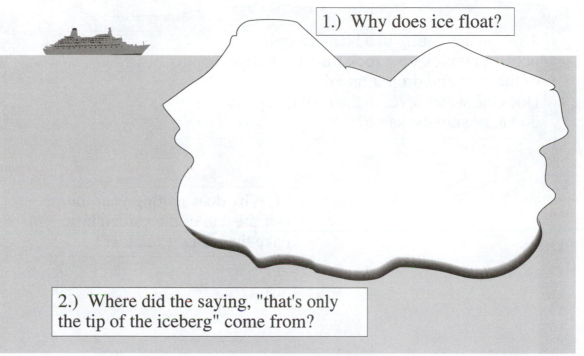

1.) Why does ice float?

2.) Where did the saying, "that's only the tip of the iceberg" come from?

3.) You are standing on a scale when all of the air is suddenly sucked out of the room. Bravely, you look at the reading on the scale before you run out of the room in search of breath. Was the reading with the air removed higher, lower or the same as that before the air was removed?

𝕸ysteries:

4.) You are sitting in a boat in a quiet pool. You pick up a rock from the bottom of the boat and drop it into the water. Does the water level in the pool go up, go down, or stay the same?

5.) Why does putting your thumb over the end of the garden hose make the water go farther?

6.) Why do golf balls have dimples?

𝔐ysteries:

7.) When water is coming out of the faucet in a nice, steady stream, the stream gets smaller at the bottom. Why?

8.) Why does smoke from a cigarette or incense stick rise smoothly and then start to swirl?

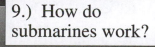

9.) How do submarines work?

𝔐agic:

The Non-Spilling Water

Cover the top of a narrow-necked bottle with cheesecloth and hold it in place with a rubber band. Now fill the bottle with water and invert it. The cheese-cloth will prevent the water from leaving the bottle!

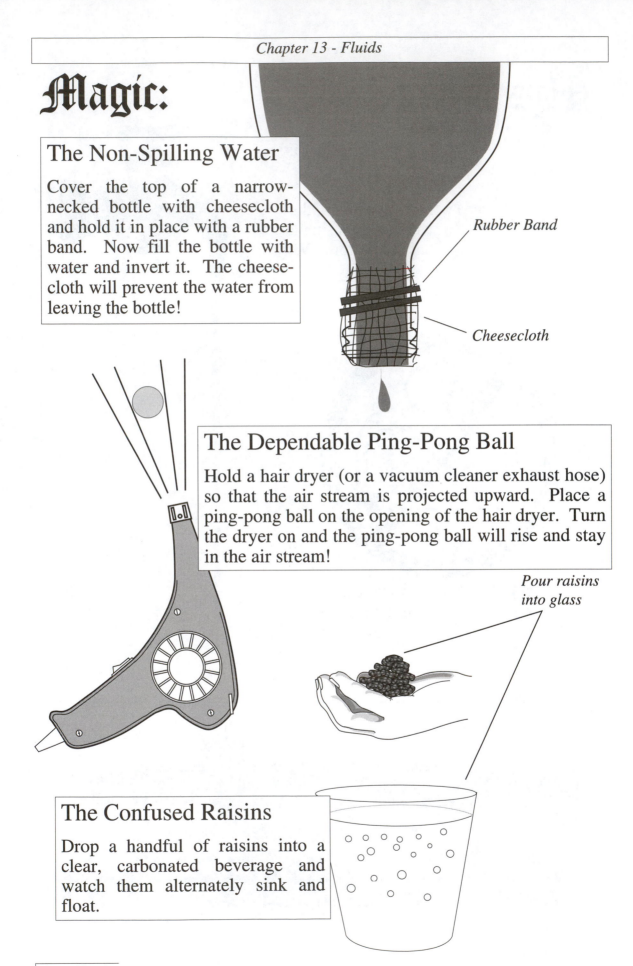

Rubber Band

Cheesecloth

The Dependable Ping-Pong Ball

Hold a hair dryer (or a vacuum cleaner exhaust hose) so that the air stream is projected upward. Place a ping-pong ball on the opening of the hair dryer. Turn the dryer on and the ping-pong ball will rise and stay in the air stream!

Pour raisins into glass

The Confused Raisins

Drop a handful of raisins into a clear, carbonated beverage and watch them alternately sink and float.

𝕸agic:

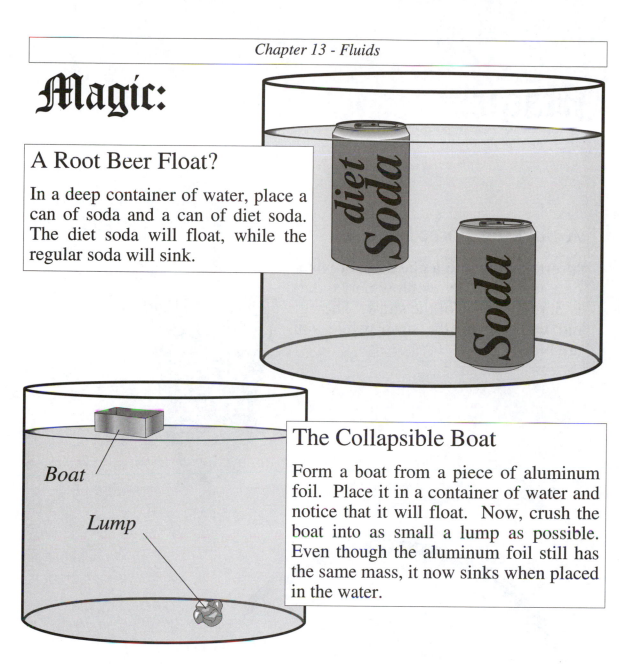

A Root Beer Float?

In a deep container of water, place a can of soda and a can of diet soda. The diet soda will float, while the regular soda will sink.

Boat

Lump

The Collapsible Boat

Form a boat from a piece of aluminum foil. Place it in a container of water and notice that it will float. Now, crush the boat into as small a lump as possible. Even though the aluminum foil still has the same mass, it now sinks when placed in the water.

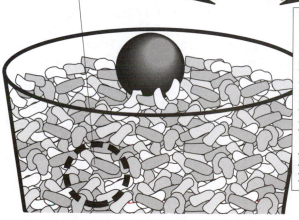

Secretly sink Ping-Pong ball beforehand

Shake back and forth

Magical Beans?

Fill a jar with dried pinto beans. Now push a ping pong ball below the surface of the beans. Place a steel ball on top of the beans and shake the jar. The pinto beans will turn a steel ball into a ping pong ball!

Magic:

Blow across straw

A Sucking Force?

Insert a straw into a glass full of water or a soft drink. Now blow horizontally across the top of the straw. The liquid will rise up the straw while you are blowing!

The Attractive Cereal Force

Place two non-flake-type pieces of cereal (loops, letters, puffs, etc.) in a bowl of water or milk. If the pieces are close to one another they will be attracted toward each other and will stick together. What causes this strange attraction?

𝔐𝔶𝔱𝔥:

Cleaning Fluid...

1.) A fluid is a liquid.

Lighter Fluid....

Transmission Fluid.....

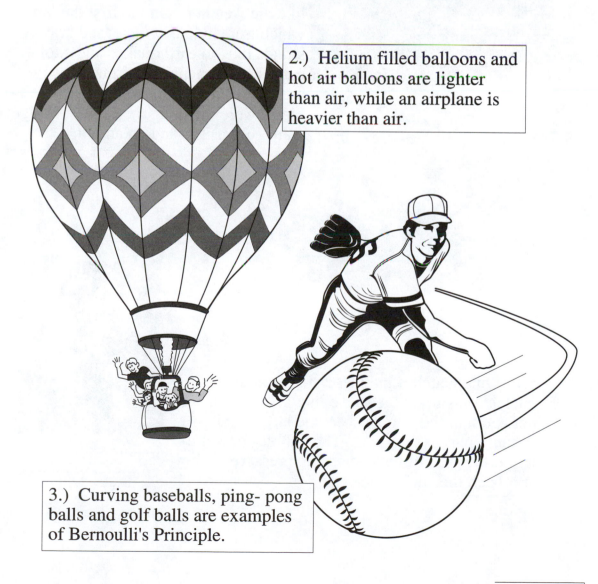

2.) Helium filled balloons and hot air balloons are lighter than air, while an airplane is heavier than air.

3.) Curving baseballs, ping-pong balls and golf balls are examples of Bernoulli's Principle.

𝕸𝖞𝖙𝖍:

4.) Kurt Vonnegut's novel, *Slapstick* (Dell Publishing Co., New York, 1976), is a curious story of an unusual man in an unusual world of the future. In this world, gravity varies from day to day like the weather. On the first day on which tremendously increased gravity strikes like an earthquake, "elevator cables were snapping, airplanes were crashing, ships were sinking, motor vehicles were breaking their axles, bridges were collapsing, and on and on".

5.) If one watches bubbles rising from a nucleation site on the side of a glass of beer, it is noticed that the bubbles become larger and farther apart as they rise. The bubbles become larger because of the decreasing pressure as they approach the surface.

𝔐𝔶𝔱𝔥:

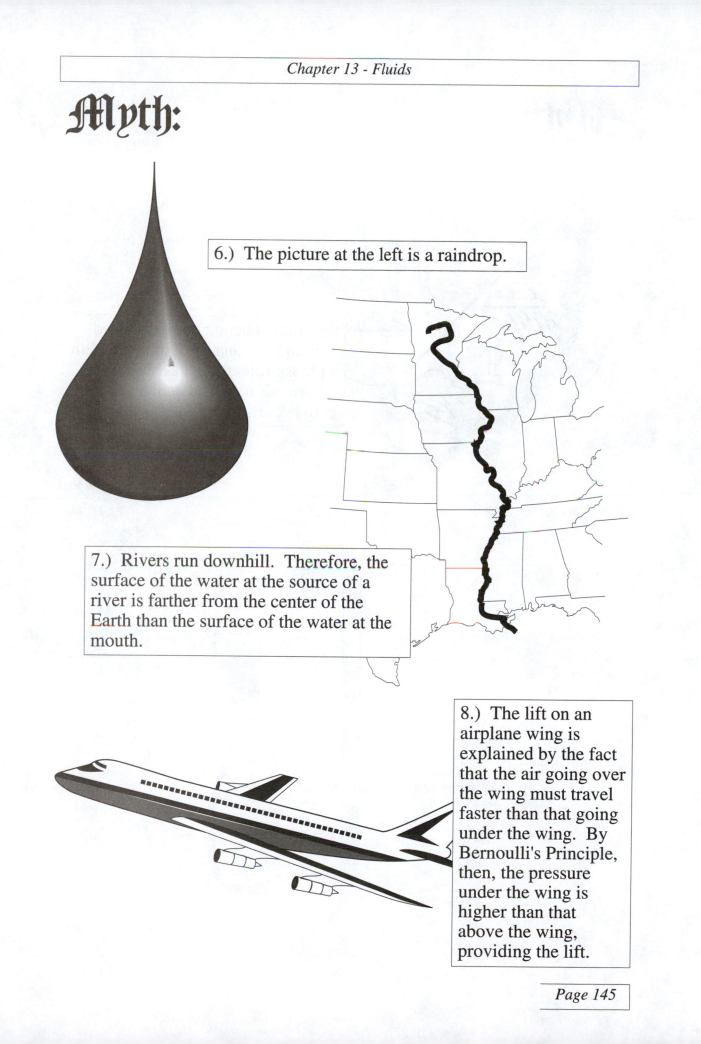

6.) The picture at the left is a raindrop.

7.) Rivers run downhill. Therefore, the surface of the water at the source of a river is farther from the center of the Earth than the surface of the water at the mouth.

8.) The lift on an airplane wing is explained by the fact that the air going over the wing must travel faster than that going under the wing. By Bernoulli's Principle, then, the pressure under the wing is higher than that above the wing, providing the lift.

Myth:

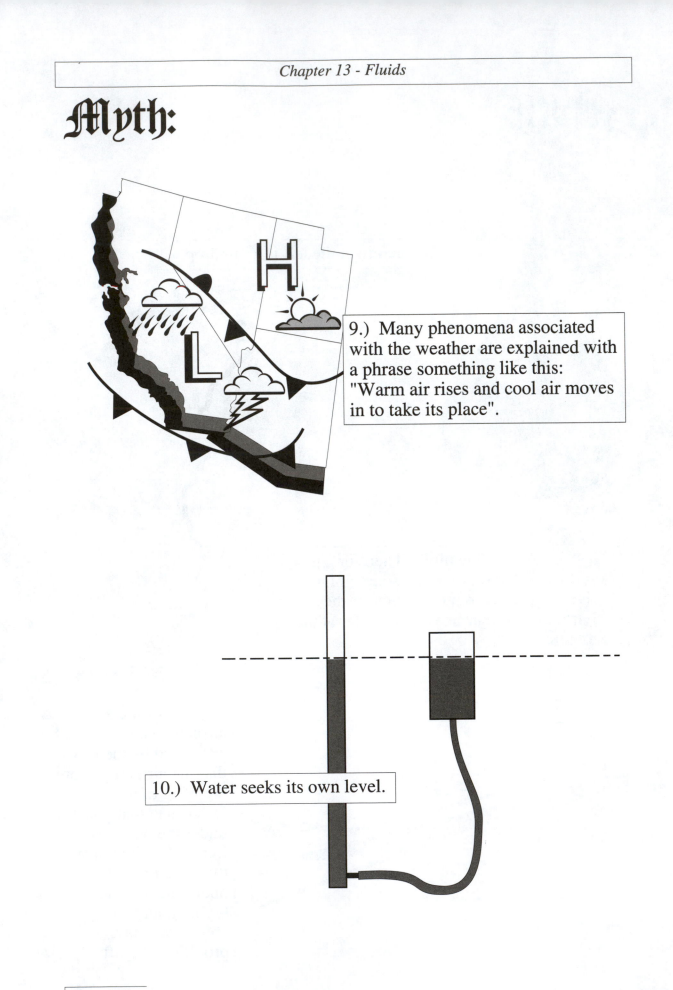

9.) Many phenomena associated with the weather are explained with a phrase something like this: "Warm air rises and cool air moves in to take its place".

10.) Water seeks its own level.

Concepts of Fluids

Fluid Statics

Following directly from the variation of pressure with depth (discussed in Chapter 12) is *Archimedes Principle:* When

> *The buoyant force is exerted in the upward direction by the fluid.*

an object is immersed in a fluid, it feels an upward (*buoyant*) force equal to the weight of the fluid displaced by the object. If the fluid is incompressible, then the buoyant force is given by,

$$F_B = \rho_{fluid} g V_{disp}$$

where ρ_{fluid} is the fluid density, g is the acceleration of gravity and V_{disp} is the volume of fluid displaced by the object. Now, the weight of the object is

$$W = mg = \rho_{obj} g V_{obj}$$

where ρ_{obj} is the object density, g is the acceleration of gravity and V_{obj} is the volume of the entire object. If an object is <u>completely immersed</u> in the fluid and <u>released</u>, then the buoyant force and the weight are the two forces that act on the object. The final disposition of the object is determined by which force is larger. Since we have completely immersed the object, the two volumes (V_{disp}, V_{obj}) are the same. The acceleration of gravity is the same in both forces. Thus, the ensuing result depends only on the relationship between the densities:

$$\rho_{obj} > \rho_{fluid} \implies W > F_B \implies \text{object sinks downward}$$
$$\rho_{obj} < \rho_{fluid} \implies W < F_B \implies \text{object floats upward}$$

The sinking object will eventually reach the bottom of the container, where the normal force from the surface of the bottom will assist the buoyant force in counteracting the object's weight. Thus, in this final situation,

$$N + F_B - W = 0$$

For the case of the object which floats to the surface, the buoyant force and the weight will continue to be the only two forces that act on the object. In the expression for the buoyant force, however, the volume V_{disp} is just that part of the volume of the object that is under the fluid as it floats on the fluid surface, which, we can show, is that fraction of the volume that is the same as the ratio of the fluid density to the object density. Thus, as it floats on the surface, the equilibrium situation is described by,

$$F_B - W = 0$$

so that the weight and the buoyant force just balance.

Fluid Dynamics

The first principle in fluid dynamics is the equation of continuity, which we discussed in Chapter 7. This principle, as applied to fluids, is fundamentally a state-

Concepts of Fluids (continued)

ment of conservation of material - <u>the fluid that flows into a region of space minus that flowing out will equal the change in the amount of fluid in the region</u>. For an incompressible fluid, the amount of material cannot be increased in a region of space, so the volume rate of flow into the region must equal the rate of flow out. Necessarily, then, if the fluid must pass through a cross sectional area of changing size, the speed will increase through smaller areas and decrease through larger areas. If the fluid is incompressible, then the mathematical statement is:

> *If the fluid must move through a constriction, it must move faster.*

$$v_1 A_1 = v_2 A_2$$

where the v's and A's correspond to the velocities and areas in two places that are being compared.

The second principle is <u>Bernoulli's Principle</u>, which states that <u>the pressure in a moving fluid decreases as the velocity increases</u> and vice versa. For an incompressible fluid, the mathematical statement of Bernoulli's Principle is the following, where we are comparing two points in the fluid flow:

> *If the fluid velocity goes up, its pressure goes down.*

$$P_1 + \rho g h_1 + \frac{1}{2}\rho v_1^2 = P_2 + \rho g h_2 + \frac{1}{2}\rho v_2^2$$

In this equation, P is the pressure in the fluid, ρ is its density, g is the gravitational field, h is the height of the fluid above some reference point and v is the velocity of the fluid. Notice that changes in height will cause a change in the pressure, but in most situations involving Bernoulli's Principle, it is the dependence of pressure on <u>velocity</u> that is of interest.

Discussions; Chapter 13- Fluids

𝔐𝔶𝔰𝔱𝔢𝔯𝔦𝔢𝔰:

1.) Why does ice float?

Water behaves anomalously in that it <u>expands</u> upon cooling between 4°C and 0°C. Thus, ice is less dense than water at 0°C. Accordingly, if the ice is completely immersed, the buoyant force is larger than the weight and the ice will rise to the surface of the water and float. This is good news for fish, since lakes will freeze from the top down and the top layers of ice represent thermal insulation against further freezing of water. Imagine the poor plight of fish if lakes froze from the bottom up. It is also good news for ice skaters, since they would like to skate on the top of the lake, not the bottom!

2.) Where did the saying, "that's only the tip of the iceberg" come from?

Following up on the answer to Mystery #1, the density of ice is about 92% of that of water. Thus, according to Archimedes' Principle, ice will float in equilibrium with 92% of its volume submerged. For ice floating in salt water, as in the oceans, 89% of the iceberg will be submerged, since ice has a density of 89% of that of seawater. Correspondingly, only 8% of the ice will be visible above the fresh water, or 11% above seawater. This is the origin of the saying, since the viewing of a small amount of ice above water represents much more mass of ice below.

3.) You are standing on a scale when all of the air is suddenly sucked out of the room. Bravely, you look at the reading on the scale before you run out of the room in search of breath. Was the reading with the air removed higher, lower or the same as that before the air was removed?

The reading on the scale will be <u>higher</u> with the air removed. When the air was in the room, you were immersed in a fluid. Thus, there was a buoyant force on you in the upward direction, which reduced the reading on the scale below your actual weight. When the air is removed, the buoyant force vanishes and the reading goes up (although the effect is quite small!).

4.) You are sitting in a boat in a quiet pool. You pick up a rock from the bottom of the boat and drop it into the water. Does the water level in the pool go up, go down, or stay the same?

The water level will go <u>down</u>. When the rock is in the boat, its weight is supported by displacing water, causing the water level to be higher than that with no rock. Since water is less dense than the rock, the volume of water displaced to support it in the boat is larger than the volume of the rock itself. As the rock is falling through the air, the water level in the pool goes down, since the rock is no longer supported. As the rock enters the water, it displaces its volume, but, as mentioned before, this is less than the volume of water displaced when the rock was in the boat. Thus, the water level does not rise back to its previous level and the net movement of the level from the original position is <u>down</u>.

5.) Why does putting your thumb over the end of the garden hose make the water go farther?

This is an example of the Equation of Continuity in action. By putting your thumb over the end of the hose, you force the water to move through a smaller cross sectional area. As a result, according to the Equation of Continuity, it must increase its speed. The water projected with a higher velocity can then be sprayed farther.

6.) Why do golf balls have dimples?

An important part of a golf drive is the lift that is provided by the spinning of the ball (see Myth #3 for the details of this effect). This lift is provided by friction between the ball and the air. The dimples provide additional friction over a smooth ball. Various dimple designs have been researched for many years. Bartlett's World Golf Encyclopedia (1975), states that "a dimple depth of 0.012 inches produces the most distance". For more information on golf ball flight, see D. Griffing, *The Dynamics of Sports*, The Dalog Company, Oxford, Ohio, 1987, H. Erlichson, "Maximum Projectile Range with Drag and Lift, with Particular Application to Golf", *American Journal of Physics*, **51**, 357 (1983) and W. M. MacDonald and S. Hanzely, "The Physics of the Drive in Golf", *American Journal of Physics*, **59**, 213 (1991).

7.) When water is coming out of the faucet in a nice, steady stream, the stream gets smaller at the bottom. Why?

This can be explained by Bernoulli's Principle. As the water falls, it accelerates. As the velocity increases, according to Bernoulli's Principle, the pressure in the stream decreases. Thus, as the water falls, the difference in pressure between the inside of the stream and the atmospheric pressure outside increases. As a result, the water is pushed into a thinner stream by the atmosphere as it falls. Eventually, the stream gets so thin that any slight variations in the diameter of the stream will allow surface tension to take over and the steady stream degenerates into individual droplets.

This effect can also be argued from the continuity equation. Since the water falls faster in the lower portions of the stream, the cross sectional area of the stream becomes smaller.

8.) Why does smoke from a cigarette or incense stick rise smoothly and then start to swirl?

This is Mystery #7 turned upside down! The hot smoke rises due to Archimedes' Principle, since it is less dense than air. As it rises it accelerates, initially in streamline motion. As the velocity increases, the smoke stream gets thinner. Eventually, the velocity is so high that turbulence sets in and the stream breaks up into swirls that are at the mercy of air currents in the room.

9.) How do submarines work?

Submarines are beautiful examples of Archimedes' Principle in action. When floating, the overall density of the submarine is less than that of water, just as for a normal ship. When the diving process is initiated, water is allowed to flood into holding tanks in the submarine. As a result, the density of the submarine becomes greater than that of water, and the submarine sinks. With fine control over the overall density, the submarine can travel at a given depth in the water.

𝔐𝔞𝔤𝔦𝔠:

The Non-Spilling Water

This is a demonstration of the <u>surface tension</u> of water. As the water wets the cheese-cloth, it fills the holes between the threads. Since each hole is fairly small, the surface tension of the water forms a "skin" over each hole that is sufficiently strong to hold back the water in the bottle. This is related to the situation with ink in a ball point pen (Mystery #13, Chapter 12). The ink must pass through a narrow opening to leave the ink supply. It is normally prevented from doing so by the surface tension of the ink, just like the water is prevented from leaving the bottle in this demonstration. When the pen is stroked across a piece of paper, the friction breaks the surface tension and the ink can flow.

The Dependable Ping-Pong Ball

This demonstrates Bernoulli's Principle. In the center of the air stream, the velocity of air is high and, therefore, the pressure is low. To the sides of the air stream, the air is still and the pressure is high. Thus, if the ball starts to drift to the side, the high pressure pushes the ball back in.

The Confused Raisins

As the raisins sit in the soft drink, bubbles of carbon dioxide will attach to them. Eventually, enough bubbles will have attached to make the overall density of the raisin-bubble system less than that of the liquid. According to Archimedes' Principle, then, the raisin will float to the surface. At the surface, the bubbles are given off to the atmosphere, the density returns to that of the raisin alone and the raisins sink to the bottom again, ready for another cycle.

A Root Beer Float?

The diet soda is sweetened with an artificial sweetener which is less dense than sugar. The overall density of the can with the artificially sweetened beverage is just less than water, so the can floats. The overall density of the can with sugar is just larger than water, so the can sinks. As a side note, if this demonstration is attempted with beer and light beer, both will float, since the density of alcohol is low enough to make both cans less dense than water.

The Collapsible Boat

This demonstration shows the dependence of the buoyant force on the volume of the water displaced by the object. When in the boat shape, the volume of the aluminum foil is large, so that a relatively large amount of water is displaced and the buoyant force is sufficient to support the boat. When the foil is crushed into a ball, the volume of the object is smaller, the buoyant force is less, and the crumpled ball sinks. You may have to pinch the foil with pliers to drive enough air out to make the overall density less than that of water. Alternately, you could float a high-density object (e.g., keys) in the boat and then wrap it up when you crush the boat. This will remove the necessity for such tight squeezing of the foil.

Magical Beans?

While the beans are shaking, two processes are occurring. The first is the tendency for large objects to move upward due to the higher probability of small openings occurring, as discussed in "The Anti-Gravity Brazil Nuts" in Chapter 6. The second is the combination of the buoyant force and the weight of the object, which will give a net force either up or down, depending on the relative density of the beans and the ball. For the steel ball, the difference in densities is very large, so that the weight is easily able to overcome the buoyant force and the upward tendency due to probability, and the ball sinks. For the ping-pong ball, the weight is so small that the buoyant force and the upward tendency due to probability are able to push the ball up to the surface. For objects which are close to the density of beans, the situation becomes murky. Winter (R. G. Winter, "On the Difference Between Fluids and Dried Beans", *The Physics Teacher*, **28**, 104 (1990)) finds that a large (relative to the beans) aluminum ball (with density about three times that of the beans) <u>rises</u> when it is "submerged" and the system is shaken. A bean-sized piece of aluminum (with the same density as the ball) rapidly <u>sinks</u>, however. This demonstrates clearly the effect of the probability discussion above on the large size ball.

A Sucking Force?

This is Bernoulli's Principle in action. The air moving across the top of the straw creates a low pressure area over the straw. Then, just as when you suck on the straw, the atmosphere is able to push the liquid up the straw. If you blow hard enough, you may be able to actually spray some liquid. This is the principle behind the "atomizer" bottle.

The Attractive Cereal Force

This is another example of the results of surface tension, which is a force pulling "inward" on a fluid surface. At a boundary between the fluid and a solid, the fluid will exert a force parallel to its surface due to surface tension. If we consider the water (or milk) surface and a floating piece of cereal, the water is flat and horizontal far away from the cereal. As we move closer, the water surface curves upward in a meniscus, since the water wets the cereal. There is a similar meniscus on the other side of the piece of cereal, so that the two forces on the cereal from the water surface balance and the piece of cereal remains in equilibrium. The first diagram below shows the Newton's Third Law forces between the water surface and the cereal as well as the two forces on the piece of cereal providing the equilibrium situation.

Force on water surface due to left side of piece of cereal

Force on water surface due to right side of piece of cereal

Piece of Cereal

Water Surface

Force on left side of piece of cereal due to water surface

Force on right side of piece of cereal due to water surface

Now imagine that two pieces of cereal come close to one another. As the menisci overlap, the water surface between them flattens and the force from the water surface on the piece of cereal near the other piece becomes more horizontal than before. Now, the horizontal components of the two forces on the two sides of the piece of cereal are unbalanced and the piece moves in the direction of the more horizontally directed force, as shown below (only the forces on the left piece of cereal are shown).

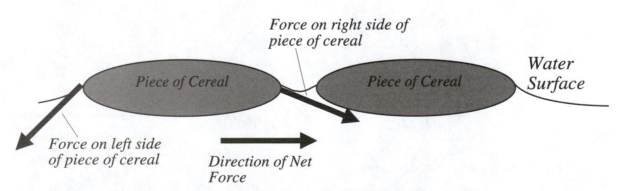

Force on right side of piece of cereal

Water Surface

Piece of Cereal

Piece of Cereal

Force on left side of piece of cereal

Direction of Net Force

The same effect can be seen if a piece of cereal arrives close to the edge of the bowl - it will stick to the edge.

𝔐𝔶𝔱𝔥:

1.) A fluid is a liquid.

Although many liquids are called fluids, the root of the word *fluid* is *flow*. Since both liquids and gases can flow, we use the word fluid to describe both liquids <u>and</u> gases.

2.) Helium filled balloons and hot air balloons are lighter than air, while an airplane is heavier than air.

The phrase, "a balloon is lighter than air" is meaningless, since it compares the weight of a definite object (the balloon) to that of a substance (air). Without specifying how much air we are talking about, we cannot assign a weight to air. The more correct statement would be "a balloon is less dense than air", although it will be difficult to convince the public to make this change! More information about the physics of the hot air balloon can be found in O. E. Haugland, "Hot-Air Ballooning in Physics Teaching", *The Physics Teacher*, **29**, 202 (1991).

3.) Curving baseballs, ping pong balls and golf balls are examples of Bernoulli's Principle.

These examples are often given by physics teachers as examples of Bernoulli's Principle. The argument is made that the spinning balls drag a layer of air around with them and the interaction of the layer of air with the outside air results in a difference of pressure that causes the ball to curve. This argument gives a <u>qualitative</u> result that describes the correct direction of the curve. But it cannot provide a correct <u>quantitative</u> description. The more correct explanation involves Newton's Third Law rather than Bernoulli's Principle. If we imagine a baseball with air flowing past it and not spinning, the path of the air is as shown below, as in D. Griffing, *The Dynamics of Sports*, The Dalog Com-

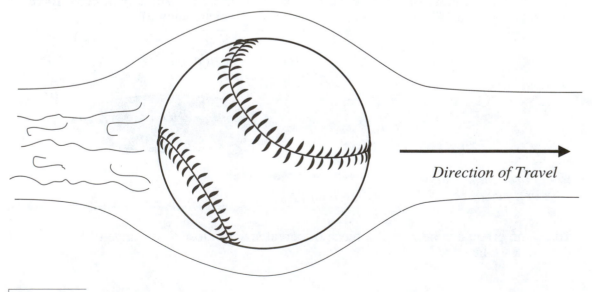

Direction of Travel

pany, Oxford, Ohio, 1987. Notice that the air flow is symmetric about the direction of travel of the ball which is from left to right in the diagram. Shown behind the ball is the turbulent wake formed by the passage of the ball through the air.

Now, imagine that the ball is spinning in the counterclockwise direction, as in the diagram below. Friction between the ball and the air will sweep air up over the top and throw it off the back, in the downward direction in the diagram. As a result, the turbulent wake will not be directly behind the ball, but will be directed downward:

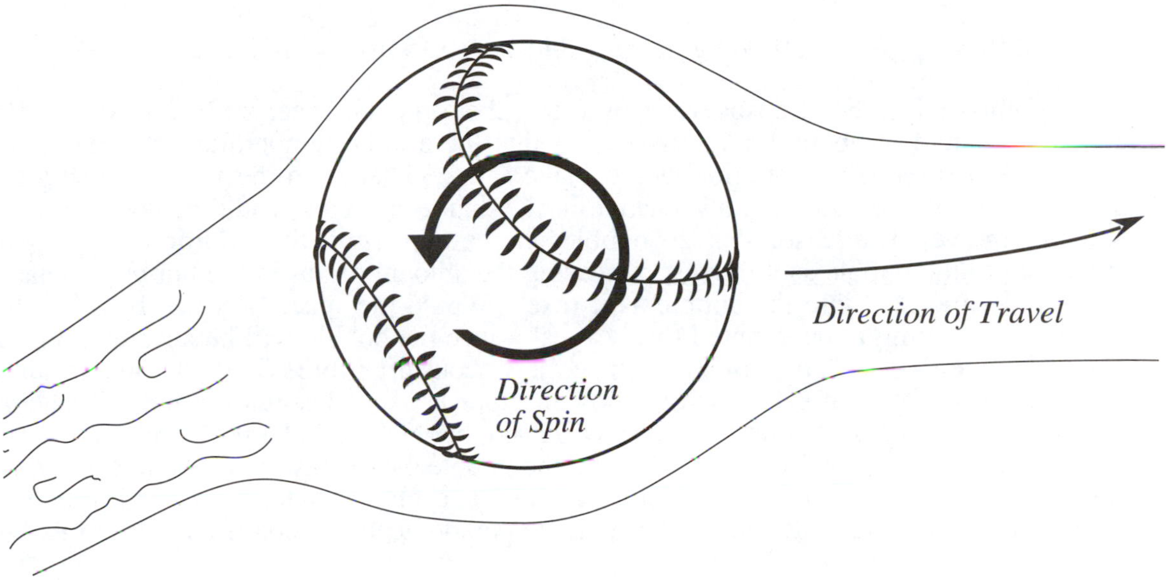

Direction of Spin

Direction of Travel

Thus, the ball exerts a downward force on the air in the above diagram. By Newton's Third Law, then, the air must exert an upward force on the ball, which causes it to curve from its original path. For more details on the interactions of sports balls and the air, see the Griffing text cited above.

4.) Kurt Vonnegut's novel, *Slapstick* (Dell Publishing Co., New York, 1976), is a curious story of an unusual man in an unusual world of the future. In this world, gravity varies from day to day like the weather. On the first day on which tremendously increased gravity strikes like an earthquake, "elevator cables were snapping, airplanes were crashing, ships were sinking, motor vehicles were breaking their axles, bridges were collapsing, and on and on".

We focus here on the phrase, "...ships were sinking...", in the list of happenings. As seen in the discussion of concepts of fluids, both the buoyant force on an object and the weight of an object are proportional to g, the acceleration of gravity. Thus, if gravity were to be increased, the weight of the ship would increase, but so would the buoyant force holding it up. Thus, ships would continue to float. People on the ships, however, would notice the difference, as it would be much harder to get up out of the deck chairs!

5.) If one watches bubbles rising from a nucleation site on the side of a glass of beer, it is noticed that the bubbles become larger and farther apart as they rise. The bubbles become larger because of the decreasing pressure as they approach the surface.

The change in pressure is not the cause of the increase in bubble radius. If we estimate the longest distance that a bubble could rise in the glass to be roughly the size of the glass, we can assign a distance of about 15 cm. Thus, the pressure change from the bottom of the glass to the top is,

$$\Delta P = \rho g h = (1000 \text{ kg·m}^{-3})(9.8 \text{ m·s}^{-2})(0.15 \text{ m}) = 1470 \text{ Pa}$$

This, compared to the atmospheric pressure at the top of the beer surface (1.013×10^5 Pa) is a change of about 1.5%. Assuming that there is no temperature change in the bubble as it rises to the top, this would make a 1.5% change in the volume (using the Ideal Gas Law), or about a 0.5% change in the diameter. This would not be visible to the human eye. The reason that the bubbles increase in size is that <u>additional liquid vaporizes into the bubble as it moves</u>, increasing the amount of gas in the bubble. What if no liquid vaporized into the bubble as it rose? Would the space between bubbles still increase? You might be tempted to say "yes", since the bubble will be accelerating upward. The answer is "no", however, since the viscous drag force on the bubble is sufficient to bring it to terminal velocity almost instantaneously. The reason that the spacing increases is related to the increase in size. As the bubble grows larger by additional vaporization, the buoyant force increases and the bubble moves faster. Of course, the drag almost immediately brings it to terminal velocity at each new size. Thus, as it grows in size it moves upward faster and faster with the paradoxical concept of an increasing, but almost always terminal, velocity!

There is a great deal of fascinating information about the physics of beer and many other everyday situations in C. F. Bohren, *Clouds in a Glass of Beer*, Wiley, New York, 1987.

6.) The picture at the left is a raindrop.

The shape shown is typical for a drawing used in layperson's publications to represent a raindrop. This is inconsistent with the effects of surface tension, however, which will reduce the energy of the drop by pulling it into a <u>spherical</u> shape, which has the smallest surface to volume ratio. This spherical shape is crucial to the explanation of the *rainbow* in Chapter 26 and the *glory* in Chapter 29. The effect of air resistance on the falling drop is not to "leave behind" a tail as in the drawing, but rather to flatten the raindrop on the bottom so that it has a shape similar to a hamburger bun. The physics by which a raindrop achieves this shape is available in J. E. MacDonald, "The Shape of Raindrops", *Scientific American*, **190(2)**, 64 (1954).

7.) Rivers run downhill. Therefore, the surface of the water at the source of a river is farther from the center of the Earth than the surface of the water at the mouth.

This seems like a perfectly reasonable statement, but let's look at some data. If we take measurements of the Mississippi River (highlighted in the diagram accompanying the Myth), we find that its source (at Lake Itasca, in northwestern Minnesota) is 5 km <u>closer</u> to the center of the Earth than the mouth (emptying into the Gulf of Mexico in south-eastern Louisiana). This is due to the fact that <u>the Earth is not a sphere but is slightly flattened</u> due to the effects of its rotation and Newton's First Law (as discussed in Myth #1, Chapter 11). Thus, the equatorial diameter is some 21 km larger than the polar diameter. Let us imagine that we slice through the Earth with a "vertical" cut (that is, the cut goes through the poles) at the longitude of the Mississippi River. Due to the equatorial bulge, the cross section of the Earth will be an <u>ellipse</u> (ignoring variations due to mountains, valleys, etc.). If we analyze this elliptical shape with a 21 km difference between the major and minor axes, we find that the difference in radial distance from the center at the latitudes of the source and mouth of the Mississippi is about 7.5 km, with the source distance smaller. Our data tells us that the actual difference is 5 km. Thus, the difference between these numbers, 2.5 km, is the "head" of water that makes the river flow in the direction that it does.

8.) The lift on an airplane wing is explained by the fact that the air going over the wing must travel faster than that going under the wing. By Bernoulli's Principle, then, the pressure under the wing is higher than that above the wing, providing the lift.

The lift on an airplane wing is a common example of the use of Bernoulli's Principle in the physics classroom. Unfortunately, the lift on an airplane wing is not due to Bernoulli's Principle. If you believe in the Bernoulli's Principle argument, ask yourself how an airplane can fly upside down. The lift is due to Newton's Third Law. The shape of the airfoil (the wing) is such that air encountering the wing is deflected downward. Since the wing applies a force downward on the air, the air exerts a force upward on the wing - hence, lift. For more details on the physics and mathematics as applied to the air streams moving past the wing, see K. Weltner, "Aerodynamic Lifting Force", *The Physics Teacher*, **28**, 78 (1990) and K. Weltner, "Bernoulli's Law and Aerodynamic Lifting Force", *The Physics Teacher*, **28**, 84 (1990).

9.) Many phenomena associated with the weather are explained with a phrase something like this: "Warm air rises and cool air moves in to take its place".

This phrase, and others that are similar, indicate that it is a natural property of warm air to rise. This supposedly leaves behind an evacuated space, into which the cool air can move. But why does the warm air rise? It does so because it expands upon heating. Thus, it is less dense than the cooler air and "floats" upward by Archimedes' Principle. The cooler air is not simply moving in to take its place, <u>the cool air is what is pushing the warm air upward</u>, by means of a buoyant force.

10.) Water seeks its own level.

It is often heard that water "seeks its own level" when we open a connection between two bodies of water. It should be made clear, however, that this is <u>not</u> a fundamental principle. When water does "seek its own level", it is because of a much more fundamental concept and, as we shall see, there are times when, because of this fundamental concept, it does not "seek its own level"!

The fundamental concept that must be applied here is this - the pressure of an incompressible fluid is a function of fluid density, the gravitational field and depth, as seen in the Concepts section. Now, when we provide a connection between two bodies of water with different heights, the water from the body with the higher level will have a larger pressure at the connection and will generally flow into the body with the lower level until the pressures equalize. At this point, there is no further force causing the water to flow. If the water in the two initial bodies is of the same type, then equalizing the pressure would be equivalent to equalizing the water level, as shown in the diagram below. This diagram shows two containers of water, separated by the black barrier and connected by a valve at the bottom. The water levels are initially unequal with the valve closed, as shown on the left. When the valve is opened, the difference in pressure causes water to flow until we have the situation shown on the right, with the water levels equal.

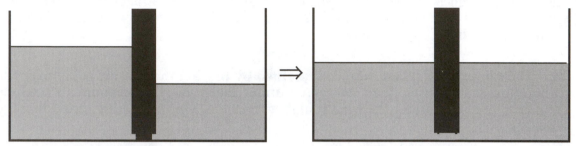

Valve Closed   *Valve Open*

But suppose we repeat this process with two initial bodies of fluids with <u>different</u> <u>densities</u>. Then, since the pressure depends on both density <u>and</u> depth, the pressure will equalize with the two types of fluid at <u>different</u> levels (as long as there is insignificant mixing of the two fluids), as shown in the following diagram:

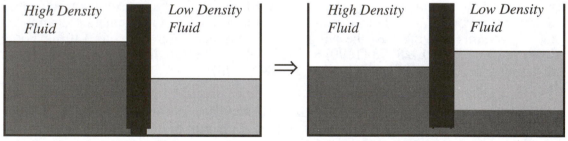

Valve Closed   *Valve Open*

Thus, the equalization of the levels is not the fundamental principle - it is the equalization of the pressures. Now, this might sound like cheating, since we have introduced <u>two</u> fluids, while the Myth referred only to <u>water</u>. But suppose the two fluids in the diagram above are fresh water and seawater. Then, the water will not seek its own level. Since salt water is denser than fresh water, performing the demonstration in the above diagram will result in the level of salt water being lower than the level of fresh water.

Just such a situation occurs in the final lock of the Panama Canal. As a ship is brought down from Gatun Lake to sea level, the last lock is filled with fresh water from the lake and a valve allows the fresh water to exit to the sea, until the equalized pressures on either side of the valve result in no further flow. Since there is fresh water on one side of the gate and salt water on the other, the level of fresh water will be higher. Thus, when the gate is opened, and the bodies of water are connected directly, the higher water in the lock will flow out to the sea, carrying the boat with it. Thus, boats receive a free ride from the last lock - no power is required!

Chapter 14
Temperature

𝕸ysteries:

1.) What is the temperature of space? Let us suppose we try to answer this question by opening up our spaceship window (take a deep breath first!) and sticking a thermometer out into space. Would its reading go up, go down or stay the same?

5 κ

0 κ

2.) What do we mean by the absolute zero of temperature?

At 30°C: *Water*

At 0°C: *Ice*

3.) Why do ice cubes rise in the tray when they freeze, as indicated in the cross sectional view of the tray above?

𝕸𝖞𝖘𝖙𝖊𝖗𝖎𝖊𝖘:

4.) When a thermometer is plunged into a hot liquid, the mercury level *drops* first and then rises. Why?

5.) When a hot liquid is poured into a cold glass container, it might break. But *Pyrex* glass will not break. Why not?

ML

6.) Why does the outside of a cold glass become wet?

Mysteries:

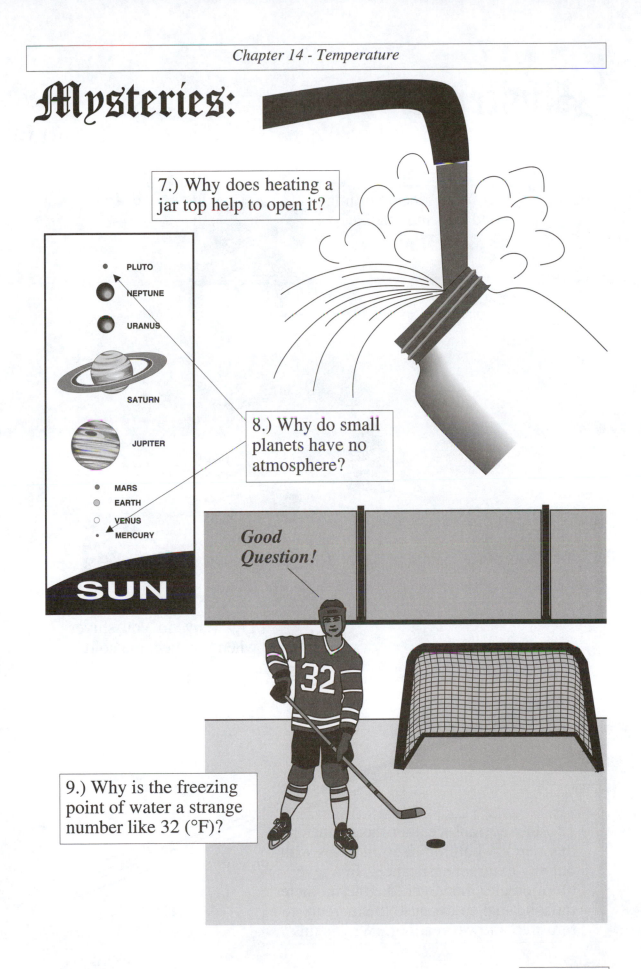

7.) Why does heating a jar top help to open it?

8.) Why do small planets have no atmosphere?

Good Question!

9.) Why is the freezing point of water a strange number like 32 (°F)?

Mysteries:

SNAP!

POP!

CRACKLE!

10.) Why does burning wood pop and crackle?

Yikes!... It's cold!

11.) Why do you shiver when you become cold?

12.) In a normal weather thermometer, the mercury column rises and falls without any external assistance. In a <u>fever</u> thermometer, however, the thermometer must be shaken to cause the mercury to return to a lower reading. Why is this?

Mysteries:

CRACK!!

13.) What causes water pipes to burst in freezing weather?

14.) Tall buildings in cities often have revolving doors, or at least double doors, at the street entrance. Why?

15.) We are told that boiling water cannot be heated to a temperature higher than 100°C. But why not? <u>What happens</u> to the energy that we transfer into boiling water?

Magic:

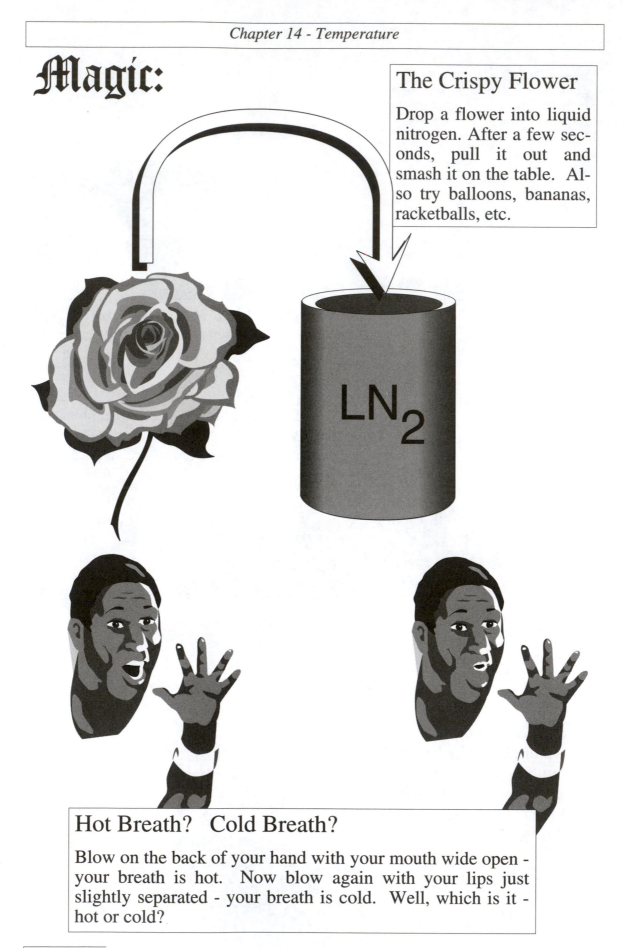

The Crispy Flower

Drop a flower into liquid nitrogen. After a few seconds, pull it out and smash it on the table. Also try balloons, bananas, racketballs, etc.

LN₂

Hot Breath? Cold Breath?

Blow on the back of your hand with your mouth wide open - your breath is hot. Now blow again with your lips just slightly separated - your breath is cold. Well, which is it - hot or cold?

𝕸𝖞𝖙𝖍:

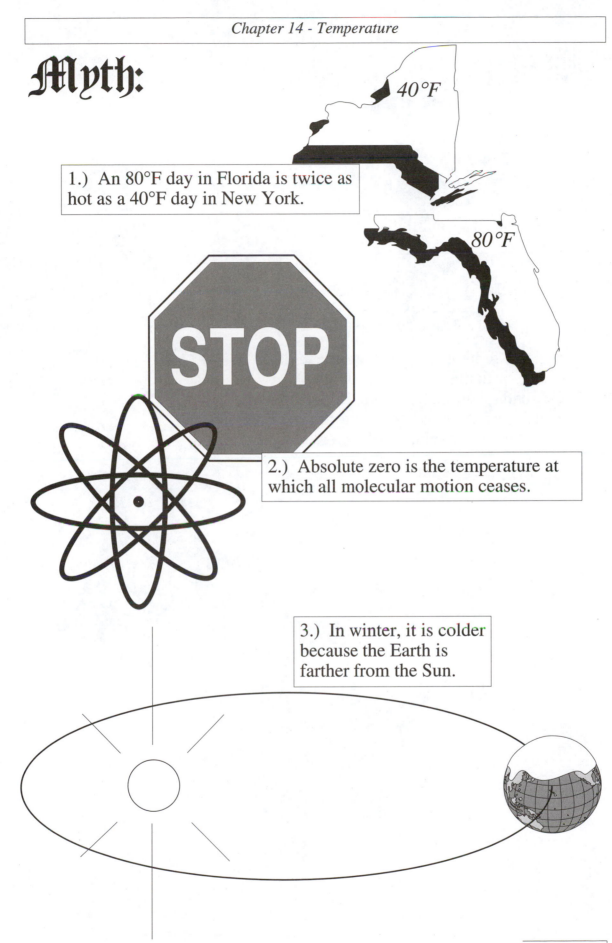

1.) An 80°F day in Florida is twice as hot as a 40°F day in New York.

2.) Absolute zero is the temperature at which all molecular motion ceases.

3.) In winter, it is colder because the Earth is farther from the Sun.

Myth:

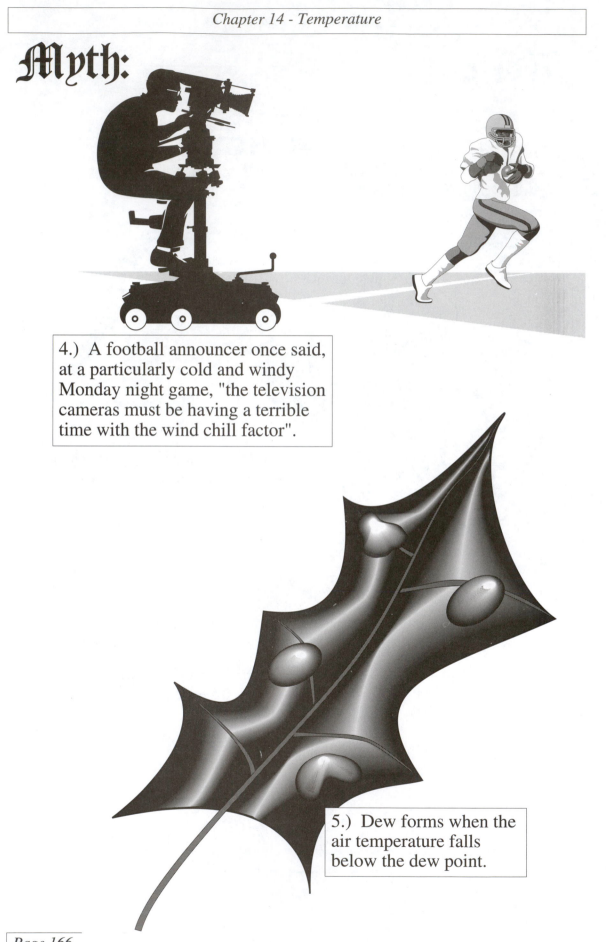

4.) A football announcer once said, at a particularly cold and windy Monday night game, "the television cameras must be having a terrible time with the wind chill factor".

5.) Dew forms when the air temperature falls below the dew point.

Myth:

6.) When boiling pasta, adding salt to the water decreases the time for the water to boil.

PHYSICS TEST

1. T = - 10 K

7.) Temperatures on a Kelvin scale cannot be negative.

Concepts of Temperature

Temperature (of an object) is a concept that we have created to describe the tendency of an object to transfer energy when placed in contact with another object. If two objects are brought into physical contact and no <u>net</u> energy is exchanged, then the objects are said to

> *Temperature - a measure of the tendency of a collection of particles to transfer energy*

be at the same temperature. We can generalize this from an object to a collection of particles. Thus, a gas has a temperature, even though we don't usually call a gas an "object". Temperature, in many cases, can be interpreted as closely related to the average energy of a collection of particles.

Culturally, in the United States, we use the *Fahrenheit* scale, on which water freezes at 32° and boils at 212°. Scientifically, we use the *Celsius* scale, where the above phenomena occur at 0° and 100°, respectively. The *absolute* scale, or *Kelvin* scale, assigns the number zero to a naturally occurring zero. This scale has the same size degree as the Celsius scale and absolute zero is at -273°C.

One effect of temperature is the changing of the size of objects as the temperature changes. Most substances expand upon heating

> *Most materials expand upon heating.*

and contract upon cooling and the fractional change in linear dimension per degree of temperature change is called the *coefficient of linear expansion*, which can be measured experimentally. Water is anomalous in that it exhibits the normal contraction upon cooling until it reaches 4°C. At that point, it begins to expand upon cooling, due to the fact that the crystal structure of ice has many open spaces and the molecules must move apart to form this open structure.

Discussions; Chapter 14 - Temperature

𝕸𝖞𝖘𝖙𝖊𝖗𝖎𝖊𝖘:

1.) What is the temperature of space? Let us suppose we try to answer this question by opening up our spaceship window (take a deep breath first!) and sticking a thermometer out into space. Would its reading go up, go down or stay the same?

As noted in the Concepts section, temperature can often be interpreted as a measure of the average energy of a *collection of particles*. In outer space, we do not have a collection of particles - we are in a vacuum (except for a few stray molecules). Therefore, the concept of temperature is not defined in space. As far as the thermometer reading on the thermometer is concerned, that will depend on our location. The thermometer will only measure the temperature of itself. If we are very far away from any stars, the thermometer will be receiving very little energy by means of radiation, but will be radiating itself. Thus, the reading will go down, as energy is radiated away. If we are close to a star, then the thermometer could absorb more radiation than it emits and the reading will go up. In principle, we could find just the right distance from a nearby star that the absorbed and emitted radiation would be balanced and the reading would be stationary. If we do not happen to be at this particular distance, then the reading will go up or down, as described above, until its temperature is such that the emission and absorption are balanced. The reading will then stabilize. Thus, at any distance from a star, the reading will eventually stabilize at a constant value.

2.) What do we mean by the absolute zero of temperature?

Absolute zero is not the temperature at which all molecular motion ceases (see Myth #2). One way to determine absolute zero is to graph the pressure of a fixed volume of gas as a function of temperature, as shown below.

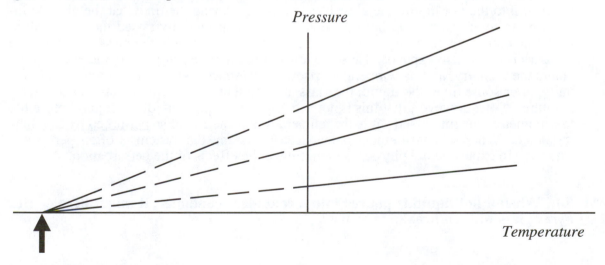

We find a linear relationship, as described by the Ideal Gas Law. If we extrapolate these lines back to zero pressure, then, <u>regardless of the type of gas and/or the initial conditions</u>, all of the lines extrapolate back through the <u>same point</u> on the temperature axis. Since there is no meaning to negative pressure, we must interpret the temperature of this extrapolation point as a lower limit of temperature for physical processes (although see Myth #7). We call it <u>absolute zero</u>.

The absolute zero of temperature can also be described as that temperature at which all of the energy that *can* be removed from the object *has* been removed.

For a method of "measuring" absolute zero in the home, see R. Otani and P. Siegel, "Determining Absolute Zero in the Kitchen Sink", *The Physics Teacher*, **29**, 316 (1991).

3.) Why do ice cubes rise in the tray when they freeze?

As noted in the Concepts section, water expands upon cooling between 4°C and 0°C. Thus, as the water freezes, it takes up more space. The only direction in which the water can expand freely in the ice cube tray is <u>up</u>, so the level of ice ends up higher than the level of water. If there is no place for the water to expand, then dire results occur, such as in the cracking of automobile engine blocks or the exploding of soft drinks in the freezer (See Mystery #13 for a discussion of freezing water in pipes.).

4.) When a thermometer is plunged into a hot liquid, the mercury level *drops* first and then rises. Why?

The explanation here is based on the fact that conduction (see Chapter 15) is not an instantaneous process - it takes time for energy to move through a material by conduction. The explanation also depends on the realization that a thermometer doesn't really measure the temperature of the liquid - it measures the temperature of *itself*. Furthermore, the thermometer doesn't even have a well-defined temperature during a time when it is not in thermal equilibrium, which is the case in this Mystery. When the thermometer is plunged into the hot liquid, the glass heats up first. During the time that the glass is increasing in temperature and before the energy has substantially raised the temperature of the mercury, the glass expands. As a result, the cavity inside the glass, in which the mercury resides, also expands. Thus, the level of the mercury <u>drops</u>. As the energy then enters the mercury, and its temperature rises, the mercury column begins to expand. Finally, after some time, the liquid, the glass and the mercury will all be at the same temperature. Note, however, that this is not actually the temperature of the liquid before the thermometer was immersed - it is the temperature of the liquid <u>as perturbed by the thermometer</u>. When a measurement is made on a system, the system is often perturbed. The trick in experimental physics is to minimize the effects of the perturbation.

5.) When a hot liquid is poured into a cold glass container, it might break. But *Pyrex* glass will not break. Why not?

In order to understand the difference between ordinary glass and Pyrex, let us list some information from standard tables of properties of materials:

Thermal Conductivity:	*Glass*:	1.32×10^5 J·m^{-1}·s^{-1}·°C^{-1}
	Pyrex:	1.09×10^5 J·m^{-1}·s^{-1}·°C^{-1}
Coefficient of Linear Expansion:	*Glass*:	9×10^{-6} °C^{-1}
	Pyrex:	3×10^{-6} °C^{-1}

Glass breaks when a hot liquid is poured in because of thermal stresses. The inside of the glass quickly becomes much hotter than the outside. In response, the inside portion of the glass expands. But if the outside is still cool (since conduction is a slow process, as described in the answer to Mystery #4) and does not expand, then there are large stresses which result in the breakage of the glass. We can avoid this in two ways. We can increase the thermal conductivity, so that the energy arrives at the outside of the glass more quickly so that there is not a large temperature difference across the glass thickness. Or, we can reduce the linear expansion so that the stresses are not so large. Notice the comparison between glass and Pyrex for thermal conductivity. Pyrex is actually about 17% worse in thermal conductivity than ordinary glass. This is in the wrong direction! But, if we look at the expansion coefficient, we see that Pyrex has only one third the expansion coefficient of glass. This is more than enough to offset the lower thermal conductivity and allows Pyrex to survive thermal stresses that glass cannot.

6.) Why does the outside of a cold glass become wet?

This is a result of the lowering of the temperature of the surface of the glass. As the surface temperature is lowered, water vapor in the air will condense to form liquid water which collects on the outside of the glass.

7.) Why does heating a jar top help to open it?

Heating a jar top to open it is taking advantage of thermal expansion. The expansion coefficients of metals are larger than that for glass. Thus, if the jar is heated, the top will expand more than the glass and, hopefully, will loosen.

8.) Why do small planets have no atmosphere?

In a gas which can be modeled with the Ideal Gas Law (planetary atmospheres can be modeled this way for the sake of this discussion), the average velocity of the molecules varies as the square root of absolute temperature and as the square root of the inverse of the molecular mass. The velocities of some gas particles can be quite high, as high as several kilometers per second. Each planet has an *escape velocity*, which is a measure of how fast an object must be moving at the surface of the planet to be able to attain a position infinitely far away from the planet (This can be calculated using the concepts in Chapter 6.). For small planets, the escape velocities are so low that gas molecules can move fast enough to leave the planet. Thus, although small planets may have started off with atmospheres, the gases have gradually seeped off into space so that no atmosphere remains. For Earth, the planet is large enough to hold onto oxygen and nitrogen, but hydrogen moves so fast that we have lost our atmospheric hydrogen. For a more detailed but quite readable discussion, see V. J. Ostdiek and D. J. Bord, *Inquiry into Physics*, 2nd edition, West Publishing Company, St. Paul, 1991, pp. 202-204.

9.) Why is the freezing point of water a strange number like 32 (°F)?

This is a somewhat long story that will be presented in short form here. In the early eighteenth century, Daniel Fahrenheit followed up on the work of Olaus Roemer. Roemer had devised a temperature scale on which the melting point of snow was 7.5°, and the boiling point of water, 60°. This scale was based on a reading of zero at the temperature of a mixture of ice and salt, which was felt to be the lowest temperature possible. Body temperature on this scale was 22.5°. Fahrenheit adopted this scale, but multiplied the readings by 4, so that ice melted at 30° and body temperature was 90°. A further modification to the scale was made by Fahrenheit by redefining the melting point of ice to be 32° and body temperature to be 96°. This modification was for ease in marking the thermometers - both the <u>value</u> of 32 and the 64 degree <u>difference</u> between 32 and 96 are powers of two. Thus, the markings on the thermometers could be easily made by dividing distances exactly in half a number of times.

So why isn't body temperature 96° anymore? Fahrenheit measured the temperature of boiling water to be 212° on his scale, but his measurement was inaccurate. Unfortunately, 212° was adopted as the official temperature of boiling water on the Fahrenheit scale, so the readings between 32° and 212° had to be adjusted to reflect the more accurate measurement, so that body temperature now falls at 98.6°.

10.) Why does burning wood pop and crackle?

Trapped in the wood are small pockets of sap. As the temperature of the wood rises, the sap in these pockets vaporizes, causing an increase in pressure. The crackling and popping sound is caused by the explosion of the sap pockets as the internal pressure causes them to burst. It is similar to the popping sound of popcorn as the corn kernel explodes due to the rising internal pressure of the heated water inside. For more information on popping corn, see R. G. Hunt, "The Physics of Popping Corn", *The Physics Teacher*, **29**, 230 (1991).

11.) Why do you shiver when you become cold?

Shivering is an involuntary physiological response to a drop in the temperature of air in contact with the body. It is a method that the body uses to increase the rate at which energy is transformed from that stored in the food that was eaten to internal energy of the body by increasing the metabolic rate. The same effect could occur by <u>exercising</u>, which would be a <u>voluntary</u> method of raising the metabolic rate. As pointed out in H. E. Landsberg, *Weather and Health*, Doubleday, Garden City, New York, 1969, shivering is not a terribly efficient process, since, although it does increase the metabolic rate, it also brings more blood to the skin, where even more energy can be transferred from the body by radiation and convection.

12.) In a normal weather thermometer, the mercury column rises and falls without any external assistance. In a <u>fever</u> thermometer, however, the thermometer must be shaken to cause the mercury to return to a lower reading. Why is this?

In a normal thermometer, the mercury column has a constant cross section. In a fever thermometer, there is a <u>constriction</u> near the low end of the column so that the mercury has to pass through one region of very small cross section. When the temperature is rising, it is easy for the mercury below the constriction to <u>push</u> the mercury through the small area. When the temperature is falling, however, the forces between molecules of mercury are not strong enough to <u>pull</u> the mercury back through the constriction. The thermometer must be <u>shaken</u> to cause the mercury to pass back through the constriction. This is an application of Newton's First Law. When the thermometer is shaken, it exhibits a motion with a very large acceleration of the glass tube in a direction away from the mercury reservoir at the end of each "shake". This acceleration causes the mercury to move through the constriction and into the reservoir because of its inertia!

Now, why is the constriction there in the first place? Once the thermometer is removed from the patient, it begins to equilibrate with the air temperature, which is lower than that of the patient. The constriction is a clever way to keep the reading of the thermometer constant at its most recent highest value, even though the temperature of the thermometer is now changing. Otherwise, the apparent temperature of the patient would change, depending on the time delay between removal of the thermometer and the actual reading.

13.) What causes water pipes to burst in freezing weather?

Energy from the pipe is transferred to the cold outside air at the surface, so the layer of water just inside the pipe is the first to freeze. Subsequently, layers closer and closer to the center of the pipe will freeze as further energy is transferred. As noted in the answer to Mystery #3, water expands when it freezes into ice. As long as the freezing does not reach the center, further freezing and the resultant expansion will simply push the liquid water down the pipe into the supply line. The problem arises when the freezing reaches the center of the pipe and there is then a solid plug of ice with water remaining between this plug and the spigot, or another plug, or any other obstruction through which the water cannot pass. The expansion resulting from the freezing of this remaining water will cause increased pressure until something has to give - often the pipe will burst under this pressure.

14.) Tall buildings in cities often have revolving doors, or at least double doors, at the street entrance. Why?

One of the advantages to revolving doors or double doors is to cut off the air path between the outside and the inside of the building, which avoids a problem caused by the <u>elevator shafts</u>. At the top of the elevator shaft is the machine room, where the motors are located. It is likely that there is some form of venting from the inside of the machine room to the outside. Suppose now that the door at the street level is <u>not</u> revolving or double but is a single door and is standing open. We now have an open air path from the outside, through the open door at street level, through the elevator doors (when open), up the elevator shaft and out to the open air again through the machine room. If the outside air is cold and the building is heated, the buoyant force of the cold air on the heated air will cause an airflow in through the open door and up the elevator shaft. The air flow of the cold air coming into the lobby can become <u>quite severe</u>. In addition to the problems associated with a violently windy lobby, this situation is not energetically wise, since much of the heated air is simply vented to the outside through the machine room. To avoid this situation, revolving doors or double doors are used, so that the amount of time when a direct air path such as that described above is zero for a revolving door (since the entryway is always sealed by one vane of the rotating door system) or very short, if it exists at all, for a double door.

15.) We are told that boiling water cannot be heated to a temperature higher than 100°C. But why not? <u>What happens</u> to the energy that we transfer into boiling water?

In order for the temperature of water to "level out" at 100°C, we must have <u>bubbling</u>. The energy that we feed into the bottom of a pan of water on a stove goes into vaporizing the water to form a gaseous bubble. This energy is carried to the top of the water by the buoyant force, where the bubble bursts and releases its steam (along with the energy) into the air. The resultant rate of cooling by energy transfer at the top surface is equal to the rate of heating at the bottom surface. If we increase the heating rate, then the rate of bubble formation increases to keep the temperature constant.

Now, this is only the case if we have bubbling, and bubbling depends on nucleation sites in the container. These sites are provided by the many extremely small cracks, scratches and nicks in ordinary household pans. If water is boiled in highly polished, smooth containers, then the water can become hotter than 100°C with minimal boiling, since the sites of bubble nucleation have been reduced in number. This kind of super-heating of water is quite dangerous since, if nucleation sites suddenly become available, the resultant boiling can become explosive.

For more details on the complexities of boiling water, see C. F. Bohren, *What Light Through Yonder Window Breaks?*, John Wiley & Sons, New York, 1991, and J. Walker, "The Amateur Scientist - What Happens when Water Boils is a Lot More Complicated than You Might Think", *Scientific American*, **247(6)**, 162 (1982).

𝕸𝖆𝖌𝖎𝖈:

The Crispy Flower

This demonstration shows that there is water in the flower that freezes in the liquid nitrogen, making the flower stiff and brittle. The balloon will shrink according to the Ideal Gas Law, since the temperature of the gas is dropping. It will return to its normal size upon reheating. The banana and the racketball will become very hard. Dropping the racketball on the floor will shatter it, resulting in a sound like an imploding light bulb.

Hot Breath? Cold Breath?

Your breath is normally warm, at about internal body temperature. When it is blown on your skin from an open mouth, it will feel warm to the skin. When the breath has to pass through a narrow opening between your lips, two effects occur. First, the breath is under pressure in your mouth and the pressure is reduced upon entering the atmosphere. As a result, the gas exhibits a rapid expansion, causing the temperature to drop. Secondly, the air is moving at higher velocity when blown between slightly separated lips than from an open mouth. Thus, internal energy in your skin is carried away by convection faster than in the open-mouth case. Both of these effects act to reduce the apparent temperature of the breath. In the first case, the feeling of coldness is due to an actual decrease in the temperature of the gas striking the skin. In the second case, in which convection plays a role, the cold feeling is due to more rapid energy transfer from the skin. This is related to Mystery #10 in Chapter 15.

Myth:

1.) An 80° F day in Florida is twice as hot as a 40°F day in New York.

Recalling that temperature is related to the tendency for an object to transfer energy, does an object at 80°F have <u>twice</u> this tendency as an object at 40°F? Although the numbers are in a 2 to 1 ratio, this statement ignores the fact that 0°F is not a naturally occurring, universally defined zero. It is an arbitrary zero, based on water. Values on a scale based on an arbitrary zero are not meaningful when entered into ratios. Only if the temperatures are expressed on a Kelvin scale, which is based on a natural zero, can we express meaningful ratios of temperatures. Thus, 80 K *is* twice as hot as 40 K. The Fahrenheit temperatures given in the statement are equivalent to 300 K and 277 K, which are certainly not in a 2 to 1 ratio.

2.) Absolute zero is the temperature at which all molecular motion ceases.

This is a commonly heard statement about absolute zero, but it is incorrect. Quantum physics tells us that the lowest energy state of a system does <u>not</u> have zero energy. There is always some energy associated with the lowest possible state of a system - it is called the <u>zero point energy</u>. Thus, as we cool a collection of molecules down, we can cool them so that they are in the lowest possible state, but they will still be moving slightly with zero point energy.

3.) In winter, it is colder because the Earth is farther from the Sun.

This can be easily seen to be wrong by making a prediction based on the hypothesis. If winter occurs because the Earth is farthest from the Sun in its orbit, then it would be winter <u>everywhere</u> on Earth at the same time. This prediction does not match observation, since it is summer in the Southern Hemisphere when it is winter in the Northern Hemisphere. In fact, during the winter in the Northern Hemisphere, the Earth is actually closest to the Sun in its slightly elliptical orbit. So we must look for another explanation.

The correct explanation is based on the tilt of the Earth's axis (see Myth #6 in Chapter 1) and a bit of geometry. During winter in the Northern Hemisphere, the Earth is in a part of its orbit such that the North Pole is tilted away from the Sun by an angle of 23.5° with respect to the plane of the Earth's orbit. As a result, the sunlight reaching the Northern Hemisphere during the daytime strikes the ground at a more glancing angle than in the summer (when the tilt is toward the Sun) and is thus spread out over a larger area. This reduces the intensity of radiation over the case in the summer, resulting in less energy transfer - hence, winter. At the same time, the South Pole is directed <u>toward</u> the Sun by 23.5°. The sunlight hits the Earth's surface here during the daytime more closely to perpendicular and we have high intensity and high energy transfer - summer. Six months later, the tilts of the Hemispheres are reversed, as are winter and summer.

4.) A football announcer once said, at a particularly cold and windy Monday night game, "the television cameras must be having a terrible time with the wind chill factor".

The wind chill factor is a combination of actual air temperature and the effects of wind. The wind will carry away energy faster from the skin by convection, as noted in the discussion of "Hot Breath? Cold Breath?". This will assist the normal evaporative cooling of perspiration from the skin surface. But television cameras do not perspire, so they cannot be affected by the wind chill factor. It is certainly true that the wind will carry away energy that is given off by the electronics of the camera, but that cannot reduce the temperature below that of the surrounding air.

5.) Dew forms when the air temperature falls below the dew point.

The offending word in this statement is *air*. Dew forms on *surfaces* and it is the temperature of the *surface* that must be lower than the dew point. The surface temperature can be lower than the air temperature due to cooling of the surface by radiant emission. A clear sky has a lower effective radiative temperature (see Chapter 15, Mysteries #3, 7, 13 and 14 for more discussions of radiation temperature) than a surface on the Earth (car, grass, fence post, etc.), so that the <u>net</u> flow of radiant energy is from the surface to the sky, allowing the surface to cool below the air temperature. On the other hand, clouds generally have a higher radiative temperature than clear sky, so dew tends to form more on clear nights than on cloudy nights. Very nice discussions of the formation of dew are given in C. F. Bohren, *What Light Through Yonder Window Breaks*, John Wiley & Sons, New York, 1991.

6.) When boiling pasta, adding salt to the water decreases the time for the water to boil.

The addition of salt to water lowers the saturation vapor pressure, thus raising the temperature at which water will boil. If this is the case, then the addition of salt <u>increases</u> the time for the water to boil. Two scientists have independently concluded that the only reason for cooks to add salt to the water is for taste and that it has nothing to do with boiling time (J. Walker, "The Amateur Scientist - What Happens when Water Boils is a Lot More Complicated than You Might Think", *Scientific American*, **247**(**6**), 162 (1982); and C. F. Bohren, *What Light Through Yonder Window Breaks*, John Wiley & Sons, New York, 1991).

7.) Temperatures on a Kelvin scale cannot be negative.

If we adopt the definition of temperature that is given in many textbooks (temperature is a measure of the average internal energy in an object), then this cannot be a myth. But notice how the phrase related to internal energy is stated in the Concepts section - "Temperature, *in many cases*, can be interpreted as closely related to the average energy of a collection of particles." This is not the correct definition in *all cases*. The more general definition involves the tendency for a collection of particles to transfer energy when brought into "contact" with another collection of particles.

Let us consider a collection of atoms making up a gas as an example. Normally, we have a distribution with fewer and fewer atoms in the higher energy states, with the number of atoms, under the right conditions, in each state described by a Boltzmann distribution:

$$\textit{Number of atoms in a state} \ \sim \ e^{-E/kT}$$

with E being the energy of the atom, T the absolute temperature and k the Boltzmann constant. Notice that, for any positive temperature, the number of atoms in increasingly higher energy states will <u>decrease</u>.

Now, consider the situation in a gas *laser* which has been pumped so that a population inversion exists. We now have more of the lasing atoms in a higher energy state than a lower energy state. The only way we can make this consistent with the Boltzmann distribution is to define the temperature as *negative*.

Now, is this negative temperature hotter or colder than a positive temperature? If we imagine bringing this collection of atoms together with a "normal" collection of atoms, energy will pass from the negative temperature collection to the positive temperature collection as the atoms decay to the ground state. Thus, according to our definition of temperature as a tendency to transfer energy, *the negative temperature is hotter than the positive temperature*! This is a general result - all negative temperatures are hotter than all positive temperatures. The hottest temperature is just below 0 K, while the coldest is very close - just above 0 K.

For another example of a negative temperature and further discussion of these ideas, see R. Baierlein, "The Meaning of Temperature", *The Physics Teacher*, **28**, 94 (1990).

Chapter 15
Heat

𝕸ysteries:

1.) Why is mountain air cold? After all, you're closer to the Sun, right?

2.) Why can't you make a three minute egg a two minute egg by turning up the heat?

3.) Why are thermos bottles silvered on the inside? Isn't silver a good conductor of heat?

Mysteries:

4.) Why doesn't dry ice *melt*?

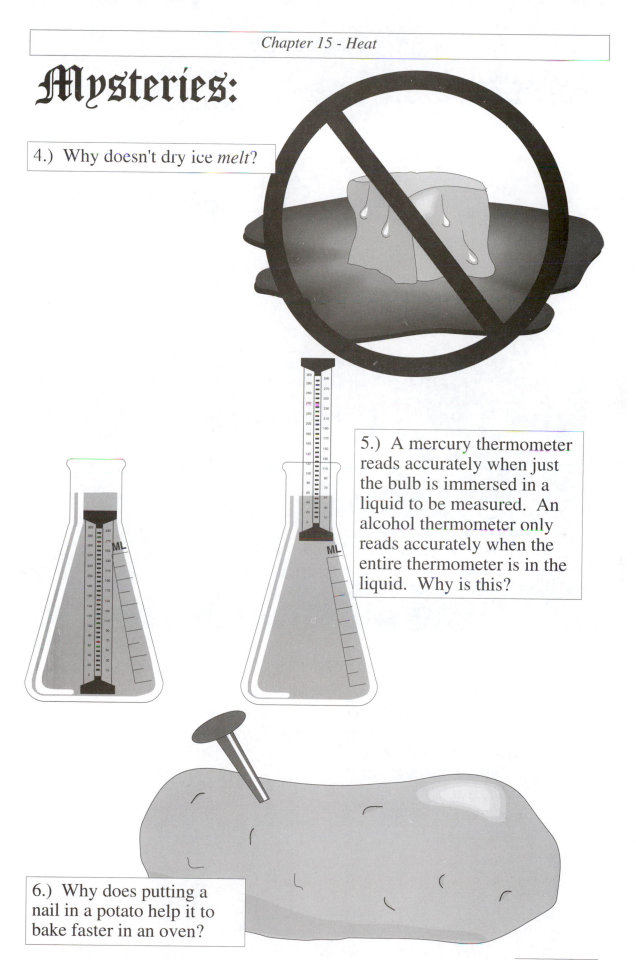

5.) A mercury thermometer reads accurately when just the bulb is immersed in a liquid to be measured. An alcohol thermometer only reads accurately when the entire thermometer is in the liquid. Why is this?

6.) Why does putting a nail in a potato help it to bake faster in an oven?

Mysteries:

BRIDGES
FREEZE
BEFORE
ROADWAY

7.) Why do bridges become covered with ice before roadways?

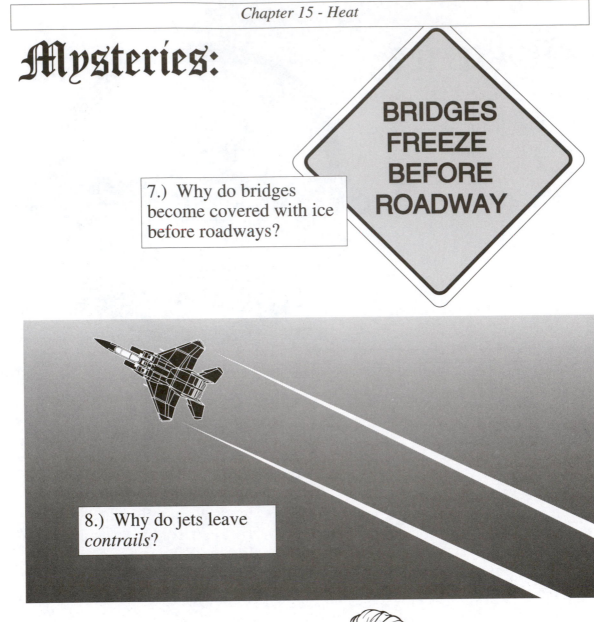

8.) Why do jets leave *contrails*?

9.) One can heat a cup of water for coffee with a small electrical coil that is immersed in the water. Why do the directions say not to operate the heater if it is not immersed in water?

Mysteries:

10.) Why does a bare floor feel colder than a carpeted floor?

11.) In Southern California, a common occurrence is a Santa Ana condition, in which warm, dry winds blow into the Los Angeles basin from the mountains. But, wait a minute. According to Mystery #1, mountain air is cold. What's going on here?

Mysteries:

12.) When you turn your car head-lights out and watch the light that they shine on a wall, they seem to go out much more slowly than normal light bulbs. Why?

13.) To keep a sandwich from a fast food restaurant warm, it is often wrapped in foil lined paper. But, wait a minute. Isn't metal foil a good <u>conductor</u> of heat?

14.) Why does frost tend to form first on the <u>tops</u> of objects (mailboxes, cars, etc.), but not on the <u>sides</u>?

𝕸agic:

*When water boils,
place cap on can and tighten*

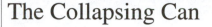

The Collapsing Can

For this activity, you will need a one gallon metal can, such as a gasoline can or a duplicating fluid can (wash it out well). Add about 1 - 2 cm of water to the can and set it on a stove or burner until the water boils. When the water starts boiling, remove the can from the heat and screw the top on securely. In a few moments, the can will crush. The crushing will be hastened by pouring cold water over the can.

This activity can also be performed by pouring boiling water into a plastic juice container and capping it tightly. It can also be done by heating a small amount of water to boiling in a soft drink can and quickly plunging the can upside down into a container of cold water.

Boiling Water in a Paper Cup

Place a small amount of water in a paper cup. Using tongs, hold the paper cup over an open flame, such as a candle or a bunsen burner. Make sure that the flame touches only the bottom of the cup. After a few moments, the water will boil without the cup catching on fire, even though the cup is in contact with the flame!

Magic:

Black and White

The effects of color on radiation absorption can be easily demonstrated with socks. Insert thermometers into two socks, one white and one black. Record the initial temperatures. Now, place the socks under an intense light source or out in the Sun, making sure that each receives the same amount of radiation. After some time, read the temperatures again.

The Paper That Would Not Burn

Place some scraps of paper in a metal kitchen strainer. Hold the strainer over a candle flame. The paper in the strainer will not burn.

𝕸𝖆𝖌𝖎𝖈:

The Glowing (?) Stove

You are probably familiar with the red-orange glow of an electric stove burner on the high setting. But does it glow on the <u>very lowest</u> setting? Turn a stove burner on the very lowest setting and take a <u>very</u> long exposure (10 - 12 <u>hours</u>) photograph of it. Use fairly high speed film and make sure that the room stays dark during the entire exposure time. Does the burner appear on the film? If you use color film, how does the color compare to the color at the high setting?

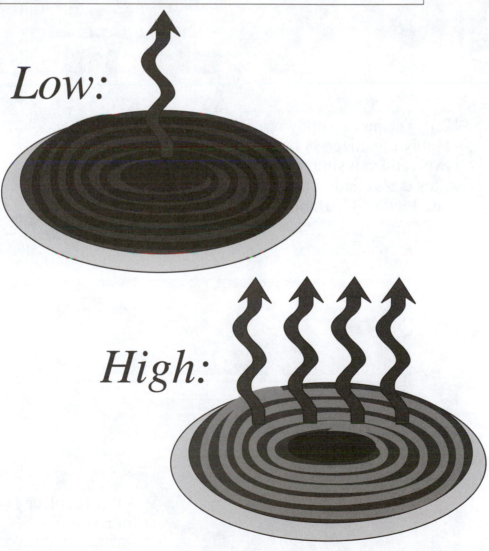

Low:

High:

Myth:

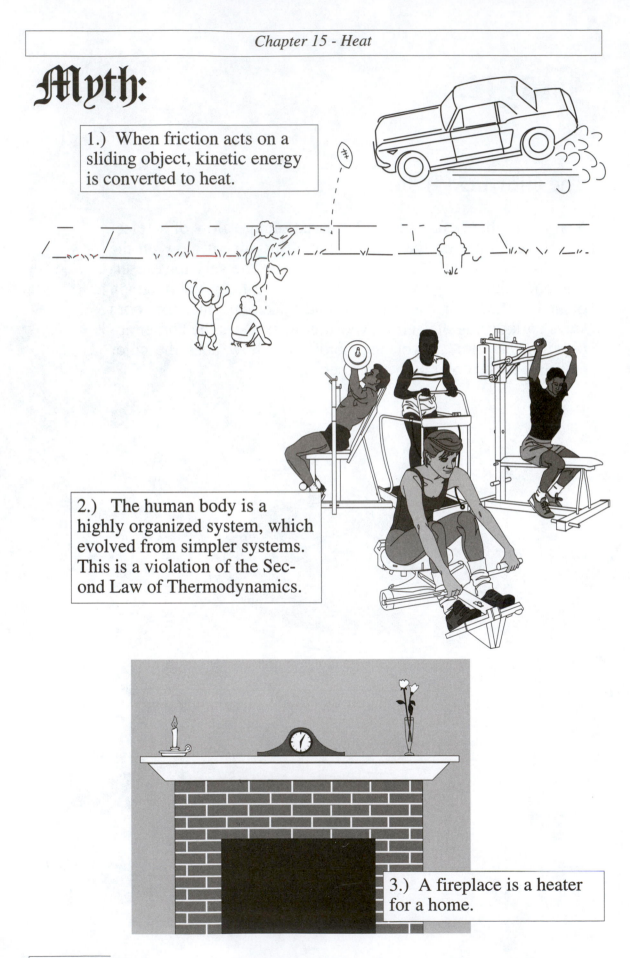

1.) When friction acts on a sliding object, kinetic energy is converted to heat.

2.) The human body is a highly organized system, which evolved from simpler systems. This is a violation of the Second Law of Thermodynamics.

3.) A fireplace is a heater for a home.

Myth:

4.) A refrigerator can be used to cool off a room in the summer - just leave the door open!

5.) At the summer solstice in June, the Sun's rays strike the Earth in the Northern Hemisphere as close to perpendicularly as they ever do. Therefore June is the hottest month. Similarly, December is the coldest month.

Myth:

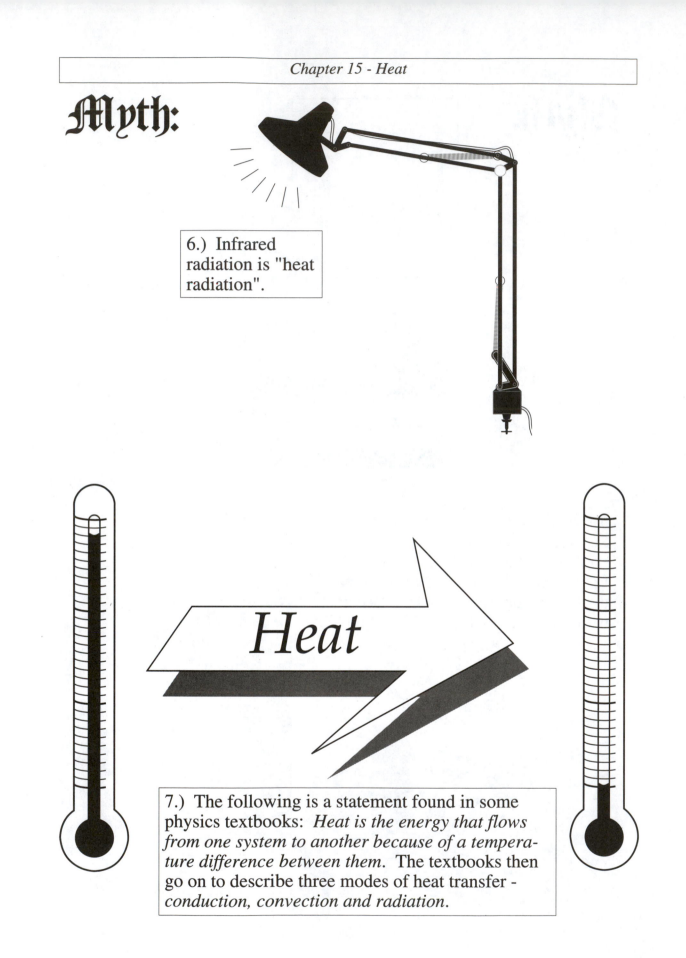

6.) Infrared radiation is "heat radiation".

Heat

7.) The following is a statement found in some physics textbooks: *Heat is the energy that flows from one system to another because of a temperature difference between them.* The textbooks then go on to describe three modes of heat transfer - *conduction, convection and radiation.*

𝕸yth:

8.) Archimedes once fought off the Romans in a ship 100 m away by having his soldiers reflect the Sun's rays from hand-held mirrors, setting the Roman ship on fire.

9.) Heat rises.

Concepts of Heat

Heat is one of the terms most commonly misused in collo- quial conversation as well as in many textbooks. Heat is often referred to as a form of stored energy, but this is not true. Heat is a means by which we can transfer energy, it is not the stored energy itself, as we discussed in Chapter 7. Thus, heat falls into the same category as *work*, which is another transfer variable. The numerical value for the heat is the amount of energy transferred.

Heat = One Way to Transfer Energy

Internal Energy = Energy associated with an object with a temperature

So if heat is not stored energy, then what is the energy contained in a warm object? The correct description is *in- ternal energy*. This is the energy as- sociated with the total kinetic and potential energy of all of the molecules of the object. If energy is transferred to an object, its internal energy rises (assuming that there is not an equal amount of energy being transferred out by some mecha- nism). BUT, the internal energy would also increase if work were done on the object (rub your hands together - they become hotter; is it work or heat?). Work and heat are two ways of transferring energy, both of which can raise the temper- ature of the object.

We have talked so far about using heat to raise the temperature of an object. There is another possibility for the result of the heat (or work, or any other energy transfer mechanism). Suppose we have an ice cube

Heat can cause changes of state as well as temperature increases.

at 0°C and we supply some energy to it. The result will not be a hotter ice cube. Rather, a change of state occurs. Instead of raising the temperature, the energy goes into breaking the molecular bonds that hold the water molecules together in- to an ice cube - the ice cube melts. Only after the bonds have been broken can the temperature rise above 0°C with the provision of more energy. If we contin- ue to provide more energy to the water which resulted from the ice cube, it will eventually reach another temperature plateau, at which the water turns into steam. After complete conversion, additional energy will cause the steam to in- crease in temperature.

The process by which the energy is transferred as heat is given as one of the fol- lowing three in most textbooks:

Convection: In this process, the temperature of a fluid is raised and the fluid is moved from one point in space to another, carrying the energy with it. Examples of convection include the water-based cooling system in an automo- bile engine and a forced air heating system in a home. *Natural* convection refers to the rising of hot fluids due to the buoyant force. Convection can also occur due to wind, such as the convective cooling of one's skin as the air blows across it. *Forced* convection re- fers to movement of energy-carrying fluid due to me- chanical means, such as in your automobile engine or your air conditioner.

Conduction: In this process, energy is transferred through a material without the actual transfer of matter. The molecules at one end of the material (in contact with a region of high temperature) start moving with increased kinetic energy. These molecules "bump" into other molecules and start them moving faster. This "bumping" moves down the material toward the cooler end. Examples include the warming up of the upper end of a spoon in a cup of coffee and the transfer of energy through the walls in a home, which can be countered by installing *insulation* in the walls.

Radiation: This process depends on electromagnetic radiation, which will be addressed in Chapter 25, the most familiar form of which is *light*. Electromagnetic radiation carries energy, which can be absorbed by materials and may result in increased temperature. Examples include the warmth one feels in front of a fireplace and the energy brought to Earth from the Sun.

It is a misconception perpetuated by textbooks that these represent processes unique to situations involving heat and temperature. Let us review these according to the taxonomy of energy transfers discussed in Chapter 7. Convection is simply one example of Mass Transfer. In Chapter 7, a bowling ball was used as an example in which the energy is stored temporarily as kinetic energy of the ball. Of course, since the ball has kinetic energy, the energy is transferred with the ball and is delivered to the pins later on. In convection, energy is stored temporarily as internal energy in the molecules of the fluid, which is then moved to a new location and the energy delivered there. As mentioned in Chapter 7, conduction can actually be considered to be Work on a microscopic level as molecules exert forces on each other. And finally, radiation is simply the transfer mode of Electromagnetic Radiation from Chapter 7, which might result in an increased temperature, but could also result in other effects such as colors from a fluorescent poster or entertainment from a television screen. Of these three modes of energy transfer, conduction is the only one that is unique to a situation involving heat.

Discussions; Chapter 15 - Heat

𝕸𝖞𝖘𝖙𝖊𝖗𝖎𝖊𝖘:

1.) Why is mountain air cold? After all, you're closer to the Sun, right?

It certainly is correct that you are closer to the sun on the top of the mountain, but a matter of a mile or two is not going to make much difference out of 93,000,000 miles to the Sun. The reason that mountain air is cold has to do with expansion of gases. As a packet of air rises up from the valley to the top of the mountain, it feels less pressure from the surrounding air, since the air becomes less dense as the altitude increases. As the pressure drops, the air expands. Thus, the packet of air is doing work on the surrounding air, which represents a transfer of internal energy out of the air mass and the temperature drops.

2.) Why can't you make a three minute egg a two minute egg by turning up the heat?

No matter how high you turn the heat up on your stove, once the water is boiling, you will not normally cause it (or the pan) to be any hotter than 100°C (unless you boil it all away, in which case the temperature of the pan could increase beyond 100°C, possibly causing it to warp!). Turning up the heat will make the water vaporize faster, but the water which remains will stay at 100°C. This is the temperature at which the egg cooks, so it will always take three minutes. See Mystery #15, Chapter 14 for more discussion about the constant temperature of boiling water.

3.) Why are thermos bottles silvered on the inside? Isn't silver a good conductor of heat?

Silver indeed is a good conductor of heat. <u>Conduction</u> of heat is countered in the thermos bottle, however, by a vacuum layer between cylinders of glass. Since conduction depends on molecules, if there are no molecules, there is no conduction! The absence of molecules also removes the possibility of <u>convection</u> as a means of energy transfer from the inner glass vessel.

With these two processes excluded, <u>radiation</u> is the primary transfer mechanism. The silver is there to counter the loss of energy by radiation. As a first guess, we might claim that the silver metal *reflects* the radiation emitted by the fluid in the bottle back into the inside of the thermos bottle. While this is consistent with our everyday experience, we must remember that most of the radiation emitted by a hot fluid is in the <u>infrared</u>, not in the visible. And is the silver a good reflector in the infrared? As it turns out, it is a good reflector in the infrared as well, so reflection <u>is</u> important.

But let's think now about another aspect - <u>emission</u>. If the silver is at the same temperature as the glass in the thermos, then wouldn't the silver emit just as effectively as the glass? The answer is <u>no</u>, because emission depends not only on the temperature but on the <u>emissivity</u> of the surface. The emissivity of a surface is the fraction of radiation emitted by the surface relative to a blackbody at the same temperature. Since silver is a good reflector (both in the visible and the infrared), it is a poor absorber and, therefore, it is a poor emitter (it has a low emissivity). Thus, the combination of high reflectance of radiation from inside the thermos and poor emission from the surface keeps the energy in and the fluid hot!

4.) Why doesn't dry ice *melt*?

It is common to observe a melting ice cube sitting in a pool of water, but a chunk of dry ice sitting in a pool of liquid carbon dioxide is not observed. Let us try to understand this with a thought experiment - we imagine that we raise the temperature of an ice cube in an environment in which we can vary the pressure. As the temperature of the ice is raised, water molecules will break free from the bonds that held them together as a solid. If the pressure is high, then these freed molecules will be "pushed back" by the gas molecules so that the water molecules spend much time close together and the intermolecular forces between the water molecules can hold them together as liquid water.

Now, suppose the pressure is low compared with the pressure necessary to assure melting. Then, the water molecules are not "pushed back" as effectively by the gas molecules. The water molecules will now be able to move far apart, so that intermolecular forces will not be able to hold them together - the molecules will simply float away as a gas.

Now, the necessary pressure above which the liquid can be maintained depends on the material. The triple point for a given material is that point on a pressure-temperature graph where all three phases can coexist and also defines the lowest pressure for which a liquid of that material can exist. Returning now to the question of dry ice, for carbon dioxide, the triple point lies at a pressure higher than atmospheric pressure. Thus, the solid carbon dioxide passes directly to the gas phase - it *sublimes*, without passing through the liquid phase.

5.) A mercury thermometer reads accurately when just the bulb is immersed in a liquid to be measured. An alcohol thermometer only reads accurately when the entire thermometer is in the liquid. Why is this?

Although mercury is a liquid, it is a metal, and, therefore, as is generally true for metals, mercury is a good thermal conductor. Thus, when the bulb of mercury is placed in a liquid, energy is conducted along the column of mercury quite quickly so that the entire column comes to a uniform temperature in a rather short time. Alcohol, however, is not nearly as good a thermal conductor as mercury. Thus, if just the bulb of an alcohol thermometer is placed in a liquid, energy takes a relatively long time to flow along the alcohol column and the column will not have a uniform temperature. Thus, the expansion of the column will not be consistent with the calibration on the thermometer. In order to take an accurate reading, the entire thermometer must be immersed so that all of the alcohol is at the temperature of the liquid.

6.) Why does putting a nail in a potato help it to bake faster in an oven?

Potato material is not a good thermal conductor. Thus, it takes time for the energy to conduct from the outside of the potato to the inside, so that it is cooked all the way through. The metal of a nail is an excellent thermal conductor. Thus, the nail provides a low resistance path for the energy to follow to the interior of the potato, resulting in a faster increase in temperature in the potato and a lower cooking time. As an electrical analogy, the nail has been put in <u>parallel</u> with the potato material. The lower resistance has increased the "current", which, in this case, is the flow of energy.

7.) Why do bridges become covered with ice before roadways?

For a roadway sitting on the ground, the Earth acts as a huge thermal mass. Water sitting on the roadway will lose energy to the surrounding cold air, but energy will be flowing in by conduction from the warmer Earth, so that the formation of ice is delayed. For the roadway over a bridge, however, the advantage of the warm Earth is not there. Underneath the roadway is more cold air, so that the roadway does not have a similar supply of energy to resist the formation of ice and the bridge roadway will become icy before the roadway on the ground.

In addition, the bridge has a large relative surface to volume ratio. For a roadway sitting on the soil, the surface is just the upper surface of the roadway, while the volume includes the ground below it. For a bridge roadway, the volume is much smaller, being just the volume of the bridge roadway itself and the surface now includes both upper and lower surfaces as well as the sides. Thus, there is less volume to cool and more area for the energy to leave, resulting in faster icing.

A third consideration is the radiation of energy into the clear night sky, such as was addressed in the discussion of Myth #5, Chapter 14. For the reasons described above, the roadway of an overpass can be brought to a temperature just above freezing. If the air currents are low, such that sufficient energy is not brought to the roadway by convection, the net energy transferred out by radiation can reduce the temperature below freezing, even though the air temperature is above freezing.

A fourth consideration is the high thermal conductivity of the metal structure of the bridge, resulting in rapid transfer of energy from these portions of the bridge.

By the way, although signs warning of this effect are common in eastern and northern regions of the United States, where the temperature often falls below freezing, the effect is not restricted to those areas. Even in the desert, where the air temperature does not fall below freezing, the effect of radiation can form frost on roadways and can even be used to make ice in the winter to be stored for use in the summer (see G. Barnes, "Jackrabbit Ears and Other Physics Problems", *The Physics Teacher*, **28**, 156 (1990)).

8.) Why do jets leave *contrails*?

Contrails from a jet aircraft are water droplets or ice crystals left in the wake of the aircraft. They are formed by a process which is called "adiabatic isobaric mixing". This refers to the mixing of combustion products from the jet engine with the atmosphere as the products are exhausted from the engine. For sufficiently low temperatures (found at the Poles or at high altitudes such as in the troposphere), the mixing of the exhaust with the atmosphere will produce a condensation which is left behind the jet in a long path and is visible from the ground as a contrail. For somewhat higher temperatures, a contrail can also be formed, but only if the ambient humidity is high enough. For even higher temperatures, the conditions are such that a contrail cannot be formed. For a mathematical treatment of contrail formation, see pp. 130 - 136 in J. V. Iribarne and W. L. Godson, *Atmospheric Thermodynamics*, 2nd ed., D. Reidel Publishing Co., Dordrecht, Holland, 1981.

For certain temperature and humidity conditions, a slight change in pressure is sufficient to cause condensation of the water vapor. Thus, it is sometimes possible for aircraft travelers to see a condensation trail left behind the tip of the aircraft wing as it perturbs the atmospheric pressure by passing through the air.

9.) One can heat a cup of water with a small electrical coil that is immersed in the water. Why do the directions say not to operate the heater if it is not immersed in water?

The design of the immersion coil depends on the high heat capacity of water. This will result in a modest increase in temperature of the water. Since the water is in contact with the coils, they will remain at a similar temperature. In addition, conduction and convection currents in the water will help to carry energy away from the coils so as to assist in keeping their temperature relatively low. If the coils were operated in air, the conduction and convection are reduced and the temperature of the coils could become quite high. What's more, since the surface to volume ratio of the coils is rather small, they are not effective at radiating energy away. Thus, when operating in air, the coils are on their way to becoming a light bulb filament! The net result of this increasing temperature is possible physical damage to the coils.

A similar situation occurs if a pan of water is left to boil too long on the stove. As long as there is water in the pan, the water can absorb energy from the bottom of the pan and, as a result, exhibit an increase in temperature or vaporize. If the water completely vaporizes, then this absorption is eliminated and the temperature of the pan can increase and cause warping of the metal.

10.) Why does a bare floor feel colder than a carpeted floor?

The sense of feel in humans is based not so much on the temperature of an object in contact with the skin as on the rate of energy flow between the object and the skin. This is why metals generally feel cold - they are good thermal conductors, so they absorb energy quickly from the skin. The bare floor and the carpet are at the same temperature, that of the room, but yet they do not feel the same temperature. The bare floor is a better thermal conductor than the carpet - hence it feels colder.

11.) In Southern California, a common occurrence is a Santa Ana condition, in which warm, dry winds blow into the Los Angeles basin from the mountains. But wait a minute. According to Mystery #1, mountain air is cold, What's going on here?

This is the reverse of Mystery #1! When air rises up the mountain, as in Mystery #1, it cools off because of a reduction in pressure and the resultant expansion. Now, think of the opposite behavior. When air moves <u>down</u> the mountain, it moves into regions of increasing pressure, which compresses the air and causes its temperature to rise. This kind of phenomenon occurs elsewhere and is called by a variety of names, such as the *Chinook* in the Rocky Mountains and the *Foehn* in Switzerland.

12.) When you turn your car headlights out and watch the light that they shine on a wall, they seem to go out much more slowly than normal light bulbs. Why?

Automobile headlights operate from 12 volt systems, unlike household bulbs, which operate at 110 volts. Thus, for a given amount of light power, an automobile headlight must draw more current than a household bulb. This translates to less electrical resistance for the automobile headlight. The resistance of a filament can be reduced by increasing the diameter. But increasing the diameter decreases the surface to volume ratio. Then, we have a similar situation to that in Mystery #7, and the automobile filament, upon cessation of the current, does not cool as quickly by radiation as the household bulb. This slower cooling is evident in the measurably longer time that it takes for the light from the automobile bulb to disappear.

13.) To keep a sandwich from a fast food restaurant warm, it is often wrapped in foil lined paper. But, wait a minute. Isn't metal foil a good <u>conductor</u> of heat?

Wrapping the sandwich in anything helps to cut down on <u>convective</u> energy transfers from the sandwich since it shields the sandwich from the air flow. In addition, the paper part of the foil lined paper provides some <u>conductive</u> insulation, along with the air trapped between the wrapper and the sandwich. Finally, the foil is there to cut down on <u>radiative</u> transfer. We have a situation similar to that in Mystery #3, where we used silver to help keep the liquid in a thermos bottle hot. Aluminum is a poor emitter. Thus, the use of aluminum foil will reduce the radiation of energy from the hot sandwich, helping to keep it warm longer.

14.) Why does frost tend to form first on the <u>tops</u> of objects (mailboxes, cars, etc.), but not on the <u>sides</u>?

The formation of frost depends on the temperature of the surface on which it forms. This temperature is the result of energy transfers by conduction, convection and radiation. The first two energy transfer processes are fairly uniform on the sides and top of an object. For the third method, radiation, the <u>emission</u> may be similar on all surfaces, but the <u>absorption</u> is not. The top surface of an object receives radiation from the sky, which, especially on clear nights, is at a fairly low effective radiative temperature (as discussed in Myth #5, Chapter 14). The sides of an object receive radiation from other objects on the surface of the Earth, which are at higher radiation temperatures than the sky. Thus, the net radiative outgoing transfer of energy is largest on the top of the object, the temperature drops fastest on the top, and this is where the frost forms first.

𝔐𝔞𝔤𝔦𝔠:

The Collapsing Can

As the water boils, the air and steam inside the can increase in temperature. When the can is capped, a fixed amount of gas is trapped inside. Then, as the gas cools, its pressure decreases. Eventually, the pressure difference between the inside of the can and the atmospheric pressure outside is enough to crush the can. Discussion of the physics of this demonstration is provided in J. E. Stewart, "The Collapsing Can Revisited", *The Physics Teacher*, **29**, 144 (1991). Photographs of a similar demonstration using a 55 gallon (200 liter) oil barrel can be found in J. O. Mattila, "Physics at the Fire Station", *The Physics Teacher*, **26**, 440 (1988).

Boiling Water in a Paper Cup

The bottom of the cup is a conductor of heat. Since it is quite thin, there will be a rather small temperature difference across the thickness of the bottom of the cup. The upper surface of the bottom of the cup is in contact with water. Despite the energy going into the water, its temperature will not increase beyond 100°C. Thus, the upper surface of the bottom of the cup is at 100°C and the lower surface is at most a few degrees higher. No part of the bottom of the cup is at a high enough temperature to ignite. If the flames are high enough that they rise up the side of the cup, the part of the cup above the water level can catch fire.

Black and White

The black sock appears to be black because it <u>absorbs</u> almost all of the visible radiation incident upon it. The white sock appears to be white because it <u>reflects</u> much of the visible radiation incident upon it. Thus, the temperature of the black sock will rise faster in the presence of visible radiation from the Sun, due to its high radiative absorption. After several minutes in the Sun, the black sock should be several degrees warmer than the white sock.

The Paper That Would Not Burn

The metal of the strainer acts as an excellent thermal conductor. Energy from the candle flame is absorbed by the metal and transferred quickly to the rest of the strainer. As a result, the energy reaching the paper is much reduced over that without the strainer in place. The paper therefore absorbs much less energy, so that it takes a long time before its kindling temperature is reached.

The Glowing (?) Stove

At the temperature of the burner at the lowest setting, the rate of emission of visible light is very low, so low that the human eye detects no radiation. The film, however, will accumulate photons from the long exposure time. Even at the low setting, some visible photons are being emitted, which will be gathered and stored on the film. You should be able to see the burner on the processed photograph. The color will be redder than the observed color at the high setting. This is due to the shift in the peak wavelength of emission farther into the infrared due to the decreased temperature.

If you are impatient to perform this long experiment or don't want to keep the room dark for so long, or don't want to use the necessary electricity, black and white photographs of the burner at the high and low settings are available on p. 74 of C. F. Bohren, *Clouds in a Glass of Beer*, John Wiley & Sons, New York, 1987.

𝕸𝖞𝖙𝖍:

1.) When friction acts on a sliding object, kinetic energy is converted to heat.

This is an example of the misuse of the word *heat*. Friction does <u>work</u> on a sliding object and transforms its kinetic energy to *internal energy*. Except for <u>subsequent</u> transfers of energy from the hotter objects to the environment, no heat is involved at all!

2.) The human body is a highly organized system, which evolved from simpler systems. This is a violation of the Second Law of Thermodynamics.

This is a common statement made by "creation scientists" as a "proof" that evolution is not possible. The Second Law of Thermodynamics states that a closed system tends toward greater <u>disorder</u> (measured by <u>entropy</u> - see Myth #1 in Chapter 7). The evolution of the complex bodies of humans is a movement toward more <u>order</u>. But the system in which humans evolved, the Earth, is not a closed system - it receives abundant continuous energy from the Sun and also (thermally) radiates energy into space. Thus, there is no requirement that organisms within this environment cannot evolve into more highly ordered organisms, as long as the entropy of the <u>universe</u> increases.

As an example of humans generating disorder, consider their appetite! Humans ingest highly complex and ordered molecules from the <u>environment</u>, such as sugars, proteins and carbohydrates and break them down into their disordered components. Thus, the environment suffers an increase in disorder, or an increase in entropy. By balancing the disorder created by the food necessary for humans, along with other processes by which disorder is generated, with the increase in order represented by their evolution, there is indeed a net increase in entropy of the universe.

For more details on evolutionists' and creationists' views on the Second Law, and other natural examples of local decreases in entropy, see J. P. Zetterberg (ed.), *Evolution versus Creationism: The Public Education Controversy*, Oryx Press, Phoenix, 1983.

3.) A fireplace is a heater for a home.

The air above the fire in a fireplace is heated. This hot air, which is less dense than warm air, is pushed up the chimney by the warm air in the room, carrying energy with it by convection. This air lost through the chimney must be replaced from somewhere. It is replaced with the warm air which enters the fireplace from the room in order to push the hot air up the chimney. Now, this air must be replaced from somewhere. It is usually replaced by <u>cold</u> air entering the house through leaks around doors and windows. The net result is that warm air from the house is going up the chimney and being replaced by cold air from outside. This is not obvious to the person sitting in front of the fire, due to the toasty feeling of the radiant heat, but, overall, the house is cooling off. This effect can be countered by putting glass doors on the fireplace, which will allow the radiant energy to pass through, but will cut off the convection flow from the room. The fire needs air, of course, so cutting off this convection flow will normally smother the fire. Allowance must be made to provide this air, which can be provided by air ducts from the outside. In this way, the fire uses cold air from outside rather than warm air from inside.

4.) A refrigerator can be used to cool off a room in the summer - just leave the door open!

While the refrigerator is very similar in design to an air conditioner, we have to look at where the exhaust energy is being dumped. An air conditioner is a heat pump which takes in energy from the inside of a house and delivers it to the warm air outside, thus reducing the temperature of the interior. The refrigerator does the same thing, pumping energy from inside the refrigerator to the air in the kitchen. Energy is absorbed by the freon from the interior of the refrigerator. To this energy is added that due to the work done by the compressor, which, in turn, was transferred in by electrical transmission. The combined energy is dumped to the air in the room. If the door of the refrigerator were left open, however, this air, and the energy it contains, could be recycled right back into the interior of the refrigerator. Thus, the energy could simply pass through the refrigeration cycle several times, with more energy added each time from electrical transmission. Thus, the net result would be to <u>heat up</u> the room, not cool it off.

5.) At the summer solstice in June, the Sun's rays strike the Earth in the Northern Hemisphere as close to perpendicularly as they ever do. Therefore June is the hottest month. Similarly, December is the coldest month.

While this seems to make sense, it does not agree with observation. In the Northern Hemisphere summer, July and August are the hottest months, even though they occur <u>after</u> the solstice. Similarly, December is not as cold on the average as January and February. This lag in the temperature is caused by the tremendous "thermal mass" of the Earth's surface. During the summer months, much energy from the Sun is stored as internal energy in the ground and the water. As the winter approaches, this stored energy is continually released and keeps the air temperature higher than one would expect. This causes the coldest months to be delayed past the winter solstice. Similarly, as the summer solstice approaches, the Earth's surface is cold from the long winter and energy from the air is absorbed, keeping the air temperature lower than expected, again causing a delay in the "appropriate" temperature.

The same effect occurs on a smaller scale each day. Even though the Sun is highest in the sky at solar noon, the hottest part of the day does not occur until later in the afternoon. This is due to the cooling effect of the ground, which is at a low temperature from the lack of solar radiation during the previous night.

6.) Infrared radiation is "heat radiation".

There is no reason to restrict the transfer of energy by radiation to infrared wavelengths. All wavelengths are able to transfer energy. Perhaps this misnomer has arisen because objects of "normal" temperatures emit their energy with an emission curve which peaks in the infrared.

7.) The following is a statement found in some physics textbooks: *Heat is the energy that flows from one system to another because of a temperature difference between them.* **The textbooks then go on to describe three modes of heat transfer - conduction, convection and radiation.**

As discussed in the Concepts section, these three processes are often discussed in the section of a textbook on heat, but can be applied to other situations that do not involve heat at all. This Myth addresses another complaint about textbooks. In reality, conduction is the only heat process depending on a temperature difference. We indeed do need a temperature gradient to drive energy flow by conduction. But let's look at radiation. Suppose we have two stars of identical temperatures and emissivities sitting near one another. They are constantly exchanging energy with one another, due to radiation, yet there is no temperature difference. We might be able to rescue the statement in this case by adding the small word *net*: *The net heat is the energy that flows from one system to another because of a temperature difference between them.* Thus, if the stars are at different temperatures, there would be a net heat flow between them.

If we look at convection, there seems to be no way to rescue the statement, unless we only include natural (buoyant) convection. If we consider forced convection, the flow of energy has nothing to do with a temperature difference. We can even think of a convective flow from low temperature to high temperature - the forced convection of energy carried by freon, for example, in your air conditioning unit from the evaporator coils inside the house to the condenser coils outside.

8.) Archimedes once fought off the Romans in a ship 100 m away by having his soldiers reflect the Sun's rays from hand-held mirrors, setting the Roman ship on fire.

The question here is that of possibility - can a group of hand held mirrors reflect enough light to raise the temperature of a wooden ship fast enough so that the Romans could not fetch buckets of water to douse the flames? If we do a back of the envelope calculation, even assuming a large number of large mirrors, we find that the Romans have plenty of time to obtain buckets of water and put out any fires which the Greeks succeed in starting. For details on one such calculation, see p. 139 in N. Thompson, *Thinking Like a Physicist*, Adam Hilger, Bristol, England, 1987.

9.) Heat rises.

This phrase indicates that heat is something tangible, as in the days of *caloric* theory. The correct statement should be that a hot fluid rises in the presence of a cooler fluid, due to the buoyant force applied by the cooler fluid.

Chapter 16
Simple Harmonic Oscillation

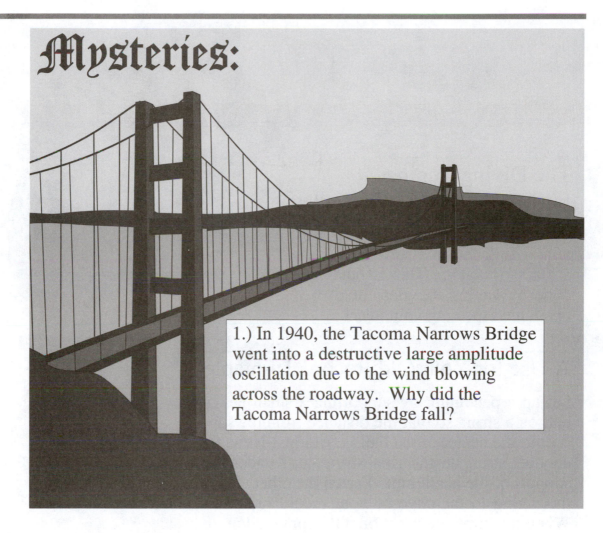

Mysteries:

1.) In 1940, the Tacoma Narrows Bridge went into a destructive large amplitude oscillation due to the wind blowing across the roadway. Why did the Tacoma Narrows Bridge fall?

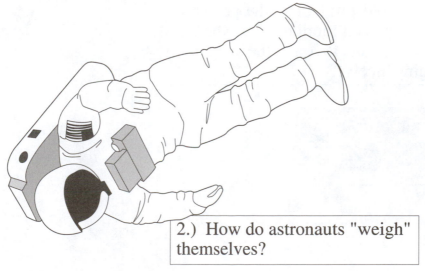

2.) How do astronauts "weigh" themselves?

Magic:

The Diving Tuning Fork

Insert the vibrating ends of a tuning fork into the surface of a glass of water. This is best done by having the glass held by an unsuspecting volunteer!

A Use for Your Old Record Turntable

Set up a pendulum by supporting a small heavy ball on a string so that the distance from the support to the center of the ball is 0.44 m. Now, tie another string to the first string, just under the support of the pendulum. Fasten the other end of this string to the edge of your record turntable. When the turntable is set on $33\frac{1}{3}$ rpm (and/or 78 rpm if it is a very old turntable!) the pendulum will show a disorganized motion. But when the turntable is operated at 45 rpm, the pendulum will start swinging nicely!

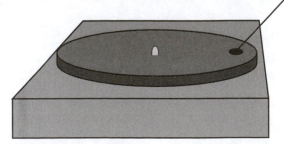

𝕸𝖆𝖌𝖎𝖈:

A New Interpretation for "Resonance in Pipes"

From the hardware store, obtain a 10 foot section of 1/2 " PVC (polyvinyl chloride) plumbing pipe. It is the standard white plastic pipe which is sold everywhere. Grasp the pipe in the center and hold it out in front of you in a horizontal position. Move the pipe slowly back and forth (perpendicular to your body and to the pipe) and watch the vibrations of the ends of the pipe. They will be very small if they exist at all. Now, move the pipe back and forth as fast as you can. The ends of the pipe will wiggle, but not very much. Now, move the pipe back and forth at an intermediate frequency. It is not too critical what this frequency is - feedback from the pipe will help you locate the correct frequency fairly easily. At this frequency, the ends of the pipe can achieve an amplitude of vibration of a meter or more, and the pipe can actually be broken in this way.

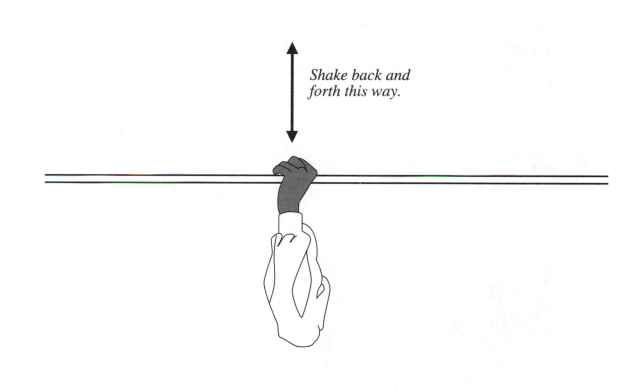

Shake back and forth this way.

Myth:

1.) An old television commercial for audio recording tape showed a singer breaking a wine glass with her voice. The question was then asked if this were actually her voice or a recording. The inference is that the tape is of such high quality that the excellent reproduction of the sound is able to break the glass.

2.) The moon always shows us the same face.

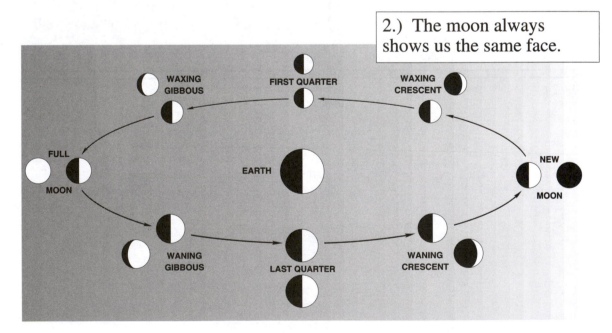

𝔐𝔶𝔱𝔥:

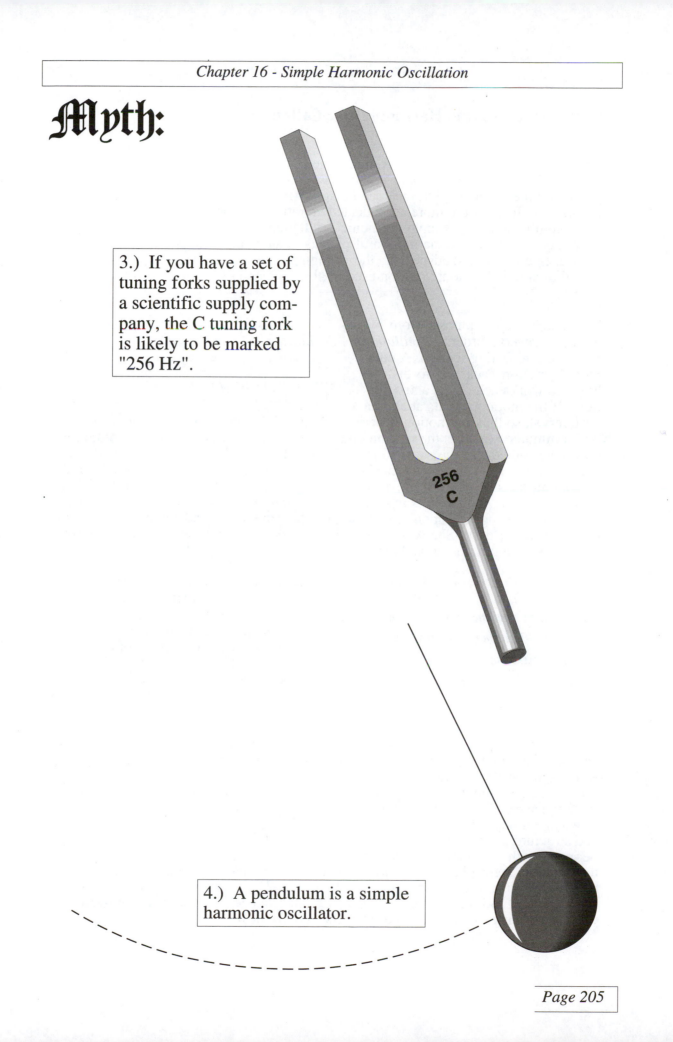

3.) If you have a set of tuning forks supplied by a scientific supply company, the C tuning fork is likely to be marked "256 Hz".

256
C

4.) A pendulum is a simple harmonic oscillator.

Concepts of Simple Harmonic Oscillation

Simple harmonic oscillation is the model used by scientists to describe a fundamental type of *vibration*. It is a model in the sense that very few vibrations in real life are purely simple harmonic, but they can be approximated very well by simple harmonic oscillation. In addition, more complicated vibrations can be expressed mathematically as combinations of simple harmonic oscillations. Examples of vibration in our everyday world include the motion of pistons in an automobile engine, vibrations of a speaker cone in a loudspeaker, the swinging of a chandelier, etc. The standard model of simple harmonic motion used in physics classes is the mass on the end of a spring.

Let us analyze the phrase, word-by-word, in reverse order. *Oscillation* refers to a repeating motion. A commuter's motion back and forth from home to work each day is an oscillation. If the motion repeats at a regular interval, so that the motion is periodic, then it is called *harmonic oscillation*. For our commuter example, the motion could be considered harmonic if the commuting took place at exactly the same time each day. This would not be true for our commuter, however, unless she traveled to work on weekends, also. If the position of the oscillating particle can be described by a single sinusoidal function of time (a sine wave), then the motion is *simple harmonic oscillation*. For our commuter, if her motion were simple harmonic, then she would move more and more slowly as she approached her desk. She would just reach the desk as zero velocity was achieved, but then she would immediately get up and start for home again!

> *Oscillation: A repeating motion*
> *Harmonic: Repeating at a regular interval*
> *Simple: Repeating in a certain way*

There are various parameters that are used to describe simple harmonic motion. The time for one complete cycle of oscillation to occur is the *period*. The inverse of this parameter is the *frequency*, which indicates the number of cycles per unit time. The maximum displacement of the particle from the equilibrium position is the *amplitude*, measured in units of length.

> *Period: Time for one complete cycle*
> *Frequency: Number of complete cycles per unit time*
> *Amplitude: Maximum displacement from equilibrium*

An important aspect of vibration is *resonance*. This refers to the large amplitude response of a vibrating system when it is driven by an external source at a frequency equal to the system's natural frequency (or an integer submultiple of the natural frequency). The natural frequency of a system depends on its shape, size and structure. Perhaps your earliest experience with resonance was learning to "pump up" on a playground swing. As a child, you learned to move your legs back and forth to increase the amplitude of the swing. The movement of your legs was a driving force and, of course, you had to pump your legs at the same frequency at which the swing oscillated, or your efforts were fruitless.

> *Resonance: The large amplitude response of an oscillating system when driven at its natural frequency.*

Discussions; Chapter 16 - Simple Harmonic Oscillation

𝔐𝔶𝔰𝔱𝔢𝔯𝔦𝔢𝔰:

1.) In 1940, the Tacoma Narrows Bridge went into a destructive large amplitude oscillation due to the wind blowing across the roadway. Why did the Tacoma Narrows Bridge fall?

This example is often used in physics classes as a demonstration of *resonance*. According to the resonance explanation, as the wind blew across the roadway of the bridge, the turbulence in the air caused the wind to "flap" across the roadway, similar to the way that a flag flaps in a strong breeze. At the particular wind speed on the day of the collapse, the flapping frequency matched the natural frequency of the torsional vibrational mode of the bridge and the flapping air thus drove the bridge into a resonance oscillation that was too large for the structure to maintain. This understanding was accepted for many years after the collapse. Current studies of the Tacoma Narrows Bridge challenge this explanation as being too simple and claim that non-linearities in the behavior of the bridge led to <u>chaotic</u> motion. It appears from these recent studies that the resonance explanation is far too simplified. One article (K. Y. Billah and R. H. Scanlan, "Resonance, Tacoma Narrows Bridge Failure, and Undergraduate Physics Textbooks", *American Journal of Physics*, **59**, 118 (1991)) convincingly argues that the observed vibrational frequency of the bridge was only 20% of that predicted by the simple resonance model. The article then proceeds to explain a newer model, known as *single-degree-of-freedom torsional flutter*. The Tacoma Narrows Bridge will continue to be studied for quite some time and it will be interesting to see the models which develop in the future. The collapse of the Tacoma Narrows Bridge is widely distributed on film loops, videotapes (such as *The Mechanical Universe*) and other audio/visual resources.

2.) How do astronauts "weigh" themselves?

The astronauts cannot weigh themselves by simply standing on a scale. This is not because they are weightless (See Myth #2, Chapter 6), but because they are in free fall - they are accelerating toward the Earth at the acceleration of gravity. The same would be true for someone standing on a bathroom scale in a freely-falling elevator (See Mystery #2, Chapter 3). The scale reading would be zero. The astronauts cannot really determine their <u>weight</u>. In order to "weigh" themselves, they need to determine their <u>mass</u>, which can be done even in accelerated motion. To do this, the astronaut makes use of simple harmonic motion. He or she sits on an oscillating machine and sets it into oscillation. The period of the motion, which depends on the mass, is then measured. The vibrating mass is then calculated from the period, the mass of the seat is subtracted and the weight of the astronaut can then be calculated.

𝕸𝖆𝖌𝖎𝖈:

The Diving Tuning Fork

It is difficult to see the vibrations of a tuning fork unless your eyes are close to the fork. It is also difficult to hear the fork, unless an efficient radiating mechanism such as a table top or a blackboard is brought into contact with the fork. Dipping the vibrating fork into a glass of water makes an effective visual display of the vibrations as the vibrating tines cause a shower of water to jump out of the glass.

A Use for Your Old Record Turntable

This is a demonstration of *resonance*. The length of the pendulum was chosen so that its natural frequency is 45 oscillations per minute. Thus, when the phonograph turntable rotates at this frequency, the pendulum receives a tug once each cycle of its natural motion and the oscillation builds up. At $33^1/_3$ rpm (or 78 rpm), the driving frequency does not match the natural frequency and the pendulum does not exhibit a resonance response.

A New Interpretation for "Resonance in Pipes"

This is another dramatic demonstration of resonance. The structure of the pipe is such that the resonance frequency for transverse vibrations is between zero and "as fast as you can shake it" for a typical human. The frequency is also close to that of walking, so that if you are walking while carrying the pipe at your side, it may go into a resonance vibration! For more information, see J. W. Jewett, "Giving New Meaning to 'Resonance in Pipes'", *The Physics Teacher*, **31**, 253 (1993).

𝕸𝖞𝖙𝖍:

1.) An old television commercial for audio recording tape showed a singer breaking a wine glass with her voice. The question was then asked if this was actually her voice or a recording. The inference is that the tape is of such high quality that the excellent reproduction of the sound is able to break the glass.

This is a demonstration of *resonance*. It is certainly possible to break a wine glass with an amplified singing voice. If the frequency of the voice is the same as the natural frequency of the glass, and the sound is loud enough, the glass can be set into a resonance vibration whose amplitude is large enough to surpass the elastic limit of the glass. But the inference that high quality reproduction is necessary is not justified. All that is important is that the <u>frequency</u> is recorded and played back correctly. The <u>waveform</u> of the sound can be altered as long as the frequency remains the same. Suppose, for example, that the singer sings a perfect <u>sine</u> wave, but the tape records it as a <u>square</u> wave. If the tape player plays the sound back at the right speed, the glass will still receive energy at the resonance frequency and will be set into vibration leading to breakage, even though the tape reproduction was <u>terrible</u>. Thus, this phenomenon does not require high quality reproduction and, thus, does <u>not</u> demonstrate the quality of the recording tape. What it does demonstrate is the quality of the tape <u>player</u>, in that it played back the tape at an accurate speed!

For a description of a classroom demonstration of breaking a glass with sound, see H. Kruglak, R. Hiltbrand and D. Kangas, "Shattering Glass with Sound Simplified", *The Physics Teacher*, **28**, 418 (1990).

2.) The moon always shows us the same face.

We discussed this aspect of the moon in Mystery #3 of Chapter 6. Why is it now presented as a myth? Well, the face isn't always <u>quite</u> the same. The moon appears to undergo a combined set of harmonic motions that are called *librations*. It should be pointed out that these are not <u>actual</u> vibrations of the moon, but <u>apparent</u> vibrations caused by the following three major factors as well as some minor ones.

1.) The moon's rotation axis is tilted by a few degrees with respect to the plane of its orbit, as shown in the diagram below:

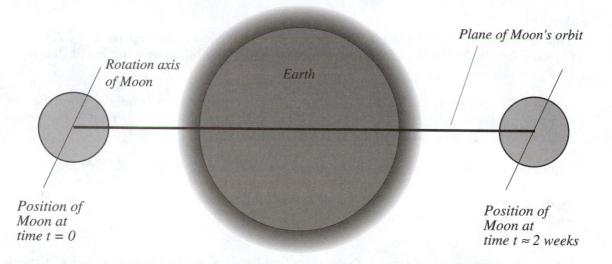

Thus, at $t = 0$, we, on the Earth, can see a little past the North Pole of the Moon (at the top of the Moon in the diagram). About two weeks later, when the Moon is on the other side of the Earth, we can see just past the South Pole. This results in an apparent North-South vibration of the Moon with a period of 29.5 days, the time for the Moon to complete one cycle of its phases around the Earth (the *synodic* period).

2.) The orbit of the Moon is slightly elliptical. According to Kepler's First Law, the Earth is at a <u>focus</u> of this ellipse, as shown in the diagram on the next page (We ignore the small motion of the Earth around the Moon-Earth center of mass and imagine the

Moon in orbit about a fixed Earth.). When the Moon is at apogee and perigee, the line of sight from the Earth is right along the major axis of the ellipse and right into the lunar tidal bulge on which the Earth has locked (See Mystery #3, Chapter 6), which is highly exaggerated in the diagram. When the Moon has completed one fourth (and three-fourths) of a rotation from perigee, it is slightly past the "top" ("bottom") of the ellipse, since, according to Kepler's Second Law, it will move faster in that part of its orbit when it is near the Earth. Considering the lines of sight from these positions in the diagram, we see that we view the Moon slightly from the side.

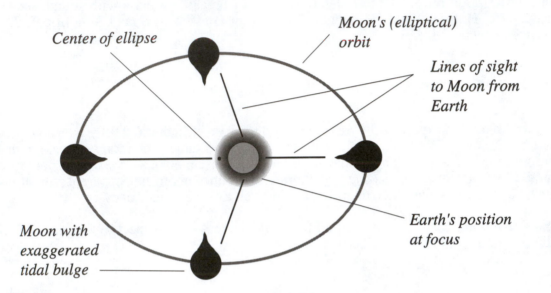

This results in an apparent East-West vibration of the Moon with a period of 27.3 days, the time for the Moon to complete one revolution in its orbit around the Earth (the *sidereal* period).

3.) The finite size of the Earth results in our viewing the Moon from slightly different vantage points at moonrise and moonset. This results in an apparent East-West vibration of the Moon with a period of about 24 hours.

These effects combine so that, although we can only see 50% of the Moon's face at any one time, we can see 59% of the face over a long period of time. Of this amount, 41% is always visible and 18% alternates in its visibility. The remaining 41% of the Moon's face is never visible from Earth.

3.) If you have a set of tuning forks supplied by a scientific company, the C tuning fork is likely to be marked "256 Hz".

The notes of the tempered musical scale are determined by setting concert A (A above middle C) at 440 Hz. The other frequencies are then established according to this reference. Octaves represent frequency ratios of 2, while each half step is a frequency ratio of the twelfth root of 2. In this scale, then, middle C is about 263 Hz. There was a brief attempt in the past by physicists to slightly adjust middle C to 256 Hz, so that all C's would be represented by frequencies which are powers of 2 (256 = 2 to the eighth power). This adjustment was not widely accepted, yet it does seem to remain by tradition on the tuning forks of scientific supply houses.

4.) A pendulum is a simple harmonic oscillator.

Simple harmonic oscillation is motion such that the position of the particle is described by a single sinusoidal function. This kind of motion results from a system that is described by <u>Hooke's Law</u>, where the force on the particle is proportional to the displacement of the particle and in the reverse direction. If the force on the pendulum is carefully analyzed, it turns out not to be a Hooke's Law force (the restoring force is proportional to the <u>sine</u> of the angular displacement, not to the displacement itself) and the motion of the particle is not a sinusoidal function. It is <u>similar</u> to sinusoidal motion, however. What's more, for small angles of oscillation, the actual motion is almost indistinguishable from simple harmonic motion, which is why the pendulum often appears as an example of this type of motion.

Chapter 17
Waves

Mysteries:

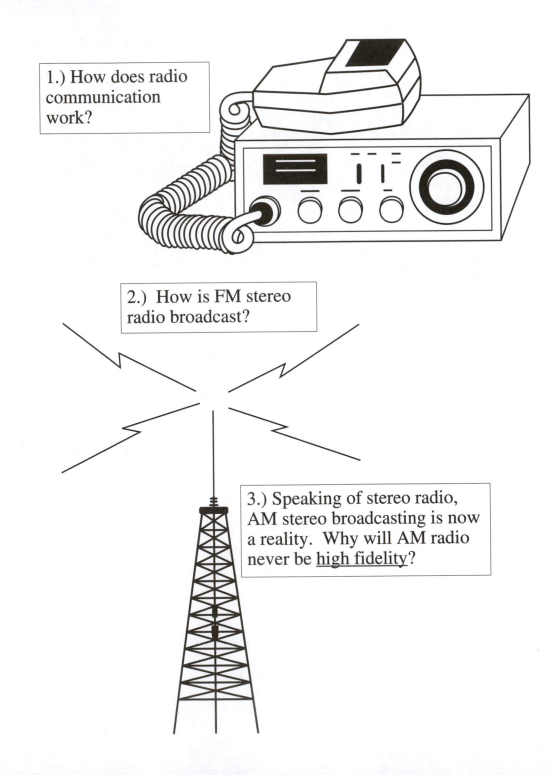

1.) How does radio communication work?

2.) How is FM stereo radio broadcast?

3.) Speaking of stereo radio, AM stereo broadcasting is now a reality. Why will AM radio never be <u>high fidelity</u>?

Mysteries:

4.) Why do ocean waves <u>break</u>?

5.) Why aren't the frets on a guitar evenly spaced?

6.) Why do people talk funny after they have breathed helium?

Waves
$f = \frac{1}{T}$

Exam
Tomorrow
Chaps 15 - 19

He

Hee, hee!

Ha, ha, ha!

Ho, Ho!

Mysteries:

7.) It is a common experience for parents to find that much of their children's bathwater is on the bathroom floor, much more than seems reasonable from simply splashing. Why is this?

8.) For a wind instrument, such as a saxophone or a trumpet, if the room temperature increases, the frequency of the sound played by the instrument increases. For a string instrument, such as a guitar or a bass fiddle, the frequency decreases if the room temperature increases. Why is this?

Magic:

The Cardboard Bass

Place a long cardboard tube, such as a carpet tube, over a wide burner flame (Be careful not to set the tube on fire!). You will hear a loud, low frequency sound.

Push string into your ear with your finger

Hanger Chimes

Tie a 2 meter long thread or string to the ends of the straight part of a wire coat hanger. Letting the hanger hang upside down from the string, knock the hanger against a table. You will hear a dull, clunking sound. Now, loop the string over your head and press the string into your ears with your fingers, letting the hanger hang below your head. Knock the hanger against a table again and listen to the difference!

Knock the hanger into a table

Magic:

Insert rolled-up paper into tube

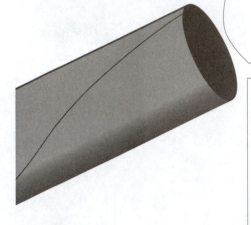

Mailing Tube Resonance

Roll up a piece of notebook paper and insert it into the end of a mailing tube with a length about 50 cm. Move the rolled-up paper back and forth in the end of the tube and you will hear a sound rising and falling in pitch. Now place your hand snugly over the opposite end of the tube. What happens to the sound?

Myth:

1.) Sometimes sudden disasters happen in the ocean: earthquakes, old horror movie monsters, meteors, etc. When this happens, a violent disturbance occurs in the water, which is the source of a *tidal wave* arriving on land somewhere.

Myth:

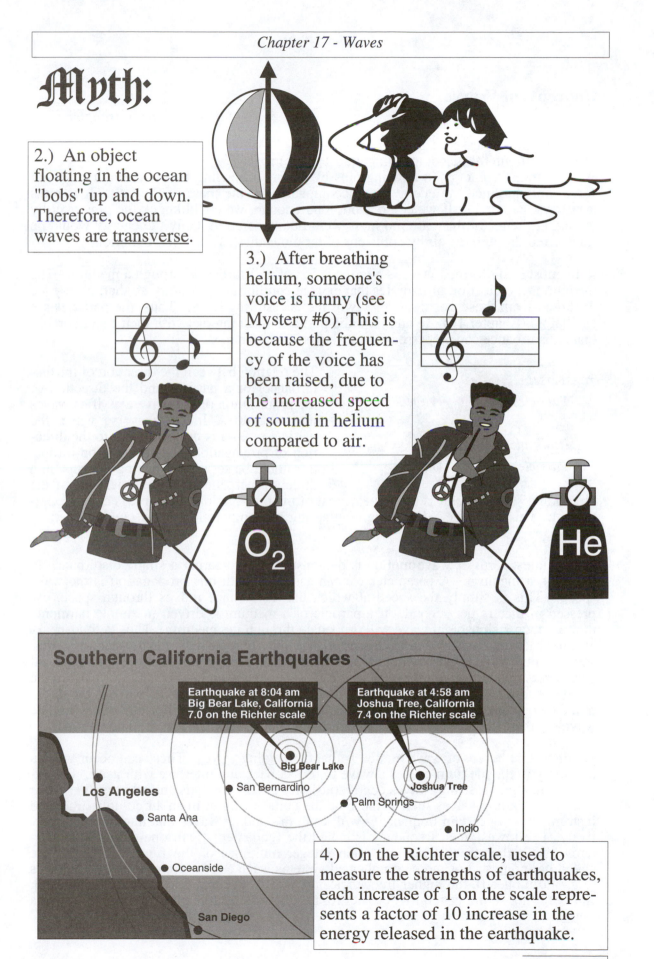

2.) An object floating in the ocean "bobs" up and down. Therefore, ocean waves are <u>transverse</u>.

3.) After breathing helium, someone's voice is funny (see Mystery #6). This is because the frequency of the voice has been raised, due to the increased speed of sound in helium compared to air.

O₂

He

Southern California Earthquakes

Earthquake at 8:04 am
Big Bear Lake, California
7.0 on the Richter scale

Earthquake at 4:58 am
Joshua Tree, California
7.4 on the Richter scale

Big Bear Lake

• San Bernardino

Joshua Tree

Los Angeles

• Palm Springs

• Santa Ana

• Indio

• Oceanside

4.) On the Richter scale, used to measure the strengths of earthquakes, each increase of 1 on the scale represents a factor of 10 increase in the energy released in the earthquake.

San Diego

Concepts of Waves

The fundamental essence of a wave is the transfer of energy without the accompanying transfer of mass.

We can transfer energy to bowling pins by rolling a bowling ball at them. This is not wave motion, since the energy was accompanied by the mass of the ball. It is mass transfer (Chapter 7). If we could shout loud enough, we could knock the pins over by yelling at them. In this case, no mass would be transferred. Only the energy would be transferred through the air by collisions of air molecules. This is wave motion.

In its most basic form, a wave is a propagation of a disturbance through a medium. The medium is a collection of particles that exist at equilibrium positions, so that, if they are disturbed from these positions, they will feel a restoring force. Thus, the particles are oscillators (Chapter 16). In addition, the particles must interact with their neighbors, so that a disturbance can be passed from one particle to the next.

Transverse:

Particles ↕   *Wave* ⟶

Longitudinal:

Particles ↔   *Wave* ⟶

The relation between the direction of the disturbance for a particle and the direction of the propagation of the wave classifies waves into categories. In a transverse wave, the particle motion is perpendicular to the direction of propagation. For a wave on a rope, this could be set up by shaking the rope in a direction perpendicular to its length. If the direction of the particles is the same as that of the propagation, the wave is called longitudinal. This could be set up on a long spring by compressing a few coils and letting them go.

The simplest form of wave motion is demonstrated by creating a single disturbance (a pulse) in a medium. A parameter we can associate with this propagation is the wave speed. This is simply the speed at which the disturbance moves through space, expressed in meters per second. If a particle of a medium is driven in simple harmonic motion, then a sinusoidal wave will propagate through the medium. Thus, a loudspeaker cone moving back and forth in simple harmonic motion will (ideally) cause a sinusoidal sound wave to propagate through the air. This type of wave allows for additional parameters in addition to the wave speed. The frequency and period of the wave are the same as those for the simple harmonic motion. The amplitude of the wave is the maximum displacement of a particle of the medium from the equilibrium position. The wavelength of the wave is the distance through which the wave moves in one period.

An important aspect of wave motion is that of standing waves. These can occur when a boundary in the medium causes a wave to reflect back and interfere with itself. For example, on a guitar string, waves continuously reflect off the ends and add together. For confined standing waves such as those on the guitar string or in an air column of a wind instrument, only certain frequencies will result in standing waves that interfere constructively. The lowest such frequency is called the fundamental. In one-dimensional systems, the higher frequencies are often integer multiples of the fundamental and are called harmonics. For more complicated systems, such as two-dimensional membranes (e.g., a drumhead), the higher frequencies are not integer multiples of the fundamental. The resulting sound generally does not result in an identifiable pitch and is thus called a noise rather than a musical sound.

Discussions; Chapter 17 - Waves

𝔐𝔶𝔰𝔱𝔢𝔯𝔦𝔢𝔰:

1.) How does radio communication work?

Radio communication depends on two waves - the *sound wave* representing the audio information (from a microphone, tape deck, phonograph or CD player at the radio station) and the *carrier wave* (Within the transmitter electronics, these are more properly described as electronic <u>oscillations</u> rather than waves, but they are popularly called waves.). The carrier wave is of higher frequency than the sound wave - on the order of 10^5 - 10^6 Hz for AM (amplitude modulation) or 10^8 Hz for FM (frequency modulation). The word *modulation* in the previous parentheses refers to the process of encoding the sound onto the carrier wave. In amplitude modulation, the pure (electronic) carrier wave is added to the (electronic) sound wave, so that the resulting combined wave has an amplitude which varies in time according to the sound wave. This combined wave is then broadcast by means of an antenna as an electromagnetic wave. In your radio receiver, another antenna detects the electromagnetic wave. The wave causes a current in the antenna which reproduces the oscillations of the wave. This current is processed by the <u>tuner</u>, which subtracts out the carrier wave, leaving the original electronic sound wave, which is then amplified and eventually passed on to the loudspeakers to be converted back into actual sound. The process for FM radio is similar, except that the sound wave is used to modulate the <u>frequency</u> of the carrier rather than the amplitude.

2.) How is FM stereo radio broadcast?

For stereo broadcasting, we need to keep two channels of information separate. This is performed by expanding the spectrum of the sound information that is used to modulate the carrier wave, as described in the discussion to Mystery #1. In normal monophonic broadcasting, the carrier wave is modulated by sound frequencies from 0 to 15 kHz, the typical range of frequencies of human hearing. For stereo broadcasting, two other modulation signals are added. A wave at 19 kHz, called the "pilot signal" is added. This is the cue for your receiver to light up the "Stereo" indicator. The second channel of information is added to a 38 kHz wave (encoding the second channel as an <u>ultrasonic</u> signal) and is then added to the original channel and the pilot signal. Since the second channel information is at such a high frequency, it does not overlap with the original information. The combination of all of these signals is now used to modulate the carrier wave and the combination is broadcast. In your receiver, the tuner demodulates the whole signal by subtracting out the carrier wave, separates the three frequency ranges (0 - 15 kHz, 19 kHz, ultrasonic), signals the Stereo light to turn on, subtracts out the 38 kHz from the ultrasonic signal and sends the sound signal on to the amplifiers. But one problem remains - what information is to be encoded onto these two signal channels? It is tempting and simple to just encode one signal channel, say the 0 - 15 kHz signal, with the left stereo channel and the other, the ultrasonic channel, with the right stereo channel information. There is one serious problem with this technique. There are a large number of monophonic radios still playing. With the scheme just outlined, these radios would only pick up the left channel, that is, only half of the information. We need a more clever design. This is performed by encoding the 0 - 15 kHz signal channel with the <u>combination</u> of left and right stereo channel information (L + R). Thus, the mono-phonic radio will pick up all of the information. The ultrasonic signal channel is encoded with the <u>difference</u> between the two stereo channel signals (L - R). Now, in the stereo receiver, we must add one more step. The <u>demultiplexer</u> circuits will <u>add</u> the two signals [(L + R) + (L - R) = 2L] and send the result to the left channel amplifier and also <u>subtract</u> the two signals [(L + R) - (L - R) = 2R] and send the result to the right channel amplifier. Now, both monophonic and stereo radio owners are happy!

3.) Speaking of stereo radio, AM stereo broadcasting is now a reality. Why will AM radio never be <u>high fidelity</u>?

The scheme outlined in the answer to Mystery #2 is independent of whether the modulation is AM or FM. Either can be broadcast as stereo and, in principle, either can be broadcast as high fidelity. But a comparison of AM and FM reception clearly demonstrates the inferior fidelity of AM radio, even if it is in stereo. Musical sounds or voices contain combinations of frequencies that are integer multiples of a fundamental frequency ("harmonics"). Fidelity depends on the accurate reproduction of the original sound, which requires the reproduction of all frequencies up to at least 15 kHz, the typical upper limit of the range of human hearing. AM radio, by law, must cut off frequencies above 5000 Hz, which necessarily removes the higher harmonics and decreases the fidelity of the reproduction. Why does this law exist? We can understand this by appealing to our understanding of <u>beats</u>. If we add a 400 Hz sound wave to a 402 Hz sound wave, the result is a sound wave of 401 Hz, beating at 2 Hz. Now, let's reverse the process - suppose we start with a steady 401 Hz wave and <u>modulate</u> it at 2 Hz. This gives us the same result as combining the two waves. Thus, <u>a 401 Hz wave amplitude modulated at 2 Hz is equivalent to a combination of waves of frequencies 400 Hz and 402 Hz</u>*. We generalize from this example to the following - if a pure wave is amplitude modulated at a frequency f (by a sound wave, say, of frequency f), the result is a spread in the frequency of the wave of half of f on either side of the original frequency. When we modulate a radio carrier wave with the sound wave information, we introduce just such a spread in the carrier wave frequencies. Thus, when a modulated carrier wave is to be received in our stereo system, we cannot just receive the carrier frequency, we must receive a range of frequencies in order to reproduce the entire sound. Now, AM radio stations are separated by only 10 kHz. If AM stations were allowed to modulate their carrier waves with sound waves up to 15 kHz, <u>the resulting spread of the carrier wave frequencies would cause stations to overlap on the dial</u>. Thus, the modulation must be cut off at a frequency level, which is chosen at 5000 Hz, to prevent this overlap. As a result, the fidelity of reproduction is necessarily low. The problem could be solved by separating stations by larger frequency differences, but this would be very difficult to do now that stations are established at their carrier frequencies. FM radio stations are separated by frequency differences of 200 kHz, so the problem does not exist for FM. It is interesting to note that the <u>entire</u> AM radio band has enough "room" for only 5 stations separated by the frequency differences in the FM band!

---

*We can relate this to the *uncertainty principle*. By introducing the modulation, we have gained information in the time domain, since the wave amplitude is no longer constant. As a result, we have lost information about the frequency domain - the single frequency has become a combination of two frequencies.

4.) Why do ocean waves <u>break</u>?

For water waves traveling in deep water, the depth of water plays little role in the wave speed. In shallow water, however, the wave speed depends on the depth of the water. When a wave approaches the shallow water of the beach, it slows down, causing the amplitude to increase. In very shallow water, then, the "top" of the wave is in deeper water than the "bottom" of the wave. As a result, since the speed depends on depth, the top of the wave is moving faster than the bottom and the wave "breaks".

5.) Why aren't the frets on a guitar evenly spaced?

The frets on a guitar are placed in the appropriate positions that will introduce standing waves on the guitar strings whose fundamental frequencies correspond to the notes in the musical tempered scale. There are two contributions to the fact that the frets are not evenly spaced. The first is a physics reason, in that the fundamental frequency of the standing wave on a string is <u>inversely proportional</u> to the length. Thus, as we move to new notes, the length of the string is determined by fractions of the length, not by differences. For example, for an octave above the open string, the fret is at the half point. The next octave is at the quarter point, etc. The second reason is a psychological-musical reason. The tempered scale that we use for our music is based on frequency <u>ratios</u> rather than differences. Each half step represents a ratio of the twelfth root of 2 while each octave represents a frequency ratio of 2. It is a useful exercise to imagine how the frets would be spaced under other imaginable possibilities - for example, frequency proportional to length and the scale based on ratios, or frequency inversely proportional to length and the scale based on differences.

6.) Why do people talk funny after they have breathed helium?

The sound of a voice depends strongly on the <u>mixture</u> of frequencies that are emitted. The driving frequency is that of the vocal folds. The sound produced by the folds is then filtered by the vocal tract. The filtering effects are determined by the shape and volume of the mouth as well as by the gas that is in it. After someone has breathed helium, their mouth is filled with a gas in which the speed of sound is higher than in air. As a result, higher frequencies in the spectrum of the sound from the vocal folds are enhanced and the voice has the familiar "funny" sound. See Myth #3.

7.) It is a common experience for parents to find that much of their children's bathwater is on the bathroom floor, much more than seems reasonable from simply splashing. Why is this?

While some of these situations may be explained by vigorous splashing, others depend on standing waves. Many children learn in the bathtub that they can cause the water to slosh back and forth violently by sliding their body back and forth at the right frequency. This is another childhood case of *resonance*, in which the driving force of the child's sliding body is in resonance with the standing wave frequency of water waves in the length of the bathtub.

8.) For a wind instrument, such as a clarinet or a trumpet, if the room temperature increases, the frequency of the sound played by the instrument <u>increases</u>. For a string instrument, such as a guitar or a violin, the frequency <u>decreases</u> if the room temperature increases. Why is this?

The frequency of a wind instrument depends on standing <u>sound</u> waves in the air column of the instrument. If the air in the instrument increases in temperature, the speed of sound increases and the standing wave frequency correspondingly increases. (It should be noted that the temperature of the gas within an air column instrument is not room temperature - it is raised above room temperature by the relatively warm breath of the player.) In a string instrument, the frequency depends on standing <u>string</u> waves on the strings. If the temperature rises, the strings expand (Chapter 14). This loosens the strings, resulting in lower tension. The wave speed on strings is proportional to the square root of the tension, so that the wave speed decreases with increasing temperature as does the standing wave frequency.

𝕸𝖆𝖌𝖎𝖈:

The Cardboard Bass

The burner heats the air inside the cardboard tube. This heated air is pushed up the tube by natural convection, resulting in a flow of air through the tube. The moving air makes a rushing noise, which contains many frequencies, including the fundamental standing wave frequency of the tube. The tube responds to this particular frequency by resonance, setting up a strong standing sound wave in the tube that emits a surprisingly loud sound.

Hanger Chimes

With the second method described, the hanger will sound like Big Ben! Since the hanger has a small surface area, it is not an efficient radiator of sound - the small surface area cannot drive many air molecules into vibration. The string provides an effective mechanism for propagating the vibrations of the hanger directly into your ears where they are converted into audible sound. Thus, more energy is delivered to your ears per unit time than when the energy travels through the air and, consequently, the sound is louder.

Mailing Tube Resonance

The sliding of the paper against the cardboard creates a sound with many frequencies, similar to the rushing air in The Cardboard Bass.   One of these frequencies will excite the fundamental standing wave frequency of the mailing tube by resonance. But the actual air column consists of the mailing tube <u>and</u> your rolled-up paper. As you move the paper back and forth, you change the length of the air column, which results in a change in the standing wave frequency. Placing your hand snugly over the opposite end converts the air column from open to closed. The fundamental frequency is lowered by an octave.

𝕸𝖞𝖙𝖍:

1.) Sometimes sudden disasters happen in the ocean: earthquakes, old horror movie monsters, meteors, etc. When this happens, a violent disturbance occurs in the water, which is the source of a *tidal wave* arriving on land somewhere.

The phrase *tidal wave* is often used colloquially to mean a large, violent wave. In reality, however, this phrase just refers to the two wavelike swellings of the water on the Earth due to the gravitational attraction of the Sun and the Moon (Mystery #2, Chapter 6), which result in the twice-daily ebb and flow of the tides. The correct term for a large violent wave in the ocean is a *tsunami*, a word borrowed from Japanese (*tsu* - harbor; *nami* - wave).

2.) An object floating in the ocean "bobs" up and down. Therefore, ocean waves are <u>transverse</u>.

Waves on the surface of water are subject to two restoring forces. For small wavelengths (ripples), the dominant restoring force is surface tension, which tends to keep the surface flat to restore equilibrium. For large wavelengths, the dominant restoring force is gravity. We are referring to gravity waves in the ocean in this discussion.

Look more carefully again at the "bobbing" object in the ocean. Careful observation shows that it is actually moving in a <u>circular</u> path as it follows the movement of the water molecules surrounding it. Water at the surface of the ocean moves with both a transverse <u>and</u> a longitudinal motion so that the resulting motion is a Lissajous figure - a circle! The diagram below indicates various parcels of water (gray circles) spread out along a wavelength and the circular paths which they follow. As the wave moves from left to right in the diagram (you would sweep your eye *from right to left* to simulate this motion!), the parcels execute a clockwise motion, as indicated by the arrows on the circles.

The radius of the circular motion is the amplitude of the wave on the surface. Water underneath the surface also moves in circular paths as the waves passes, but the radii of the circles decrease with depth.

3.) After breathing helium, people talk funny. This is because the frequency of the voice has been raised, due to the increased speed of sound in helium compared to air.

The first sentence is true. The second sentence is a common, but erroneous, explanation, even appearing in some physics textbooks. The fundamental frequency of the voice is determined by the vibration frequency of the vocal folds, <u>which is unaffected by the gas in the mouth</u>. The funny voice is caused by the enhancement of higher frequencies in the voice spectrum, as explained in the discussion to Mystery #6. Thus, it is the <u>quality</u> of the voice that is changed, not the fundamental <u>frequency</u>.

4.) On the Richter scale, used to measure the strengths of earthquakes, each increase of 1 on the scale represents a factor of 10 increase in the energy released in the earthquake.

Before a standard scale of earthquakes existed, earthquakes were "rated" primarily by the amount of shaking and resultant damage. In this way, however, a relatively small earthquake near a population center could be rated as stronger than a major quake in the middle of the ocean. The Richter scale was devised in order to standardize earthquake ratings. The Richter scale was actually developed from a scale originated by K. Wadati in Japan in 1931. For a description of the factors to consider in rating earthquakes by magnitude, see C. F. Richter, *Elementary Seismology*, W. H. Freeman & Co., San Francisco, 1958.

When an earthquake strikes, waves spread out from the epicenter. These waves move both through the bulk of the Earth and along the surface of the Earth. Measurements of these waves are used to estimate the energy released in the earthquake. One relationship between the Richter magnitude and the energy released (several varying relationships can be found in different texts*) is given by the following expressions:

$$E = 2.5 \times 10^{(4.4 + 1.5M)}$$

$$or \ \ \log E = 4.8 + 1.5M$$

where E is the energy released (in Joules) and M is the Richter magnitude. Thus, a difference of 1 on the scale corresponds to an increase in energy released by a factor of $10^{1.5} = 31.6$, rather than 10. The statement in the myth is commonly heard in everyday conversation (at least in places such as Southern California!). The origin of the erroneous energy factor of 10 may be confused with the fact that a difference of 1 on the scale corresponds to a 10-fold increase in the pen deflection (in thousandths of millimeters) on a standard seismograph at a standard distance (100 km) from an earthquake. The pen deflection is proportional to the <u>amplitude</u> of the earthquake waves, not the energy carried by them. For a method of converting the earthquake magnitude from the standard distance to other distances, see B. A. Bolt, *Earthquakes: A Primer*, W. H. Freeman & Co., San Francisco, 1978.

---

*For example:

C. F. Richter, *Elementary Seismology*, W. H. Freeman & Co., San Francisco, 1958:	log E = 4.4 + 1.5M
B. A. Bolt, *Earthquakes: A Primer*, W. H. Freeman & Co., San Francisco, 1978:	log E = 4.8 + 1.5M
D. Halliday & R. Resnick, *Fundamentals of Physics*, 3rd Ed., Wiley, New York, 1988:	log E = 4.4 + 1.5M
D. Halliday, R. Resnick & J. Walker, *Fundamentals of Physics*, 4th Ed., Wiley, New York, 1993:	log E = 5.24 + 1.44M

Chapter 18
Sound

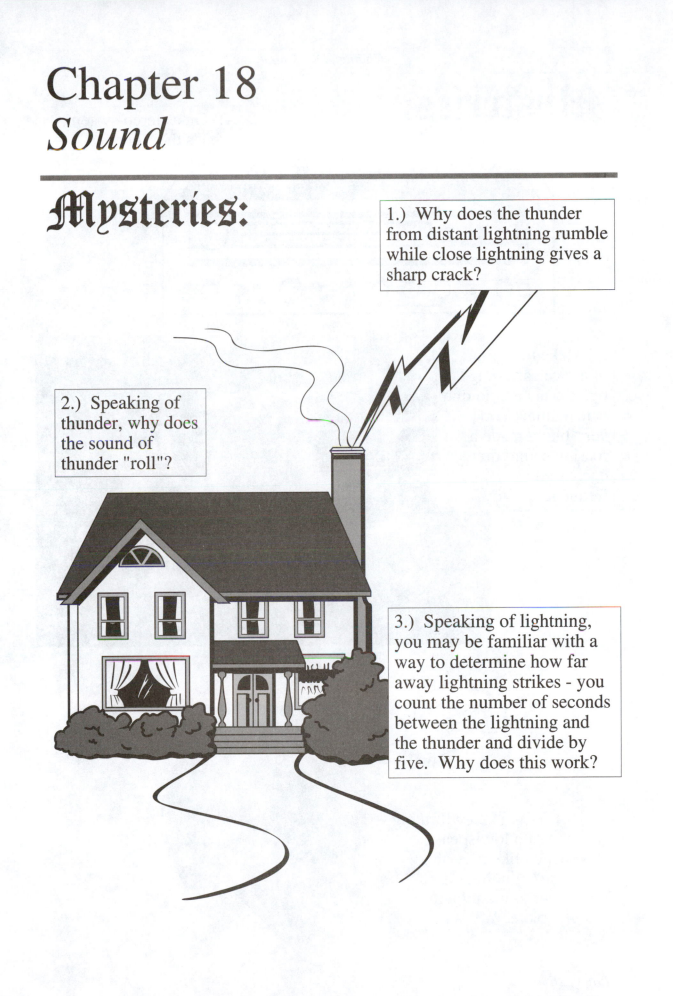

𝔐ysteries:

1.) Why does the thunder from distant lightning rumble while close lightning gives a sharp crack?

2.) Speaking of thunder, why does the sound of thunder "roll"?

3.) Speaking of lightning, you may be familiar with a way to determine how far away lightning strikes - you count the number of seconds between the lightning and the thunder and divide by five. Why does this work?

Mysteries:

4.) On a stereo system, what's the "loudness" switch for?

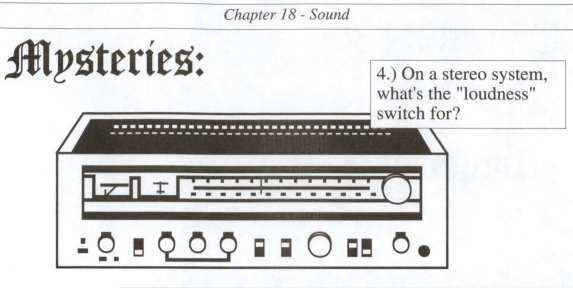

5.) Make the sound of a race car going by. Go ahead, no one is listening. You very likely made a roaring sound that <u>decreased</u> in frequency. Why?

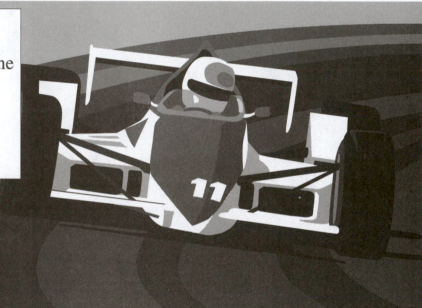

6.) Why is the vibrating part of a loudspeaker shaped like part of a *cone*? Wouldn't it be easier to manufacture a flat plate?

Mysteries:

7.) In a satellite-sub-woofer loudspeaker system for a stereo, why do you only need <u>one</u> subwoofer?

8.) Have you ever been in a boat in the middle of a lake and been able to hear sounds from the shore? Why does sound seem to travel so far and so well over water?

And then, she said that.....

9.) If you are fortunate enough to be able to visit a number of water-falls, listen to the sense of pitch from the sound of the falling water. How does this pitch vary with the parameters of the waterfall (height, width, etc.)?

Mysteries:

10.) Why do you hear the ocean when you hold a seashell up to your ear?

11.) Why does a teakettle make noise as the water approaches a boil? What determines the frequencies of the sound?

12.) When you turn on your electronic flash unit for your camera, you hear a high-frequency squeal which rises in frequency. What causes this sound?

𝔐𝔞𝔤𝔦𝔠:

The "Amplifying" Poster

Strike a tuning fork to start its oscillation and notice that it is not very loud. Now, in a piece of poster board or construction paper, cut a strip out about half an inch wide and approximately the length of a tine of the tuning fork. Strike the tuning fork again and hold it near the hole in the poster board. The sound becomes louder. Why?

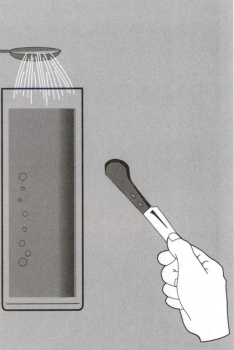

The Frequency of Water

Heat up a cup of water in a microwave oven or on a stovetop to a typical coffee or tea temperature, but <u>not to boiling</u>. Pour the hot water into a tall glass and tap the glass with the wooden handle of a knife or other kitchen utensil. While tapping, pour a teaspoon of salt into the water. Listen for the change in frequency of the sound from your tapping.

Myth:

1.) In a science fiction movie in which planets or spaceships explode in space, have you ever noticed that you <u>hear</u> the explosion? What's more, even if the scene suggests that you are observing the explosion from a very large distance, you see *and* hear the explosion at the <u>same time</u>.

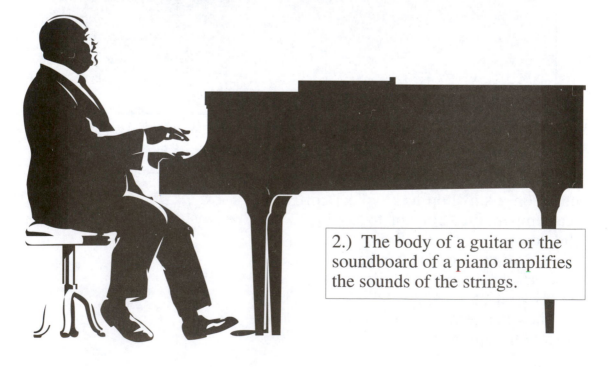

2.) The body of a guitar or the soundboard of a piano amplifies the sounds of the strings.

Myth:

3.) Sound waves are longitudinal.

4.) If a tree falls
in the forest, there
is no sound.

Myth:

5.) Choirs often use a pitch pipe before beginning to sing. The pitch pipe emits the correct pitch at which they should start singing.

6.) A common demonstration involves hanging a ringing bell in a bell jar and evacuating the air. The bell becomes silent, indicating that the sound does not propagate through the very low pressure air.

Q = 0 !!!!

7.) Sound waves are adiabatic.

Concepts of Sound

In Chapter 17, we discussed wave motion as a disturbance propagating through a medium. For the type of sound that we hear, the energy is carried by a mechanical disturbance propagating through the medium of air. This is our usual interpretation of sound, although sound can also travel through liquids (you can hear when you are under the water in a swimming pool!) and solids (think about Native Americans listening to the ground for approaching horses in movies). The source of sound is a vibrating object - tuning fork, guitar string, loudspeaker cone, piano sounding board, etc.

> *Sound - a mechanical disturbance propagating through air*

The speed of sound in air at room temperature is about 345 m·s⁻¹ or 1100 ft·s⁻¹. This is significantly slower than the speed of light, although fast enough that for nearby sources of sound, we hear the sound at very nearly the same time that we see it.

> *Speed of Sound ≈ 345 m/s ≈ 1100 ft/s*

There are a number of parameters related to sound, both in terms of its physical nature and in terms of our psychological reaction. As a physical wave, a sinusoidal sound wave has a <u>frequency</u>. Our psychological reaction that is most closely related to frequency is <u>pitch</u> - the highness or lowness, or "bassness" or "trebleness" of our reaction. The range of frequencies to which our ears are sensitive is about 20 Hz to 15,000 Hz. Frequency is the primary stimulus for pitch although many people hear a small pitch change when the intensity of a constant-frequency sound is changed. Speaking of <u>intensity</u>, this is a physical parameter associated with sound that is related to the sound energy crossing a region in space per unit area per unit time. The psychological reaction to intensity is <u>loudness</u>. If the intensity of a sound is increased, the sound is perceived as louder. Loudness, however, is strongly dependent on frequency. If the physical intensity of a sound is kept constant and the frequency is changed, the resulting psychological loudness varies significantly. As an extreme example, consider an ultrasonic sound of frequency, say, 30 kHz. No matter how intense this sound is made, it will result in no loudness.

> *Frequency ⟷ Pitch*
> *Intensity ⟷ Loudness*
> *Spectrum ⟷ Quality (Timbre)*

The parameters mentioned above exist for pure tones of a single frequency. If we have a mixture of frequencies in a sound, then we have another pair of physical-psychological parameters. The <u>spectrum</u> of the sound is the recipe for the particular mixture of frequencies. The resulting psychological reaction is called <u>quality</u> or <u>timbre</u>. As an extreme example, consider two sounds, one a mixture of harmonics (frequencies related by integer ratios) and the other a mixture of frequencies with no integer relationship among them. The first sound will result in an identifiable pitch, that of the fundamental frequency, and is called a *musical* sound. The second sound, which we identify as a *noise*, will have a much different quality, so different that it may not even have an identifiable pitch. Thus, the difference between music and noise is a gross example of quality. If

Concepts of Sound (continued)

we compare two musical sounds, differences in quality are harder to describe in writing. But imagine listening to an oboe and a trumpet playing the same note. There is something different about their sounds - we have learned through a lifetime of practice to tell which sound comes from which instrument. Part of the identification procedure that we use is to differentiate the <u>quality</u> of the sound from the two instruments which, if the sounds are steady and continuous, is closely related to the spectrum, the physical mixture of frequencies.

Discussions; Chapter 18 - Sound

𝔐ysteries:

1.) Why does the thunder from distant lightning rumble while close lightning gives a sharp crack?

As sound travels through air, the energy carried by the wave is transformed by various means. One way in which this occurs is by absorption by the molecules of the air. As the molecules collide, the energy of the wave can be transformed into rotational or vibrational energy of the oxygen and nitrogen molecules. This absorption is particularly effective at higher frequencies. Thus, if sound travels through air for a long distance, more of the energy associated with the higher frequencies is absorbed than that associated with the lower frequencies. Now, let's apply this to thunder. When lightning strikes close by, the sound travels through a short distance in the air and little appreciable absorption takes place. Thus, you hear all frequencies, in particular, the very strong mixture of high frequencies which give the typical sound of "crackling" electricity. For lightning striking far away, the sound travels through a long distance in the air and the energy associated with the high frequencies is very effectively absorbed. What is left by the time the sound reaches you is a mixture dominated by low frequencies, which has the quality of a "rumble". For more information about thunder, see A. A. Few, "Thunder", *Scientific American*, **233(1)**, 80 (1975).

2.) Speaking of thunder, why does the sound of thunder "roll"?

When lightning strikes, a column of air surrounding the path of the lightning is very rapidly heated and expands explosively. This results in a sound pulse which moves away from the lightning path. Since the current in a bolt of lightning moves very quickly (as much as half of the speed of light), sound is created all along the path of the lightning at virtually the same time. Thus, the lightning acts as a *line source* of sound, rather than a point source such as a loudspeaker (although the line is not straight - the path of a lightning bolt is quite tortuous). Now, as an observer, you are close to the bottom of the path and far away from the top of the path, which is in the clouds. Thus, although the sound leaves every point on the path at the same time, it arrives at your ear at different times, since it has to travel different distances. You will hear the sound from the points on the path near the ground first and then from points higher up. The result is a prolonged rumbling of the thunder as sound from different parts of the path arrive at your ears. The arrival of the sounds is made even more complicated by the tortuous shape of the path.

3.) Speaking of lightning, you may be familiar with a way to determine how far away lightning strikes - you count the number of seconds between the lightning and the thunder and divide by five. Why does this work?

Sound travels at approximately 1100 ft·s^{-1} in air. If we compare this to the number of feet in a mile, 5280 ft·mile^{-1}, we see that sound travels approximately $^1/_5$ of a mile in a second, or one mile every five seconds. Thus, if we count the seconds and divide by 5 (seconds·mile^{-1}), we obtain an approximation of the distance in miles. You can also count the time between lightning and thunder and divide by 3 to obtain the distance in kilometers. This is because the speed of sound is about 345 m·s^{-1}, which is close to $^1/_3$ of a kilometer!

4.) On a stereo system, what's the "loudness" switch for?

The human ear is not equally sensitive to all frequencies, as mentioned in the Concepts section. If we provide a human ear with sounds of equal <u>physical</u> intensity and covering a range of frequencies, we find that the ear/brain combination does not have the same <u>psychological</u> response to the sounds in terms of their strength. In particular, the ear is very sensitive to sounds in the range of 3000 - 4000 Hz, but the sensitivity decreases above and below this. This effect is not very pronounced at high intensities - the ear has a somewhat "flat" response for loud sounds. The effect is very pronounced at low intensities, however, especially at low frequencies. For example, a low-intensity pure tone of 50 Hz has to be <u>1000 times as intense physically</u> to be psychologically as loud as a 1000 Hz tone! This is the reason for the loudness switch on a stereo, which is designed only to be used at low listening levels. For these low intensity sounds, the ear is not sensitive to low and high frequencies. Activating the loudness switch electronically increases the physical intensity of the low and high frequencies in order to help the ear to hear them as loudly as the midrange frequencies and maintain the balance of the sound. If the loudness switch is left activated as the volume is turned up, the sound can be heard to be extremely heavy in the bass.

A similar variation in response is evident for light. If we present violet light, green light and red light to the eye such that all three lights have the same physical intensity (*irradiance* is the optical term), the psychological response will be very different. The green light will appear brighter than the red or violet.

5.) Make the sound of a race car going by. Go ahead, no one is listening. You very likely made a roaring sound that <u>decreased</u> in frequency. Why?

You made this type of sound due to your experience in watching race cars on television or in person. What you are demonstrating in making this sound is the <u>Doppler Effect</u>. When a wave source is moving toward you, the wavelength of the wave is compressed, resulting in an increase in the apparent frequency. Similarly, when a source of a wave moves away from you, the wavelength is stretched, resulting in a drop in the frequency. Thus, as a sound source approaches you and passes, the frequency that you hear will start high and drop to a lower value. This is exactly what you hear for the roar of the race car engine as it approaches you and passes.

It should be pointed out that, theoretically, a mixture of equal intensities of all frequencies (*white* noise) would exhibit no Doppler Effect, since the same mixture would exist whether the car is moving toward you or away from you or not even moving at all. White noise has no discernible pitch anyway, so the Doppler Effect is meaningless. The automobile engine, however, does not generate white noise. Although it is noisy in character, it engenders a discernible pitch, associated with the rotation frequency of the engine.

6.) Why is the vibrating part of a loudspeaker shaped like part of a *cone*? Wouldn't it be easier to manufacture a flat plate?

As we discussed during our treatment of vibration (Chapter 16), oscillating systems exhibit *resonance*, whereby they will exhibit a large response if driven at their natural frequency. Imagine a flat plate vibrating as a source of sound. This would be a perfectly reasonable loudspeaker until we realize that the plate has natural frequencies associated with standing waves established across the surface of the plate. Thus, the flat plate does not <u>stay</u> flat when it vibrates. When excited into one of its resonance modes, the plate will possess adjacent regions that are moving in different directions on either side of a nodal line (*Chladni plates* are a popular demonstration of this effect.). As a result, a higher pressure area of air in front of a region of the plate at a standing wave "crest" will just slosh over into the lower pressure area in front of a part of the plate that is at a "trough". Thus, the plate is very inefficient as a radiator of sound, since the energy just sloshes back and forth in front of the plate instead of moving away to your ear. We could eliminate this problem by making the plate stiffer by making it thicker. Using this technique, we could move the resonance frequencies up above the range of human hearing, so that resonance would not occur. The problem with this, however, is that the more massive plate now represents more resistance to changes in motion and it is very hard to cause this plate to vibrate over a significant distance at audio frequencies - it is too massive! What we need is a vibrating system that has the required stiffness without much mass. A perfect candidate is a paper or plastic *cone*, as we see in present-day loudspeaker systems. The structure of the cone gives the system the necessary stiffness, while the mass of the paper or plastic is quite small.

7.) In a satellite-subwoofer loudspeaker system for a stereo, why do you only need <u>one</u> subwoofer?

The ability of the ear to determine the direction of the source of a sound is dependent on the frequency of the sound. For low-frequency sounds such as bass notes in music, the directional ability of the ear is relatively poor. Therefore, it is not necessary to provide two speakers for the bass notes, since you can't tell where they are coming from anyway. For high frequency sounds, the ear is quite good at directionality, so we do want to provide two separated stereo speakers for the high frequencies.

8.) Have you ever been in a boat in the middle of a lake and been able to hear sounds from the shore? Why does sound seem to travel so far and so well over water?

Part of the answer here is reflection off the water, particularly if the water is calm. If you imagine listening to a sound over water compared to listening to it over a grassy field, the water will tend to reflect the sound, while the grass tends to absorb it. The more important reason, though, has to do with <u>refraction</u> (see Chapter 26). Due to the high heat capacity of water, the water in a lake tends to stay cooler than the air. It is often possible, then, that a <u>temperature inversion</u> occurs over water, that is, the air above the lake is cool and the air higher up is warm. Since sound travels faster in warm air than cool air, this has the effect of refracting upward traveling sounds back toward the Earth. Thus, sound that would normally have been lost into the atmosphere is directed back downward to combine with the direct sound, so that sounds from distant sources are surprisingly loud. This type of temperature inversion also occurs at night as the ground cools, so that sounds from distant sources seem louder in the dark than in the daytime.

9.) If you are fortunate enough to be able to visit a number of waterfalls, listen to the sense of pitch from the sound of the falling water. How does this pitch vary with the parameters of the waterfall (height, width, etc.)?

The frequency associated with a waterfall seems to be associated with only its height. In J. S. Rinehart, "Waterfall-Generated Earth Vibrations", *Science*, **164**, 1513 (1969), a graph is presented, showing frequency against inverse height. The data suggests a straight line dependence, with a slope equal to one fourth of the velocity of sound in water. The author of this article suggests that the entire waterfall is vibrating in a standing sound wave with a node at the brink of the waterfall and an antinode at the base.

10.) Why do you hear the ocean when you hold a seashell up to your ear?

This phenomenon is a demonstration of the extreme sensitivity of the ear. The threshold pressure of hearing for the human ear is equivalent to moving the eardrum a distance of about one tenth of the diameter of an atom! The pressure difference needed at the eardrum to surpass the threshold of hearing is the same as that between two points in the atmosphere at sea level separated vertically by *one third of an inch*! The seashell is acting as a *Helmholtz resonator*. Soft sounds from the environment enter the seashell. Some of the sound is at resonant frequencies of the shell. These set up standing waves, which are intense enough to surpass the threshold of hearing when the shell is held close to the ear. The mixture of resonant frequencies is not harmonic, due to the complicated shape of the interior of the shell. Thus, the resultant sound is something like "white noise", which is similar to the sound of the ocean at the seashore.

11.) Why does a teakettle make noise as the water approaches a boil? What determines the frequencies of the sound?

The source of the sound in a teakettle is the bursting of bubbles during the heating of the water. This sound results in the excitation of certain resonant modes in the kettle. One of these modes is the standing wave mode in the air above the water, which is very similar to the situation with an organ pipe. The frequency of this mode is dependent on the height of the water in the kettle. Another standing wave mode is associated with the radial direction across the diameter of the kettle and is unaffected by the height of the water. Additional standing wave modes in the column of water contribute to the sound. Also, vibrational modes of the kettle itself can color the sound. A detailed study of the sound from a teakettle, along with a number of spectra can be found in S. Aljishi and J. Tatarkiewicz, "Why does Heating Water in a Kettle Produce Sound?", *American Journal of Physics*, **59**, 628 (1991).

12.) When you turn on your electronic flash unit for your camera, you hear a high-frequency squeal which rises in frequency. What causes this sound?

Within an electronic flash unit, high voltage is necessary to charge a capacitor. The energy stored in the capacitor is suddenly released into the flash tube, resulting in the very bright light. A transformer is used to step up the voltage to that necessary to achieve the desired energy storage. If the flash unit is powered by a battery, however, the DC voltage from the battery must be converted to AC for use by the transformer. This is performed with an <u>oscillator</u> circuit within the flash unit.

Now, the transformer used in flash units is an inexpensive variety, with coils of wire that are not perfectly solid in their mounting on the transformer core. Thus, as the AC current from the oscillator passes through the transformer coils, adjacent coils exert forces on each other and the coils vibrate. It is the vibration of all of these coils of the transformer that causes the whining sound. As the capacitor charges, the characteristics of the oscillator-capacitor circuit are such that the oscillator frequency increases, so that the whining sound appears to rise in pitch.

𝕸agic:

The "Amplifying" Poster

The explanation of this effect is similar in some ways to the discussion of Mystery #6 above. During a single sweep of a vibration of the tine of a tuning fork, a high pressure area is created in the air in front of the tine, while a low pressure area occurs behind it. The high pressure can be relieved easily by the movement of the air from the front of the tine to the back. Thus, much of the vibration energy of the tine is transformed to energy of the air simply sloshing back and forth around the tine. As a result, the efficiency of sound radiation is low and the sound is quiet. By placing the tuning fork near the opening of the poster board, the poster board acts as a barrier for the air to move from front to back, cutting off the "leakage path". As a result, the cancellation of pressure is reduced and the fork becomes a more efficient radiator. As a result, the sound is louder. This is the same reason that loudspeakers are placed in speaker cabinets - the wood at the edge of the speaker cuts off the leakage path so that air does not slosh back and forth around the edge of the speaker. Notice that the word *Amplifying* is in quotes in the title of this *M*. Amplification requires a boost in power from some external source - your stereo amplifier requires electrical energy, for example. In the tuning fork trick, the sound is louder, but it is <u>not</u> amplified - there is no external power source (see Myth #2). The sound is louder because more of the energy is radiated as sound and less is stored in "sloshing".

The Frequency of Water

This is a phenomenon called the "hot-chocolate effect" (F. S. Crawford, "The Hot Chocolate Effect", *American Journal of Physics*, **50**, 398 (1982)). When the salt is added to the hot water, nucleation sites for bubble formation are provided and bubbles appear in the water. This lowers the speed of sound through the water, which reduces the frequency of standing waves established in the water column. Crawford claims that the sound can be reduced in frequency by about 3 octaves with this technique (F. S. Crawford, "Hot Water, Fresh Beer, and Salt", *American Journal of Physics*, **58**, 1033 (1990)). This reference also gives details on reproducing the effect in a tavern with beer.

𝔐𝔶𝔱𝔥:

1.) In a science fiction movie in which planets or spaceships explode in space, have you ever noticed that you <u>hear</u> the explosion? What's more, even if the scene suggests that you are observing the explosion from a very large distance, you see *and* hear the explosion at the <u>same time</u>.

Sound requires a medium for propagation. Often, this medium is air for the normal sounds that we hear. In space, there is no air or other medium between an exploding planet and an observer to carry the sound waves. Thus, the explosion should be silent. This would not be dramatic or exciting in the movies, however, so a little cinematic license is taken!

As a mechanical wave, sound must necessarily travel slower than light in any medium. Thus, if a source emits both sound and light, the light will reach an observer before the sound. Thus, if it were possible to hear an explosion in space, the sound would be *heard* some time after the explosion was *seen*. In the movies, this would create confusion in the viewer and also slow the movie down while continuation of the action waited until the sound wave arrived. Thus, more cinematic license is taken.

2.) The body of a guitar or the soundboard of a piano amplifies the sounds of the strings.

The body of a guitar and the soundboard of a piano certainly do make the sound of the instrument louder, but they do not <u>amplify</u> the sound. As mentioned in the discussion of The "Amplifying" Poster, amplification requires adding energy to the sound signal from an external source. In an electric piano or an electric guitar, the sound is definitely amplified, with the use of external electrical energy transmission. But in the acoustic guitar or piano, there is no amplification. The guitar body or soundboard make the sound louder by creating a more efficient coupling with the air than that provided by the vibrating string alone. This is primarily due to the large area of the vibrating body or soundboard. Thus, the energy is radiated into the air <u>faster</u> than is the case with the string alone. As a secondary result, the vibration does not last as long as that of a string vibration in air without an attached soundboard.

The intensity of a sound is defined as,

$$\text{Intensity} = \frac{\text{Energy}}{(\text{Area})(\text{Time})}$$

Thus, the intensity can be changed (and the loudness thereby changed) by adjusting any one of the three parameters on the right side of the equation. In amplification, the energy, on the top of the fraction, is increased, resulting in a louder sound. The use of a soundboard reduces the time for a given amount of energy to be radiated in the above expression, resulting in a louder sound. Is there a way to reduce the area for a given amount of energy so as to increase the intensity? Yes - remember when people who were hard of hearing used horns in their ears? The function of these horns was to take energy arriving over the large area at the end of the horn and compress it down to the smaller area of the tip, thus increasing the intensity of the sound.

3.) Sound waves are longitudinal.

This is the case in air, where we hear most of our sounds, but, if we consider sound moving through a solid, it can have both a longitudinal and a transverse component. The nature of the wave depends on the strength of the interaction between molecules of the medium. In a gas, the interactions are weak, and only longitudinal propagation can be maintained over a long distance. For a weak interaction with a neighboring molecule located in the x-direction, a small movement in the y-direction will be almost undetectable. In a solid, the intermolecular interactions are much stronger, allowing for transverse waves to propagate.

4.) If a tree falls in the forest, there is no sound.

This is a popular conundrum which depends on the fact that the word *sound* is used to mean two different things. The inference is that there are no creatures nearby to hear the sound of the falling tree. If you are defining sound as the psychological response of a creature to this disturbance, then there is no sound of this type. If you are defining sound as a physical disturbance propagating through a medium, then there is definitely a sound. The falling tree disturbs the air and the wave propagates, regardless of the presence or absence of an ear.

5.) Choirs often use a pitch pipe before beginning to sing. The pitch pipe emits the correct pitch at which they should start singing.

This is a misuse of the word *pitch*. The pitch pipe cannot emit the right pitch, since pitch is only a psychological reaction. What the pitch pipe emits is a sound of the correct *frequency*.

A similar situation exists for light. Phrases such as "a red light", "the color of the light", etc., are used often. In reality, however, the light has no color. Color is a psychological response of the eye-brain system of the human. The parameter of the light that is most closely related to the sensation of color is the frequency, but the light has no color of its own.

6.) A common demonstration involves hanging a ringing bell in a bell jar and evacuating the air. The bell becomes silent, indicating that the sound does not propagate through the low pressure air.

The propagation of sound in a gas depends on the relative size of the mean free path between molecules and the wavelength. As long as the mean free path is small compared to the wavelength, sound will propagate just fine. At the pressures attainable by a classroom pumping system, say a few millimeters of mercury, the mean free path is a fraction of a millimeter. This is still far smaller than the wavelength, which is on the order of meters. Thus, <u>there is no barrier to sound propagation through the air</u>.

The silence of the bell is due to the acoustic impedance mismatch at the air-glass boundary in the bell jar. Acoustic impedance is similar to electrical impedance. In electricity, we try to drive a current through a material by applying a voltage difference. The opposition to this current flow is measured by the <u>electrical impedance</u>, which takes into account resistance as well as capacitive and inductive reactance. In acoustics, we are trying to drive a sound wave through air by applying a pressure difference. The opposition to this sound wave is measured by the <u>acoustic impedance</u> of the medium.

As one might guess, solids offer much larger acoustic impedances than gases, since it is hard to cause the tightly bound molecules in a solid to move much from their equilibrium positions. Thus, there is already a significant mismatch in impedances between the glass of the bell jar and the enclosed air at normal pressure. In general, the worse the impedance mismatch between two media, the more energy is reflected and the less is transmitted. Thus, it is good news that air and glass have very different acoustic impedances, since, if they were similar, we would be able to hear quite well through windows!

While the acoustic mismatch between the glass of the bell jar and the enclosed air is large, it is not so large that we cannot hear the bell faintly when the air inside is at atmospheric pressure. Now, the acoustic impedance of a gas is given by,

$$Z_{acoustic} = \sqrt{\gamma P \rho}$$

where γ is the ratio of molar heat capacities, P is the pressure and ρ is the density. As the bell jar is evacuated, both the pressure and density drop, so that the impedance drops quickly. The impedance of the glass, of course, remains constant. Thus, the impedance <u>mismatch</u> becomes great.

Now, the transmission coefficient for sound going from air through a solid and back to air depends on the <u>square</u> of the ratio of acoustic impedances of the air and the solid. Thus, as the impedance of the air in the bell jar drops quickly, the transmission coefficient drops very quickly. As a result, very little energy crosses the boundary between the air and the glass and then only a fraction of that is transmitted from the glass to the outside air. As a result, the bell is essentially silent.

7.) Sound waves are adiabatic.

Statements in many textbooks argue that sound waves are adiabatic because the variations in pressure occur so quickly that there is insufficient time for any appreciable heat to flow so as to re-establish thermal equilibrium. But what about low-frequency waves, in which the processes occur more slowly? Here, we find that the wavelength is so long that the distance between regions of different temperatures precludes the flow of significant heat. Thus, we have two contributions - the time during which the process takes place and the distance between regions of different temperatures. As the frequency of the sound wave increases, the process occurs faster, but the distance shrinks. Now, which is more important - time or distance? A thermodynamic analysis shows that sound waves in the audio range, 20 Hz - 15,000 Hz, can be modeled very successfully as adiabatic. But as the frequency increases, the process tends to be more and more isothermal. Thus, for very high frequency ultrasonic signals, the increasing isothermal nature of the sound must be taken into account, since the speed of sound depends on the type of thermodynamic process which occurs. For ultrasonic signals, then, audio dispersion (variation of wave speed with frequency) is a serious consideration. We will see more examples of dispersion, applied to light, in Mysteries #1 and #8 in Chapter 26.

Chapter 19
Electric Fields & Forces

𝔐𝔶𝔰𝔱𝔢𝔯𝔦𝔢𝔰:

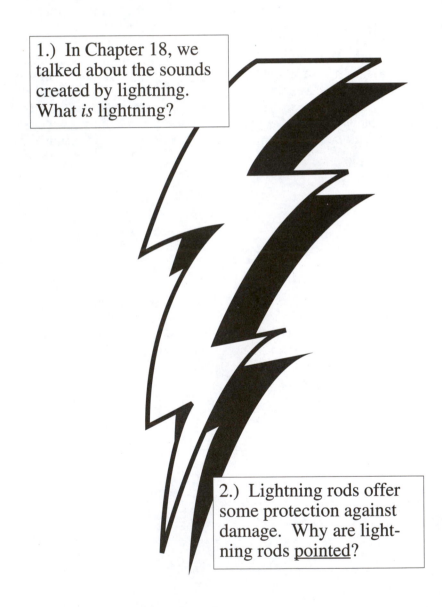

1.) In Chapter 18, we talked about the sounds created by lightning. What *is* lightning?

2.) Lightning rods offer some protection against damage. Why are lightning rods <u>pointed</u>?

𝕸ysteries:

3.) Inkjet printers work by spraying the ink onto the paper as it passes through the printer. How is the ink spray controlled so carefully?

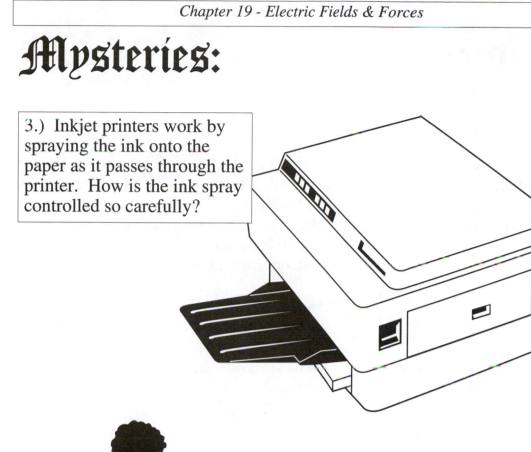

4.) How does a photocopy machine work?

𝕸𝖆𝖌𝖎𝖈:

The Plastic Wrap Hand Grabber

Lay a piece of kitchen plastic wrap against a wall and rub it briskly with your fingers. As you rub, it will adhere to the wall, since it is becoming charged. Peel it off the wall and lay it across a horizontally held pencil. The two sides of the plastic will stand out at an angle from the vertical. Now, place your hand between the two sides of the plastic. The plastic wrap will collapse on your hand!

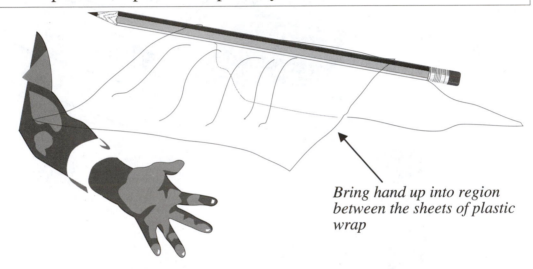

*Bring hand up into region
between the sheets of plastic
wrap*

Lighting a Bulb with No Wires

On a dry day, rub a balloon on your hair. Hold the charged balloon near a fluorescent light bulb and a small spark will jump, making the light bulb flash!

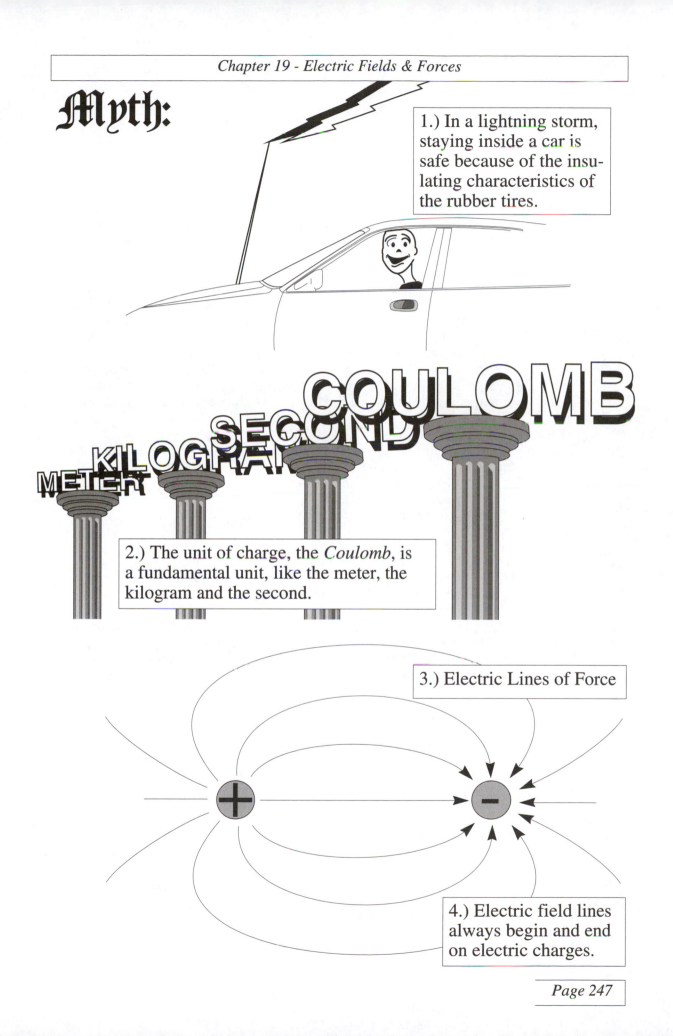

Myth:

1.) In a lightning storm, staying inside a car is safe because of the insulating characteristics of the rubber tires.

2.) The unit of charge, the *Coulomb*, is a fundamental unit, like the meter, the kilogram and the second.

3.) Electric Lines of Force

4.) Electric field lines always begin and end on electric charges.

Concepts of Electric Fields & Forces

In Chapter 6, we discussed the force of gravity. This is one of what we presently believe to be the four fundamental forces of nature*. It exists between objects that have the quality that we call *mass*. We do not understand the basic nature or origin of mass, but we know how it behaves, in terms of its role in determining the strength of the gravitational force (*gravitational mass*) and as the resistance of objects to changes in their motion (*inertial mass*).

> *Charge is similar to __mass__ in some ways.*

Now, why are we talking about mass here in a chapter on electricity? The reason is that we now look at another quality of some objects in the universe, *electric charge*, for which we also have no understanding of its nature or origin. Thus, it is similar to mass in this way. It is also similar in that charge is the basis of another fundamental force of nature, electromagnetism. We will look only at the electrical characteristics of this force in this chapter and save the magnetic effects until Chapter 22.

Not only is charge similar to mass in that it is the basis for a force, but if we look at the mathematical description of the force between point charges charges q_1 and q_2, as given by *Coulomb's Law*, in the box to the right, we see that this Law has exactly the same mathematical form as Newton's Law of Universal Gravitation in Chapter 6. The constant ε_o is called the *permittivity of free space*.

> *Coulomb's Law:*
>
> $$F = \frac{1}{4\pi\varepsilon_o}\frac{q_1 q_2}{r^2}$$

Since the days of Benjamin Franklin, we have used the notation *positive* and *negative* to describe the two types of charge possessed by protons and electrons, respectively. Experimentation shows us that the electric force between like charges is repulsive and that between unlike charges is attractive.

The point of view suggested by Coulomb's Law is a one-step process for describing a force which exists between charges. In this model, the two charges interact directly with each other. It is useful to imagine an alternative two-step process as follows. We imagine one charge sitting by itself in the universe. This charge somehow alters space around itself due to the fact that it possesses charge. This alteration of space is given the name of an *electric field*. Now, if a sec-

> *An electric __field__ is established by a single charge.*
>
> *An electric __force__ requires two charges.*

---

*We identify four fundamental forces (gravity, electromagnetic, nuclear strong, weak), although intense efforts are being made to unify the four forces into manifestations of a single concept. Success has already been achieved in unifying the electromagnetic and weak forces into the *electroweak* force.

ond charge enters the same region of the universe, we do not imagine it interacting directly with the first charge, as in Coulomb's Law. We imagine that it interacts with the altered space, with the electric field, in such a way that it feels a resultant force, either toward or away from the first charge. The overall result is the same as that from Coulomb's Law, but the concept of an electric field set up by a single charge allows us some insights into electrical behavior that are not possible from the one-step point of view.

We compared the gravitational force to the electric force and discussed the fact that both had similar mathematical behavior. We now look at another comparison between gravity and electricity which will lead to a very useful concept for electrical situations. As we saw in the Concepts section of Chapter 6, the gravitational potential energy for a system of two masses is given by,

$$PE_G = -G \frac{m_1 m_2}{r}$$

This expression comes from the equation for the gravitational force. Since the electric force has exactly the same mathematical form, we should be able to define an electric potential energy in the same way. Indeed, we find,

$$PE_E = \frac{1}{4\pi\varepsilon_o} \frac{q_1 q_2}{r}$$

An immediately apparent difference between these two equations is that the gravitational equation has a negative sign, while the electric equation does not. A negative potential energy represents an attractive force situation, since these equations are defined such that the zero of potential energy corresponds to separation of the masses or charges by an infinite distance. In the gravitational situation, we only have positive mass, so substitution for the two masses into the equation leads to a negative potential energy, as is appropriate, since the gravitational force is always attractive.

An attractive electric force, however, arises between two <u>unlike</u> charges. Thus, in the electric equation, when charges of opposite sign are substituted, the negative potential arises that corresponds to the attractive force. Substitution of two charges of like sign leads to a positive potential energy, corresponding to the resulting repulsive force.

Now, we take an action similar to that which we took in going from the electric force to the electric field. The <u>force</u> is only defined as existing between two charges. The <u>field</u>, however, is something which is established by a single charge. The electric potential energy is defined between two charges. Is there a corresponding variable that is established by a single charge? We can certainly define one, and it turns out to have great usefulness for us. We defined the field as the force per unit charge and we define

An electric <u>potential</u> is established by a single charge.

An electric <u>potential energy</u> requires two charges.

Concepts of Electric Fields & Forces (continued)

our new variable in a similar way. We calculate the potential energy per unit charge (letting $q_1 = q$, a source charge and $q_2 = q'$, a test charge) and call it the *electric potential V*:

$$V = \frac{PE_E}{q'} = \frac{1}{4\pi\varepsilon_o} \frac{q}{r}$$

The unit of electric potential is that of energy per charge, or Joules per Coulomb. We call this combination a *volt*. This is the origin of the symbol V for electric potential and leads to the popular description of electric potential as a *voltage*. We will have more to say about voltage in Chapter 20.

Discussions; Chapter 19 - Electric Fields & Forces

𝕸𝖞𝖘𝖙𝖊𝖗𝖎𝖊𝖘:

1.) In Chapter 18, we talked about the sounds created by lightning. What *is* lightning?

Lightning is the electrical discharge that occurs when air molecules are ionized in a large electric field, often between the Earth and a thundercloud.   The phenomenon of lightning is very common, with about 100 lightning flashes around the world every second, from some 2,000 thunderstorms always in progress somewhere on the globe. A typical lightning bolt arises from a potential difference of hundreds of millions of volts and can carry currents up to 10,000 amperes. A complete discussion of lightning would require more space than is appropriate for this book. For detailed descriptions of lightning, there are a number of reference books, including R. H. Golde (Ed.), *Lightning*, Academic Press, London, 1977 and P. E. Viemeister, *The Lightning Book*, MIT Press, Cambridge, Mass, 1972.   A nice concise description of lightning is available as an "interlude" in H. C. Ohanian, *Physics*, 2nd. ed., Norton & Company, New York, 1989. Related phenomena are covered in A. A. Few, "Thunder", *Scientific American*, **233(1)**, 80 (1975) and H. W. Lewis, "Ball Lightning", *Scientific American*, **208(3)**, 106 (1963). The "current" understanding (pun intended!) of the mechanism by which clouds become charged is discussed in E. R. Williams, "The Electrification of Thunderstorms", *Scientific American*, **259(5)**, 88 (1988).

2.) Lightning rods offer some protection against damage. Why are lightning rods <u>pointed</u>?

The intent of a lightning rod is to "attract" the lightning to it rather than have the lightning strike houses or trees or other objects. We can do this if we can establish a very large surface charge density around some metal object.   This large surface charge density will result in a large electric field around the object, which will ionize the air and act as the source of a return stroke from the ground (while a "leader" comes down from the cloud, a "return stroke" jumps up from the ground to meet it - see the references listed in Mystery #1 for more information) when a lightning bolt is released from a cloud. Now, why does a point give us such a large surface charge density? Imagine that we have two charged, equal-radius metal spheres attached by a wire. We now start shrinking one of the spheres until we have the situation shown below.   As the right-hand

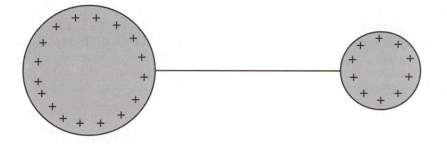

sphere shrinks in the figure, charge will flow from the shrinking sphere to the constant-radius sphere through the wire, in order to keep the electric <u>potential</u> constant everywhere on the surface.

The <u>amount</u> of charge on each sphere will be proportional to the radius of the sphere. But the area of the sphere is proportional to the <u>square</u> of the radius. Thus, even though the smaller sphere has less charge, it has a <u>larger charge density</u> than the stationary sphere. Now, if we imagine shrinking the sphere to a very tiny radius, it becomes a <u>point</u>, which will have the very large surface charge density that we discussed needing above.

3.) Inkjet printers work by spraying the ink onto the paper as it passes through the printer. How is the ink spray controlled so carefully?

An inkjet printer is a technological application of the force on a charged particle moving through an electric field. In one type of printer, nozzles spray drops of ink through a charging electrode, which is controlled by the computer. The drops then pass through a pair of deflection plates with an electric field between them. If the printhead is passing over a region of the paper where ink is to be deposited (as for a part of a letter or a graphic image), the computer turns off the charging electrode and the drops pass through the deflection plates undeflected and strike the paper. If the printhead is over a region of the paper where no ink is to be deposited, the computer turns on the charging electrodes, giving the drops a charge. They are then deflected by the electric force as they pass through the deflection plates into a gutter that collects the unused ink.

4.) How does a photocopy machine work?

This is another device that uses the electric force in a technological application. This process of making copies is known as <u>xerography</u>, from the Greek *xeros*, meaning <u>dry</u> and the Greek *graphia*, meaning, among other things, <u>recording</u>. The central part of the copier is the drum, which is a metal cylinder, usually aluminum, covered with a <u>photoconductive</u> material, often selenium. A photoconductive material is one that will not pass current in the absence of light, but becomes a conductor when it is irradiated. At the beginning of the copying process, the selenium surface is given a positive charge. The drum remains out of the light, so that the charges remain in their original positions on the non-conducting surface. A light is then turned on the paper to be copied and an image of the paper is formed by a series of optical elements (lenses, mirrors, etc.) on the drum. Areas where there is no printing on the original will be lit up on the drum. These areas will become conducting, and electrons from the aluminum will pass through the selenium and neutralize the charge. Thus, only the image of the printing will retain charge on the selenium surface. The drum is now spread with particles of toner that are given a negative charge. These negatively-charged particles stick to the positively-charged areas on the drum, attracted by the electric force. A piece of blank paper is given a larger positive charge than that on the drum. Thus, when the paper rolls across the drum, the toner is transferred to the paper. Finally, the paper passes through a heater which melts the toner into the paper, and the process is complete.

𝔐𝔞𝔤𝔦𝔠:

The Plastic Wrap Hand Grabber

Rubbing the plastic wrap on the wall gives it a net charge. When it is laid over the pencil, the two halves begin to droop on either side, due to the gravitational force, but are repelled by the electric force, since both halves carry the same type of charge. The result is that the halves of the plastic stand out at an angle. Now, when your hand is placed between the two halves, free charges in your hand like the ones on the plastic are driven away from the surface of your skin. In addition, polarized molecules in your hand will align with the charge unlike that on the plastic wrap near the outside surfaces of your hand. Thus, opposite types of charge are on the surface of your hand and the plastic wrap, resulting in an attractive force between your hand and the plastic wrap, which causes the plastic to move toward and grab your hand.

Lighting a Bulb with No Wires

Rubbing the balloon gives it a net charge. When the balloon is then brought near a fluorescent light bulb, there is a spark representing a transfer of charge to one end of the bulb. There now exists a voltage difference between the ends of the bulb, which generates a brief current through the gas of the bulb, causing it to emit a flash of light.

𝔐𝔶𝔱𝔥:

1.) In a lightning storm, staying inside a car is safe because of the insulating characteristics of the rubber tires.

It is true that staying inside a car in a lightning storm is safe. This is not because of the tires, however. A current which has passed through a long distance in air is not going to think twice about a couple of inches of rubber. The protection offered by a car is due to the fact that the electric field inside of a conducting shell (like the body of the car) is zero, which can be shown by Gauss' Law. Thus, even if lightning were to strike the car and result in a deposit of charge on the metal, the electric field inside is still zero.

2.) The unit of charge, the *Coulomb*, is a fundamental unit, like the meter, the kilogram and the second.

This seems like it would make sense, especially since charge is so similar to mass, whose unit is one of the fundamental group. But it is not so in the SI system (Le Système International d'Unites). The main reason is that <u>it is experimentally difficult to measure charge with great accuracy</u>. The preferred procedure is to measure the force between wires carrying a given <u>current</u>. Both force and current can be measured very accurately. This procedure is used to define the *Ampere* as a unit of current. Once this has been established, the Coulomb is defined as the charge that flows past a given point in one second when the current is one Ampere.

3.) Electric Lines of Force

Electric field lines are discussed in many textbooks as if they are real - that is, that there are definite points in space through which a line passes and other points through which no lines pass. The truth is that <u>electric field lines are only a useful visual construction</u> to enable us to imagine an electric field. If electric field lines are to be used, it should be made clear that we are <u>arbitrarily</u> defining how many lines should be drawn from a given charge and that the number of lines should be proportional to the charge - larger charges should have more field lines connected. Drawings of field lines carry explanations that the density of lines is proportional to the field strength, but it needs to be remembered that the density of field lines depends on the original number that we choose to draw. Finally, we need to address the electric flux, which is defined in a typical physics text as follows: "we define the electric flux to be the number of field lines that pass through a given surface". Similar definitions appear in many texts. Once again, this is ambiguous, since the number of lines passing through the surface depends on how many we choose to draw to represent the field in the first place.

4.) Electric field lines always begin and end on electric charges.

This statement is true for <u>electrostatic</u> fields - the kind created by combinations of stationary point charges. But for electric fields created by changing magnetic fields, there are no stationary charges present - the electric field is induced by the changes in the magnetic field. Thus, these types of fields can be represented by field lines which are closed paths, which have no beginning and no end.

Chapter 20
Voltage, Current and Resistance

𝕸ysteries:

1.) When a light bulb fails, why does it almost always do so just as you turn it on?

2.) As an incandescent light bulb grows older, it glows more dimly. Why?

3.) Electrical energy is transmitted over power lines at hundreds of thousands of volts. Why is such high, dangerous voltage used?

𝕸ysteries:

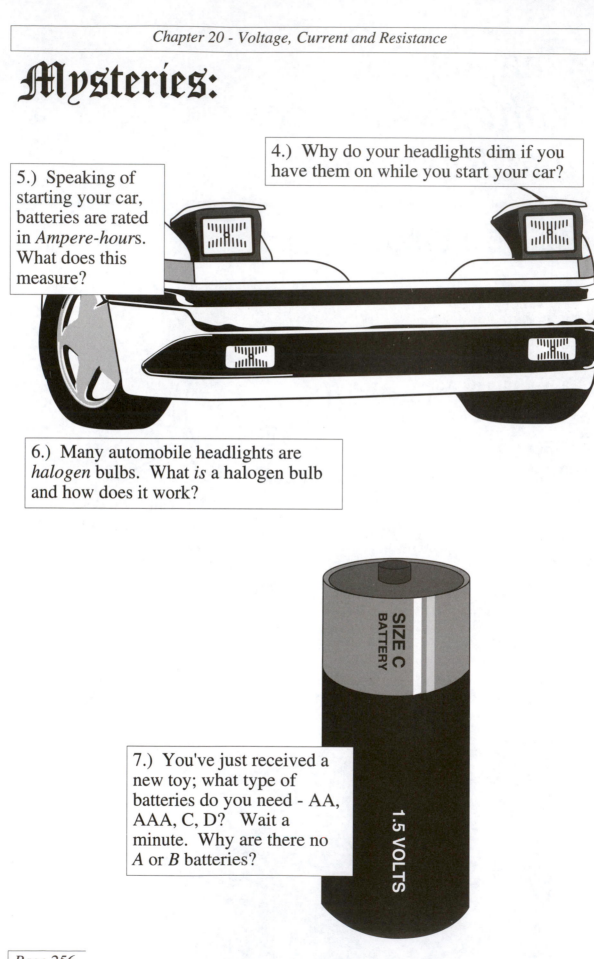

4.) Why do your headlights dim if you have them on while you start your car?

5.) Speaking of starting your car, batteries are rated in *Ampere-hours*. What does this measure?

6.) Many automobile headlights are *halogen* bulbs. What *is* a halogen bulb and how does it work?

7.) You've just received a new toy; what type of batteries do you need - AA, AAA, C, D? Wait a minute. Why are there no *A* or *B* batteries?

SIZE C
BATTERY

1.5 VOLTS

Mysteries:

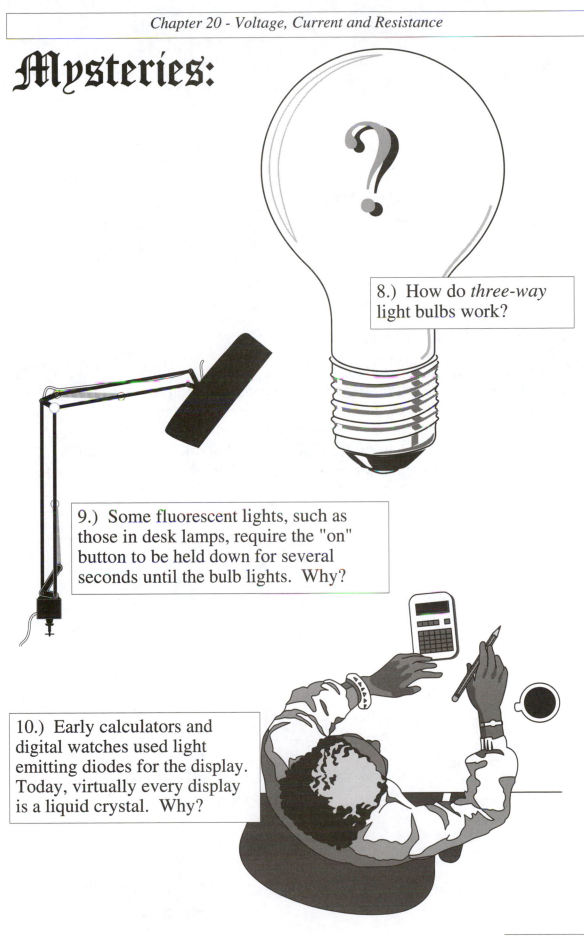

8.) How do *three-way* light bulbs work?

9.) Some fluorescent lights, such as those in desk lamps, require the "on" button to be held down for several seconds until the bulb lights. Why?

10.) Early calculators and digital watches used light emitting diodes for the display. Today, virtually every display is a liquid crystal. Why?

𝕸𝖆𝖌𝖎𝖈:

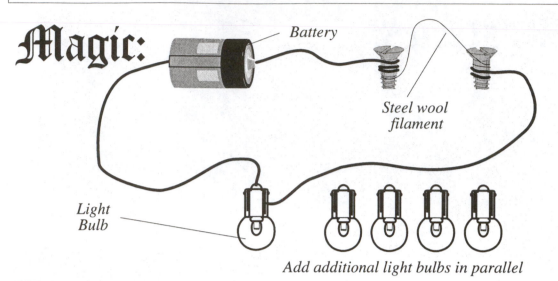

Battery

Steel wool
filament

Light
Bulb

Add additional light bulbs in parallel

The Steel Wool Fuse

Create a circuit on a wooden board using a battery holder and a flashlight bulb. Leave a gap in the circuit by wrapping the circuit wire around two screws, as shown in the diagram above. Complete the circuit by spanning the gap with one filament of very thin steel wool. Place the battery in the holder so that the light bulb glows. Now add additional light bulbs *in parallel with the first*. What happens to the steel wool?

𝕸𝖞𝖙𝖍:

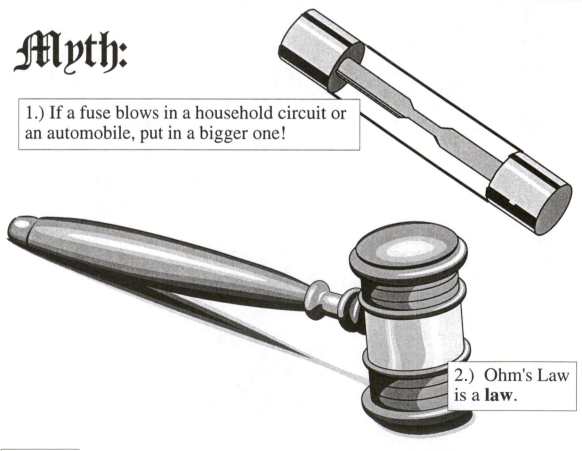

1.) If a fuse blows in a household circuit or an automobile, put in a bigger one!

2.) Ohm's Law is a **law**.

Myth:

3.) Big electrical items like televisions use lots of electrical power.

DANGER LOW VOLTAGE

4.) Low voltage is not dangerous.

5.) Electromotive <u>force</u>

EM......F?

Concepts of Voltage, Current and Resistance

The three physical quantities in the title to this chapter are the basics of electrical circuits. *Current* is a measure of how much charge passes a given point in space per unit time, measured in *Amperes*.

> *Voltage: what's pushing*
> *Resistance: what's pushing back*
> *Current: how well did we do?*

Now, what causes the current to flow? We know that electric fields apply forces on charges, and it is indeed the electric field that causes the charges to move. We represent the electric field a little differently here, though, by using the *electric potential*, discussed in the Concepts section of Chapter 19, and measured in *Volts*. Voltage (electric potential) is similar to potential energy in that the voltage at a particular point is arbitrary - it is only <u>voltage differences</u> that are well-defined. The relationship between the voltage and the electric field is that <u>the electric field is the rate of change of the voltage with spatial displacement</u>. A small electric field is reflected by a slow change in voltage from point A to point B. If the field between A and B is increased, then the voltage between A and B changes more rapidly through space. Now, as charges move through space, they will encounter some opposition. Often, the current is represented by movement of electrons through a metal, in which case the opposition is provided by the electrons scattering from the crystal lattice ions of the metal. A measure of this opposition is called the *resistance* of the piece of material, measured in *Ohms*.

These three variables are related for a voltage difference ΔV applied across a piece of material of resistance R, resulting in a current I. It should be noted that an unfortunate tradition in physics textbooks results in using the symbol V <u>both</u> for a voltage <u>and</u> for a voltage *difference*. It is more correct to call a voltage difference ΔV, but this is not done very often. The relationship among the variables is:

> *The variables are related by*
> $$I = \frac{\Delta V}{R}$$

$$I = \frac{\Delta V}{R} \qquad (I = \frac{V}{R}, \text{ in many textbooks})$$

which is often incorrectly called Ohm's Law (See Myth #2). The equation above has the form,

$$\text{time rate of flow} = \frac{\text{amount of drive}}{\text{amount of opposition}}$$

which can also be interpreted, in more simple terms, as

$$\text{effect} = \frac{\text{cause}}{\text{opposition}}$$

which is a general form of a <u>transport equation</u>. The "drive", or "cause", in the numerator of the right side is a difference in voltage for the electrical case and, in general, is a difference in some physical variable for analogous situations. Other problems involving transport have exactly the same form of equation. For example, conductive flow of heat through a piece of material can be described with an equation of the form,

$$\text{time rate of heat flow} = \frac{\text{difference in temperature}}{\text{thermal resistance*}}$$

For water flowing in a pipe, we can express the relationship as,

$$\text{time rate of water flow} = \frac{\text{difference in pressure}}{\text{viscous and frictional resistance}}$$

Even Newton's Second Law can be expressed in a similar form:

$$F = ma \implies a = \frac{F}{m} \implies$$

$$\text{time rate of change in velocity} = \frac{\text{net force}}{\text{resistance to changes in motion}}$$

Electrical Power:
$P = I\Delta V$

In cases where a voltage difference is driving a current flow, energy is being transferred and we can measure the rate of transfer by the variable <u>power</u>. The power in this case is given by**,

$$P = I\Delta V \qquad (P = IV, \text{ in many textbooks})$$

If the current is passing through a resistive load, then we can rewrite the expression for power by using the relationship between voltage, current and resistance, resulting in two additional expressions:

$$P = I\Delta V = I(IR) = I^2 R$$

$$P = I\Delta V = (\frac{\Delta V}{R})(\Delta V) = \frac{(\Delta V)^2}{R} \qquad (P = \frac{V^2}{R}, \text{ in many textbooks})$$

---

*Electrical resistance of a piece of material has the form $R = \frac{1}{\sigma}\frac{L}{A}$, where σ is the electrical conductivity, L is the length of the material through which current will flow and A is the cross sectional area through which the current will flow.

Thermal resistance of a piece of material has the form $R = \frac{1}{k}\frac{L}{A}$, where k is the thermal conductivity, L is the length of the material through which heat will flow and A is the cross sectional area through which the heat will flow.

**But see Myth # 2, Chapter 24.

Discussions; Chapter 20 - Voltage, Current and Resistance

𝔐𝔶𝔰𝔱𝔢𝔯𝔦𝔢𝔰:

1.) When a light bulb fails, why does it almost always do so just as you turn it on?

Note first that we have not used the phrase, the light bulb "burns out". The use of the word "burn" applied to a light bulb is inappropriate, since the word implies some form of <u>combustion</u> process, which is not occurring in a light bulb.

As the tungsten filament in a light bulb glows (not <u>burns</u>!), tungsten atoms vaporize from the surface. Since the filament is a <u>coil</u> (or even a coiled coil!) rather than a straight wire (as in earlier bulbs), there may be some thinner areas of the filament resulting from the manufacturing process. Even if the coiling process is perfect, the fact that it is a coil means that some areas of the filament will receive more radiation from neighboring areas (and, thus, be hotter) than other areas (and, thus, will vaporize faster). Thus, at some point in its life, the filament will either have and/or develop some regions that are thinner than others. Now, since the resistance of a wire depends on the cross sectional area, these thin regions will exhibit more resistance than the thicker regions. As a result, the voltage drop per distance along the filament is larger here than elsewhere. This results in more power being delivered to the thin regions and a correspondingly higher temperature. The higher temperature results in faster vaporization from these regions. Thus, any thin regions which do exist and/or develop become even thinner due to these effects.

The second idea that we will need to explain this mystery is the following. The resistance of tungsten, along with most materials, increases with temperature. The easiest way to understand this is to realize that the crystal lattice ions will vibrate with larger amplitudes at higher temperatures, representing a larger cross section and a higher probability of scattering electrons. Thus, when the light bulb is not operating, the resistance of the cold filament is low. When the bulb is first turned on, the filament offers relatively little opposition to electrical flow, and the current grows to a very high value rapidly. The current maximizes and falls as the temperature of the filament rises and offers more resistance. Thus, there is a spike in current right at the beginning of operation.

Now, let us combine these two ideas. When the bulb is turned on and the very large current spike grows, the adjacent coils exert forces on each other (similar to the forces between the coils of the transformer in the electronic flash unit of Mystery #12, Chapter 18). Thus, there are strong mechanical stresses on the coils at the time of the large current spike. If an area on the filament has become sufficiently thin, due to the vaporization of tungsten, it will break under this mechanical stress. The result is a voltage difference across a gap between two pointed ends of the filament at the break. This results in a very large electric field at these pointed ends, as in the pointed lightning rods of Mystery #2, Chapter 19. As a result, a small bolt of lightning passes between these points as the gas in the bulb is ionized by the electric field. This results in the flash of light that is seen as the light bulb fails. Since the current spike occurs just after the light bulb is turned on, that is most often when the bulb fails.

2.) As an incandescent light bulb grows older, it glows more dimly. Why?

As mentioned in the discussion of Mystery #1, as a bulb glows, tungsten atoms vaporize from the surface and collect on the interior surface of the glass. As a result, the radius of the filament slowly decreases with time. This causes the resistance of the bulb to increase, which lowers the current. This lowers the power delivered, which lowers the operating temperature of the filament and, in turn, lowers the amount of light radiated by the filament. The combination of less light from the filament and the increased opacity of the tungsten-coated glass results in a dimmer bulb.

3.) Electrical energy is transmitted over power lines at hundreds of thousands of volts. Why is such high, dangerous voltage used?

If we consider the power delivered to a resistance, we can combine the expression for electrical power with that relating voltage, current and resistance as discussed in the Concepts section. We obtain the result,

$$P \;=\; I\Delta V \;=\; I\,(IR) \;=\; I^2 R$$

where we consider R to be the resistance of the transmission wires. Now, imagine that we transmit electrical energy at low voltage. For a given desired power transmission to the load (the system of users - homes, businesses, etc.), the current would be high. According to the above expression, the power delivered to the resistance of the wires would be large. This would result in raising the temperature of the wires, so that electrical energy would be transferred by conductive and convective processes from the wire to the internal energy of the air and the supports for the wires. This effect of increasing temperature due to resistance is called *Joule heating* (but is it *heating*?! See Chapter 15). If the energy transmission is at high voltage, however, then the current is relatively low and the power delivered to the resistance of the wires will be much smaller, since it depends on the square of the current. This will result in less thermal transformation and more electrical power transmitted through the network of wires to the load.

Now, it might be argued that high voltage would lead to high energy loss, since, as discussed in the Concepts section, the power can also be expressed as,

$$P \;=\; \frac{(\Delta V)^2}{R}$$

This argument can be countered by noting that the high voltage that we are discussing in electrical transmission is that between the transmission wires. In the expression for power above, however, the voltage difference is that across the resistance in the denominator, which is the voltage between the ends of the wire. This voltage is given by $\Delta V = IR$, so that reducing the current through the wires (by increasing the voltage between the wires) actually reduces the voltage across the resistance of a wire. This leads to agreement that reduction of the current results in reduction of energy transformations.

In our homes, of course, we don't use devices rated at hundreds of thousands of volts. Transformers are used (see Mystery #3, Chapter 23) to step down this high voltage to the household voltage that is appropriate for devices in the home.

4.) Why do your headlights dim if you have them on while you start your car?

An automobile battery, just like any battery, has internal resistance. When current is flowing through the battery, there is a voltage drop across this resistance, and some energy is transformed by this resistance into internal energy - the battery becomes hotter. Thus, the voltage difference between the terminals drops as the current rises. The starter is an electrical device that requires a large amount of current. When this current is being delivered by the battery, the voltage between the battery terminals is significantly lowered. Since the headlights are connected across the battery terminals, they will dim in response to the lower voltage.

5.) Speaking of starting your car, batteries are rated in *Ampere-hour*s. What does this measure?

Since an Ampere is a measure of charge per unit time, the product of an Ampere and a time unit will be a charge unit. Thus, the number of Ampere-hours on a battery represents the total estimated charge that the battery can deliver. A battery with a higher rating will be able to crank an engine longer than one with a lower rating.

6.) Many automobile headlights are *halogen* bulbs. What *is* a halogen bulb and how does it work?

A major contribution to the failure of a light bulb is the continuous transfer of tungsten from the filament to the inner walls of the light bulb, as described in the discussion of Mystery #2. When a failed bulb is removed from a lamp, the coating of tungsten can be easily seen on the inside of the bulb. The halogen bulb is designed to slow this transfer process. The halogen gas, often iodine (or bromine), works in a cyclic process that removes tungsten deposited on the wall of the bulb and returns it to the filament. Let us suppose some tungsten has been transferred to the wall. For wall temperatures above about 250°C, the tungsten will react with the iodine to form WI_2, which will not adhere to the wall, but will drift off into the interior of the bulb. When one of these molecules encounters the <u>very</u> high temperature of the filament, it dissociates. The tungsten atom adheres to the filament, while the iodine moves away. The iodine is then free to drift to the wall again and start another cycle of carrying the tungsten back to the filament.

Because of the high temperatures at the bulb wall, halogen bulbs are often made of quartz, with a melting point of over 1600°C. In order to achieve the high wall temperatures, the wall must be close to the filament, so that the bulb package is small, on the order of 1 - 2 cm. It is often sold as a unit with a parabolic reflector, both to protect the fragile quartz bulb and wires and to prevent oils from human fingers on the quartz, which could lead to a cracked region on the surface of the bulb.

7.) You've just received a new toy; what type of batteries do you need - AA, AAA, C, D? Wait a minute. Why are there no A or B batteries?

There used to be A and B batteries, which were used to provide power for crank telephones and early radios. They are no longer manufactured as they have no use in to-day's devices.

As an example, B batteries were used to power separate loudspeakers for home radio sets. It is interesting to look in old magazines from the 1920's to see examples of the use of these batteries. For example, on pages 108-109 of the *Saturday Evening Post*, January 8, 1927, there is a two-page advertisement for the Eveready Layerbilt "B" battery, with reference to its use in loudspeaker sets. Another example is on page 167 of the *Saturday Evening Post*, October 17, 1925, in which the Jewett Radio & Phonograph Co. (now, why do you suppose this example was chosen?) mentions the B-battery in the speaker circuit.

8.) How do *three-way* light bulbs work?

Three-way light bulbs have two filaments which can be operated independently or to-gether, to give three different lighting levels. The connections to these filaments are shown in the diagram below for a 50-100-150 watt three-way bulb.

The bottom part of the diagram shows the bottom of the bulb. If you look at the base, you will see a central contact (the small white circle in the middle of the base in the diagram), a ring contact (the middle white annular region) and the outer shell (the outer white annular region), which is the screw base for the bulb. The dashed lines show the connections to the two filaments. At the lowest light level, the shell and the ring are connected to the line voltage and the 50 watt filament glows. At the next highest set-

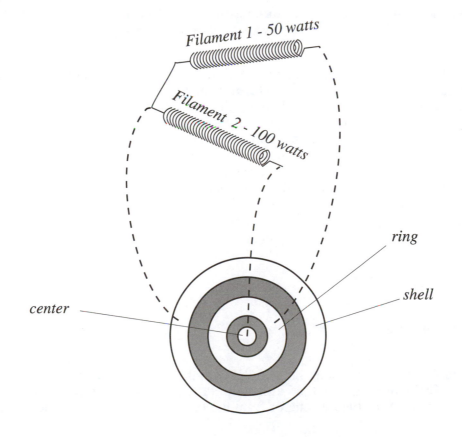

ting, the center and the shell receive the line voltage and the 100 watt filament glows. At the highest setting, the center and the ring are shorted, with the line voltage then applied between this combination and the shell. Thus, the filaments are connected in *parallel* (see Chapter 21) and both glow. The electrical power consumed in this situation is 50 watts + 100 watts = 150 watts. For more information on the physics of incandescent light bulbs, see the articles by H. S. Leff, "Illuminating Physics with Light Bulbs", *The Physics Teacher*, **28**, 30 (1990) and W. S. Wagner, "Temperature and Color of Incandescent Lamps", *The Physics Teacher*, **29**, 176 (1991).

9.) Some fluorescent lights, such as those in desk lamps, require the "on" button to be held down for several seconds until the bulb lights. Why?

A fluorescent bulb requires a current to flow through the gas inside the bulb. When the bulb is first turned on, a large voltage is needed to create the first arc across the bulb by ionizing the gas. In "instant-on" type bulbs, this large voltage can be provided by the starting system. A lower starting voltage, however, is needed if the electrodes are heated so that some electrons are present in the tube by means of thermionic emission. Some "instant start" systems provide an automatic pre-heating system for the electrodes so that the electrons are present within a second or two of turning on the bulb. In many desk lamps, this system is not present. By holding the button down, you manually provide a heating voltage to the electrodes, which causes the electrons to be emitted. After a few seconds, enough electrons are present that the button can be released and the current will be established in the bulb.

10.) Early calculators and digital watches used light emitting diodes for the display. Today, virtually every display is a liquid crystal. Why?

A light emitting diode is a source of light and therefore requires an energy conversion process (electrical potential energy to electromagnetic radiation). A battery-powered device operating with light emitting diodes will draw a large amount of energy, causing the batteries to run down quickly. Watches with light emitting diode displays did not display continuously - you needed to push a button to see the time. Otherwise, the battery life would be extremely short. A liquid crystal display (See Mystery #3, Chapter 28) is not a source of light - you can only see it due to the reflection of ambient light in your environment. The liquid crystal display draws an extremely small amount of current and, thus, requires very little energy. The battery life for this type of display is much longer.

𝕸𝖆𝖌𝖎𝖈:

The Steel Wool Fuse

As additional light bulbs are added in parallel with the first, more current is drawn from the battery. This can be understood either from the point of view that more paths are being made available for the current, or from the fact that the total resistance of the circuit is decreasing. Eventually, the current will be high enough to raise the temperature of the steel wool filament high enough for it to burn.

If the filament burns with only one light bulb in the circuit, try using a thicker strand of steel wool or multiple strands.

𝔐𝔶𝔱𝔥:

1.) If a fuse blows in a household circuit or an automobile, put in a bigger one!

Although fuses have been replaced in new construction by circuit breakers, there are still old houses (and automobiles) with fuseboxes. A fuse is designed to "blow" if the current in a given circuit becomes too high. This is intended as protection against a fire started by an overheated electrical circuit. If the current rises above the rated value of the fuse, the metal in the fuse will melt and break the circuit. Putting in a bigger fuse (higher current rating) may stop it from blowing, but it defeats the purpose, which is <u>protection</u>. Thus, the circuit with the higher rated fuse may overheat and cause a fire.

2.) Ohm's Law is a law.

As stated in the Concepts section, the equation $\Delta V = IR$ is often called Ohm's Law. Well, not only is this not Ohm's Law, but Ohm's Law is not even a law. It should more properly be called *ohmic behavior*. <u>A device obeys Ohm's Law, or exhibits ohmic behavior, if a graph of the current through the device against the applied voltage is a straight line.</u> If the graph is not a straight line, then the device is <u>non-ohmic</u>. As examples, consider the current-voltage graphs of two devices shown below. The first device is a simple electrical resistor, while the second is a semiconductor diode.

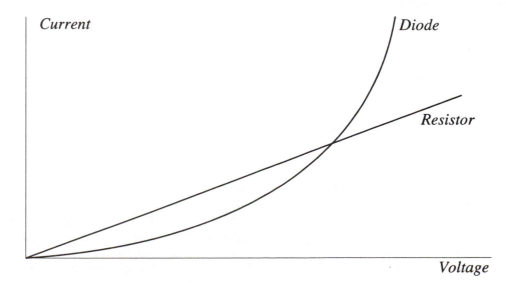

The graph for the resistor is linear, so it is an ohmic device, obeying Ohm's Law. The graph for the diode is not linear, so it does not obey Ohm's Law. For both devices, however, the relationship between the voltage, current and resistance anywhere on the graph is given by $\Delta V = IR$. The resistance R is constant for the resistor, but decreases with voltage for the diode.

3.) Big electrical items like televisions use lots of electrical power.

The amount of power consumed by an electrical device is not related to its physical size, but to its resistance. In reality, a television draws relatively little power. A typical power rating for a 27" color television is 160 watts. The types of devices that draw the most power are those which require electrical "heating" for their operation. A typical iron, for example draws 1100 watts, while a four-slice toaster may draw 1600 watts. Devices with motors are also large users of electrical power. For example, a typical upright vacuum cleaner draws about 850 watts.

4.) Low voltage is not dangerous.

While high voltage is clearly dangerous, proper care should be paid to low voltages also. Humans have received fatal shocks from voltages as low as 50 V. The important parameter is current, not voltage, and an important consideration is the path taken by the current. If you make contact with an electrical circuit so that current flows from your forearm to your fingers, you may feel a painful shock. If this same current flowed between your two hands, however, it likely will also flow through your heart and could be fatal. Currents as low as 1 mA can cause a tingling sensation, while currents in the range of 10 mA can cause painful shocks. Muscle paralysis and breathing difficulties can occur at 20 mA. Breathing may not be possible at 75 mA, and ventricular fibrillation of the heart occurs at 100 mA. Above 200 mA, the heart may stop completely and the body may receive severe burns. When electricians need to move live wires with their hands, they touch them with the backs of their hands, so that an unexpected shock will not cause a muscular contraction that will keep their hands gripping the wire.

While the current through the body will be determined by the voltage, it will also depend on the resistance. Thus, if household voltage of 120 volts is applied across your body and your electrical resistance is low, a serious shock could result. The resistance of your body depends mostly on the situation at the contact points with the voltage supply. Resistance will vary with the area of contact, as well as with the wetness of the skin.

5.) Electromotive force

Despite the word *force* in the name of this quantity, it is not a force, but is a voltage difference. It is the highest voltage difference that a battery can provide. It is not the voltage measured across the battery terminals when it is operating a device. As we mentioned in the discussion of Mystery #4, the internal resistance of a battery causes the terminal voltage to decrease as the current output increases. If the current were to be extrapolated to zero, the resulting terminal voltage would be the electromotive force.

Chapter 21
Electrical Circuits

𝕸ysteries:

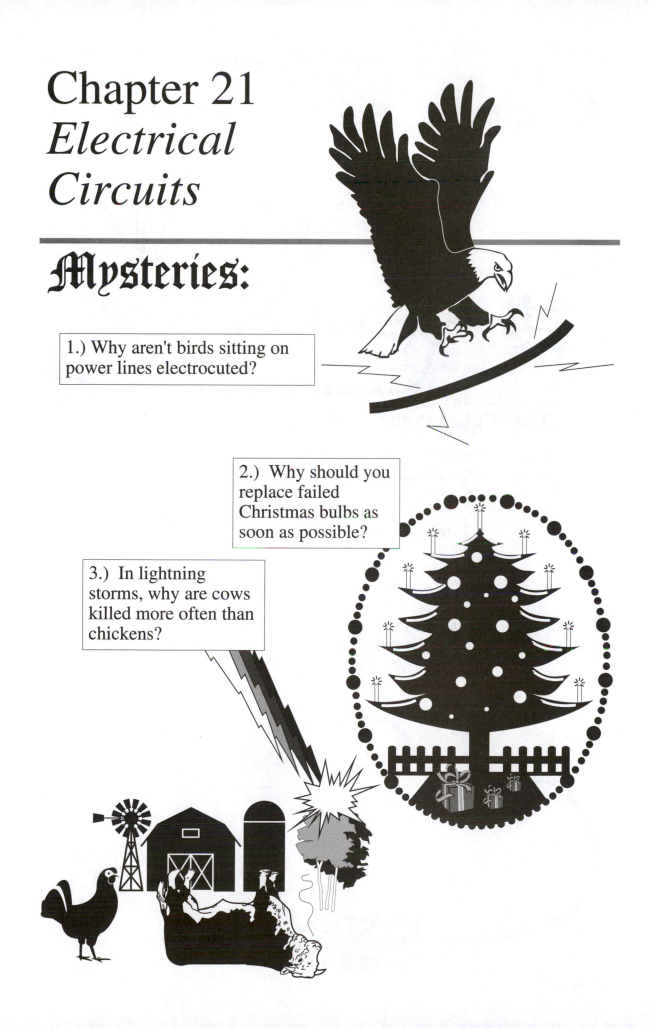

1.) Why aren't birds sitting on power lines electrocuted?

2.) Why should you replace failed Christmas bulbs as soon as possible?

3.) In lightning storms, why are cows killed more often than chickens?

Magic:

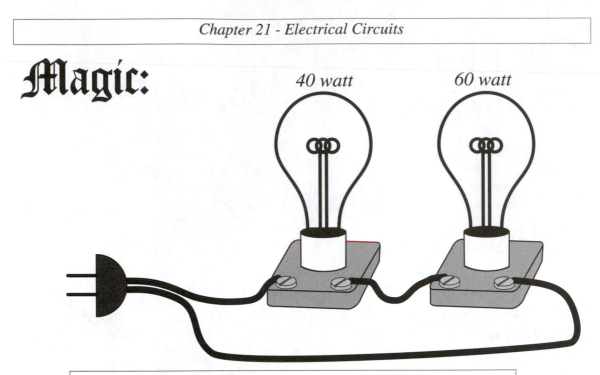

40 watt   60 watt

The Higher Powered Bulb?

Carefully connect a 40 watt and a 60 watt bulb in <u>series</u> across a 110 volt supply. Which is the brighter bulb?

The Magic of Salt

Set up a circuit with a light bulb and a battery. Now cut and strip the wires at some point in the circuit and dip the ends into a small container of distilled water. Does the light bulb light up? Now add some salt to the water. Does the light bulb light up? Look carefully at the region of the water around the wires.

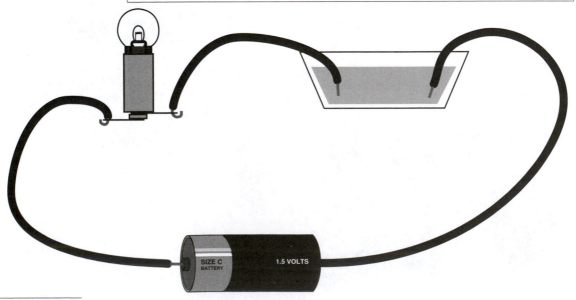

SIZE C BATTERY   1.5 VOLTS

Myth:

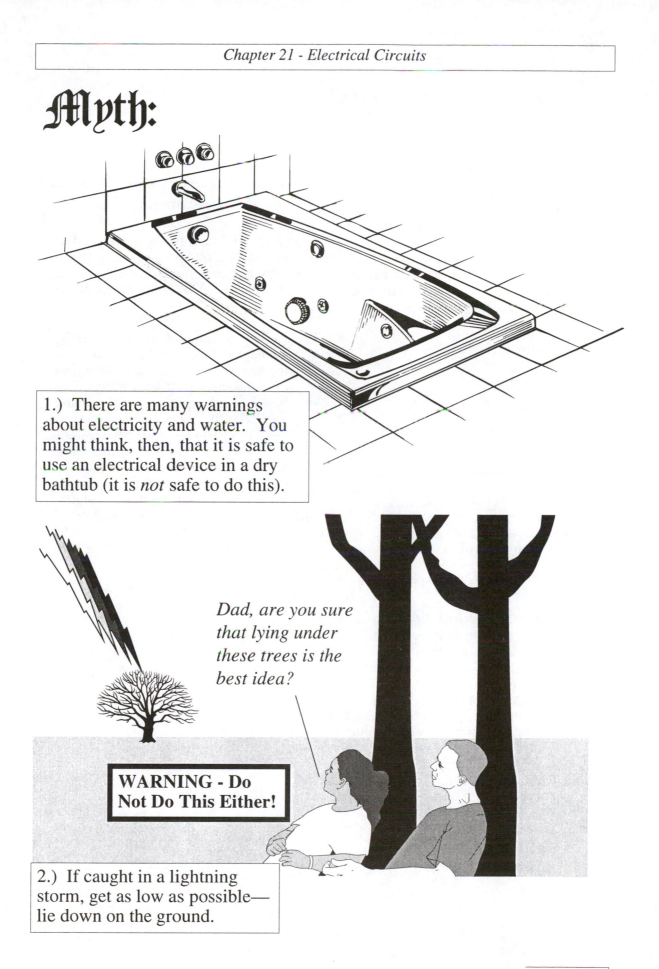

1.) There are many warnings about electricity and water. You might think, then, that it is safe to use an electrical device in a dry bathtub (it is *not* safe to do this).

Dad, are you sure that lying under these trees is the best idea?

WARNING - Do Not Do This Either!

2.) If caught in a lightning storm, get as low as possible— lie down on the ground.

Concepts of Electrical Circuits

The major concept of an electrical *circuit* is that a <u>complete</u> circuit must exist before electrical current can flow. In general, there is a voltage source (battery or power supply), wires to carry the current and some <u>load</u> in the circuit. There must be a continuous connection

> *A complete circuit must exist in order for electrical current to flow.*

from the battery through the wires to the load and back to the battery. The load is generally an electrical device of some sort - a motor, a heater, a stereo system, a television, etc. In operation, energy is transferred from the voltage source to the load, according to the power equation, $P = I\Delta V$, given in the previous chapter.

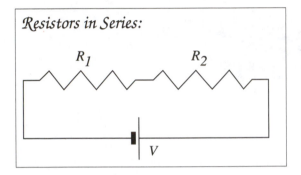

Resistors in Series:

If two or more devices are to be connected to the voltage source at the same time, there are two types of combinations that can be analyzed in a simple way - <u>series</u> and <u>parallel</u>. In a *series* circuit, two or more devices are connected "end-to-end", so that the same current must pass through each device. Thus, the current through each device is the same as that through the battery. Furthermore, the sum of the voltage differences across the devices will equal the terminal voltage difference of the battery. The equivalent resistance of the combination of two (or more) resistive devices connected in series is,

$$R_{eq} = R_1 + R_2\ (+\ R_3\ +\ R_4\ +\ .....)$$

The second possibility is to connect the devices in *parallel*. In this case, the devices can be envisioned as being side-by-side, rather than end-to-end. When connected in this way to a voltage source, the <u>same</u> voltage difference is applied to all of the devices. The current divides, so that the sum of the currents in the devices equals the current through the battery. In the case of resistive devices connected in a parallel circuit, the <u>inverses</u> of the resistances add up to equal the inverse of the equivalent resistance:

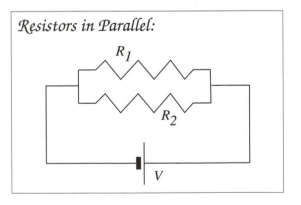

Resistors in Parallel:

$$\frac{1}{R_{eq}} = \frac{1}{R_1} + \frac{1}{R_2}\ (\ +\ \frac{1}{R_3} + \frac{1}{R_4} + \ .....)$$

Discussions; Chapter 21 - Electrical Circuits

𝕸𝖞𝖘𝖙𝖊𝖗𝖎𝖊𝖘:

1.) Why aren't birds sitting on power lines electrocuted?

We can identify three contributions to the non-electrocution of birds sitting on power lines: 1.) The power lines are covered with insulation so that the birds do not make contact with the current-carrying metal; 2.) A given power line will have a certain voltage with respect to the ground voltage. The bird is not spanning a voltage difference from a wire to the ground; it is just in contact with one wire. This is similar to trying to develop a current by making contact with only one terminal of a battery; and 3.) It would be possible to cause a significant current to pass through a bird's body if it were to touch two points widely separated on a <u>highly resistive</u> material carrying a <u>large current</u>. In this case, there would be a large voltage difference between the two points of contact, the feet of the bird. Then, we have essentially put the bird in parallel with another resistor, so that current could pass through the bird. But this is not the case with the bird sitting on the wire. We have the bird sitting on a metal conductor with <u>low</u> resistance, carrying <u>low</u> current (See Mystery #3 in Chapter 20). Thus, there is a very small voltage difference between the feet of the bird, regardless of the magnitude of the voltage on the wire with respect to the ground.

A similar argument would hold if you find yourself hanging from a live power line for some reason. You have no problem as long as the wire is the only thing you are touching. If you are close enough to jump or step onto the ground, be sure to let go of the wire before you make contact with the ground. If you are in contact with both at the same time, you have completed the circuit!

2.) Why should you replace failed Christmas bulbs as soon as possible?

Most strings of Christmas tree lights are connected in series. If a light bulb is removed, the circuit is no longer complete and the whole string of lights goes out. Many bulbs are equipped with a shunting device, so that, if the bulb fails, the current still passes through the failed bulb, and the other bulbs will still light. This is good news and bad news. The good news is that you don't have to hunt for the bad bulb - it is simply the one that is not lit up. The bad news is that there are now fewer bulbs in series across the voltage source. The source voltage is now divided over fewer bulbs, so that each bulb operates at a slightly higher voltage (and increased power delivered to the bulb), increasing the chances that another bulb will fail. If a second bulb fails, the voltage (and power) across each bulb rises a bit more. If a number of bulbs have failed, the voltage across each bulb could be dangerously high, causing the bulbs to operate at a very high temperature, which could cause a fire.

For more information on Christmas tree lights, see W. DeBuvitz, "Christmas Tree Lights - A Continuing Series?", *The Physics Teacher*, **30**, 530 (1992).

3.) In lightning storms, why are cows killed more often than chickens?

In some ways, this is the opposite of Mystery #1, at least for the cows. When lightning strikes, say, a tree, current passes through the tree and then spreads through the ground radially away from the tree. The ground has a fairly high resistance and, under these conditions, is carrying a large current. Thus, if we consider that part of the current moving at the ground surface, there is a significant voltage difference between points on the ground separated radially with respect to the tree. (We normally consider the ground to be an equipotential, but in this extreme situation of very high currents from lightning, the non-zero resistance of the ground results in substantial potential differences between different points.) Now, imagine placing a cow in a position facing the tree. We have put the cow in parallel with the current running through the ground. There is a significant voltage difference between the front and rear legs of the cow, so that a dangerous current can pass through the heart of the cow, possibly killing it. With chickens, the two legs are much closer together than the cow's front and rear legs, so that the voltage difference and, hence, the current is much smaller and less likely to be lethal. This is similar to contribution #3 to the safety of birds on power lines, as described in Mystery #1.

𝔐𝔞𝔤𝔦𝔠:

The Higher Powered Bulb?

The result is that the 40 watt bulb glows more brightly than the 60 watt bulb. The ratings of light bulbs refer to the power delivered to them at the standard voltage of 110 volts from the house voltage. Of course, all devices in a home are connected in parallel with the power source, so each device has the full voltage applied to it. If we consider the expression for power delivered to a load, $P = I\Delta V$ and express it in terms of voltage difference and resistance, as was done in Chapter 20, we have

$$P = I\Delta V = (\frac{\Delta V}{R})\Delta V = \frac{(\Delta V)^2}{R}$$

Thus, remembering that we hold the voltage difference constant for bulbs of different wattages, we see that higher powered bulbs must have lower resistance. Now, imagine that we place bulbs of different wattages in series. Both bulbs now carry the same current. The higher power, lower resistance bulb (60 watts, in the example) will have a lower voltage across it than the lower powered, higher resistance bulb (the 40 watt). Thus, in the expression for power,

$$P = I\Delta V,$$

the current is the same in both bulbs, so that bulb with the higher voltage (the 40 watt) will have more power delivered to it. Thus, it will glow more brightly than the 60 watt bulb.

The Magic of Salt

With the ends of the wire in distilled water, the bulb will either not light up or it may glow dimly, due to any impurities which may have been introduced into the water. Adding the salt results in ions of sodium and chlorine in the water which can carry a current, and the bulb will light up. The existence of the ions in the water essentially completes the circuit, as it provides charge carriers, just as a wire does. At each wire end in the water, you should see bubbles rising. You are performing <u>electrolysis</u>, and liberating hydrogen and oxygen molecules from the water.

𝕸𝖞𝖙𝖍:

1.) There are many warnings about electricity and water. You might think, then, that it is safe to use an electrical device in a dry bathtub (it is *not* safe to do this).

It is common to hear warnings about operating electrical devices while taking a bath or when water is nearby. And these warnings are perfectly valid - <u>extreme care should be taken when operating electrical devices near water, if the use cannot be avoided altogether</u>. The reason that this statement appears in the Myth section is that it is not actually the water that is the direct cause of the problem - it is only an intermediary. The main problem with operating electrical equipment in or near water is the existence of metal plumbing lines which are in intimate contact with the ground. Imagine the metal drain of your bathtub, for example. It is connected to metal plumbing pipes that go directly into the ground. Thus, <u>the drain is at the ground potential</u>. If your hands should come in contact with a high voltage in the electrical device while your foot is in contact with the metal drain, electric current will pass through your body. Thus, you can be electrocuted in a perfectly dry bathtub!

If your foot is <u>not</u> in contact with the drain, but there is water in the tub, then <u>the water provides charge carriers which can carry current between your foot and the drain</u>. (Pure water is a poor conductor, but dirty tap water contains ions and is thus a much better conductor.) If your hands should come in contact with a high voltage in the electrical device or if the device is dropped into the water, your body can become part of a circuit, along with the water, between the high voltage point on the device and the drain. Thus, the water is dangerous, in that it carries the current, but it is the metal drain connected to the ground that is the real culprit.

2.) If caught in a lightning storm, get as low as possible - lie down on the ground.

The advice to get as low as possible is good advice - you don't want to be the highest object around. But <u>do not lie on the ground</u>. If you do, you become a "cow", as in Mystery #3. By this, it is meant that you have two contact points on the ground, at your head and your feet, that are separated by a large distance on the ground. If current from a lightning stroke should pass through the ground parallel to your body, there could be a large voltage between your head and your feet, resulting in a large current passing through your heart. The best position is to stoop, so that you lower your height, but to remain in a squatting position, so that you are more like a "chicken" than a "cow".

Chapter 22
Magnetic Fields & Forces

𝕸ysteries:

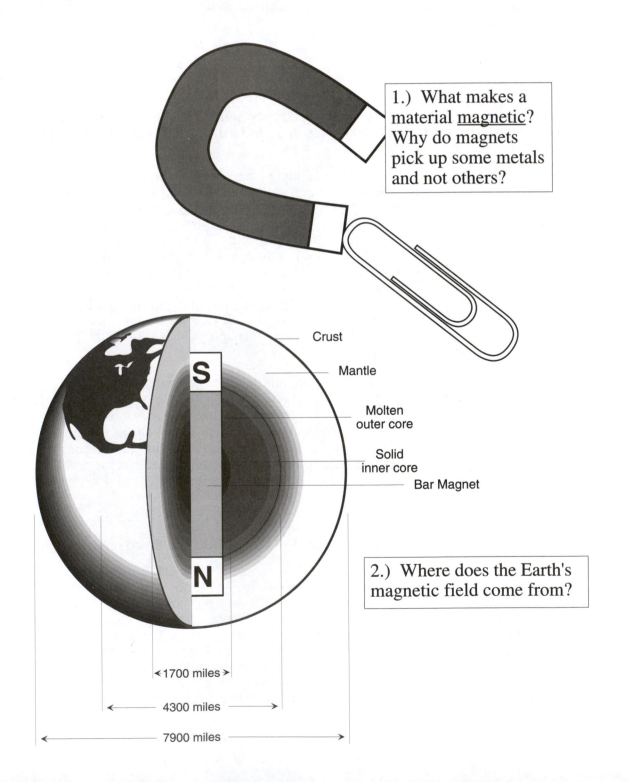

1.) What makes a material <u>magnetic</u>? Why do magnets pick up some metals and not others?

Crust

Mantle

Molten outer core

Solid inner core

Bar Magnet

S

N

◄ 1700 miles ►

◄ 4300 miles ►

◄ 7900 miles ►

2.) Where does the Earth's magnetic field come from?

Mysteries:

3.) In an episode ("Nightmares") of the television series *MacGyver* (Paramount Pictures, 1985-1992), MacGyver was poisoned. Toward the end of the show, the antidote, in an iron canister, fell into a storm drain. To extract it, MacGyver picked up an iron rod and struck a fire hydrant with it several times. In doing so, the rod became magnetized and he was able to pick up the canister. What's going on here?

Magic:

Hold magnet close to corner of bill.

An Attraction for Dollars

Strong neodymium magnets are available from scientific supply houses. These magnets provide a very strong magnetic field in a small volume. If a neodymium magnet is brought near the corner of an American dollar bill, the bill will be attracted! Try it!

The Amazing Balancing Act

Remove the magnet from a loudspeaker that you no longer need. Hide it under a small cardboard box, cut so that the top rests on the magnet. Now, amaze some observers as you balance razor blades and paper clips on the box!

Box to hide magnet

Magnet

Magic:

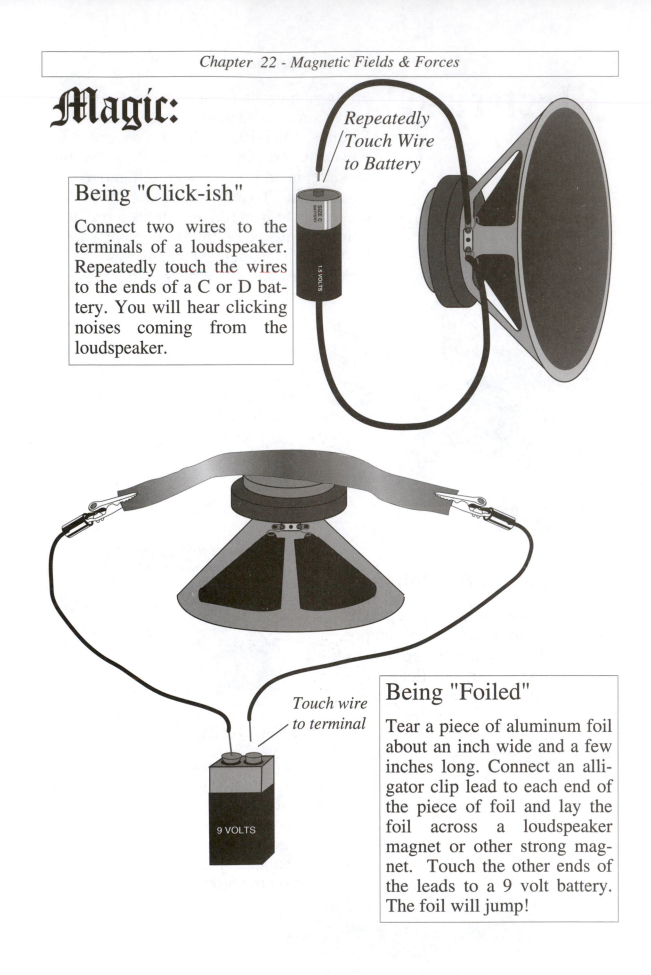

Being "Click-ish"

Connect two wires to the terminals of a loudspeaker. Repeatedly touch the wires to the ends of a C or D battery. You will hear clicking noises coming from the loudspeaker.

Repeatedly Touch Wire to Battery

Touch wire to terminal

Being "Foiled"

Tear a piece of aluminum foil about an inch wide and a few inches long. Connect an alligator clip lead to each end of the piece of foil and lay the foil across a loudspeaker magnet or other strong magnet. Touch the other ends of the leads to a 9 volt battery. The foil will jump!

Myth:

1.) The General *Motors* Corporation, the Ford *Motor* Company, Chrysler *Motors*, American *Motors*, Bavarian *Motor* Works (BMW), etc.

2.) Compasses point toward the North magnetic pole.

3.) Compasses point toward the North.

4.) Newton's Third Law - if A exerts a force on B, then B exerts an equal force on A in the opposite direction.

Concepts of Magnetic Fields & Forces

The science of magnetism developed separately from that of electricity for centuries. The word *magnetism* comes from *Magnesia*, the name of a region of Asia Minor where large deposits of lodestone, a magnetic mineral, were found.

The properties of magnets that are most readily observed are their attraction (or repulsion) for each other, their attraction for certain types of metal and their ability to point in a given direction relative to the Earth when allowed to hang freely.

Our model of magnetism recognizes two poles on a given magnet, which we name *North* and *South*. In some ways, these poles are similar to electric charges, but there are some important differences. One way in which poles behave like charges is in their attraction or repulsion. We have experimentally observed forces between magnets and determined that the

> *Like Poles Repel.*
> *Unlike Poles Attract.*

magnetic force between like poles is repulsive and that between unlike poles attractive. Poles are different from charges in that there are no isolated magnetic poles such as there are isolated charges - poles always seem to exist in pairs*.

> *A magnetic field is established by a single magnet.*
>
> *An magnetic force requires two magnets.*

When we considered electricity (Chapter 19), we imagined a two-step process, in which one charge establishes an electric field and the other charge then enters that field and, as a result, feels a force. We can develop a similar model for magnetism. We imagine one magnet to alter space around itself in a way that we describe as a *magnetic field*. When a second magnet is brought into the field, it experiences an attractive or repulsive force due to its interaction with the field.

We can also set up a situation in which a single charge is immersed in a magnetic field. In this case, we find that there is no magnetic force unless the charge is moving. If it is moving with a component of its velocity perpendicular to the magnetic field, the magnetic force on the particle is proportional to the charge, the magnetic field and the velocity. We also find that the direction of the force is perpendicular to both the field and the velocity vectors. This magnetic force is the basis of particle accelerators that use magnetic fields to guide particles along planned paths.

An electric charge is the origin of the electric field, as discussed in Chapter 19. It has been known since the nineteenth century that the origin of a magnetic field is a current, that is, moving charges. Thus, both electric and magnetic fields have charges as their origin. This discovery is an early example of unification. What were initially considered to be separate phenomena, electricity and magnetism, were unified into a single area of study called *electromagnetism*.

This unification has led to much of our modern technology. For example, the understanding that a wire moving in a magnetic field results in a current (Chapter 23) is the basis of electrical power generation in power stations. Similarly, the understanding that a current-carrying wire feels a force in a magnetic field is the basis for motors, upon which we depend greatly.

---

*Although some modern theories predict the existence of *magnetic monopoles*.

Discussions; Chapter 22 - Magnetic Fields & Forces

𝕸𝖞𝖘𝖙𝖊𝖗𝖎𝖊𝖘:

1.) What makes a material <u>magnetic</u>? Why do magnets pick up some metals and not others?

The magnetic properties of materials is an interesting area of study. The first question could actually be interpreted in two different ways. First, it might mean, "What is special about those materials from which magnets can be made?". Or, it might mean, "What is special about those materials that are not magnets, but that will be attracted to magnets?". For both questions, it is important to realize that even individual atoms act as tiny magnets. They have a North and a South pole and establish a magnetic field about themselves. The behavior of these atomic magnets is the basis for the magnetic properties of materials.

For the first question, a material that can maintain a magnetic field, after an external field is applied and then removed, is called *ferromagnetic*. In such materials, strong interatomic forces result in the "lining up" of the magnetic fields of adjacent atoms. Areas in which the atomic magnetic fields are in the same direction in a material are called *domains*.   In an unmagnetized piece of ferromagnetic material, the random directions of the many domains cancel out, so that there is no net magnetism. If the material is placed in a magnetic field, however, those domains whose magnetic fields are in the same direction as the applied field will grow in size, while other domains will shrink. When the external field is removed, the material will maintain the magnetic field - it will be a "permanent" magnet. This can be demonstrated by "magnetizing" a sewing needle by letting it sit on a magnet for a long time or by stroking the magnet with the needle. At high enough temperatures, the thermal jostling of the molecules in a ferromagnet is sufficient to overcome the interatomic forces responsible for the ferromagnetism and the material will lose its ability to remain magnetized. The *Curie Temperature* is the temperature at which this transition occurs.

Now, for the second question. Since all materials are composed of atoms, all materials exhibit magnetic properties. For reasons "beyond the scope of this book", the behavior of atoms is such that <u>all</u> materials exhibit *diamagnetism*. This is the establishment of a magnetic field <u>opposite</u> to an applied magnetic field. Thus, a diamagnetic material is <u>repelled</u> from the magnet providing the applied field. Some materials exhibit a second effect, in addition to the diamagnetism, called *paramagnetism*. Again, for reasons "beyond the scope of this book", the behavior of these atoms is such that an additional magnetic field is established in the <u>same</u> direction as the applied field. This is stronger than the diamagnetic field, so the overall effect is paramagnetic. These materials are weakly <u>attracted</u> when brought close to a magnet. In both diamagnetism and paramagnetism, the atoms act independently - there is no "coupling" as there is in ferromagnetism.

Now, suppose a magnet is brought close to a <u>ferromagnetic</u> material, such as your refrigerator door. As discussed above, the domains in the material will line up with the applied field, creating a <u>strong</u> attraction between the magnet and the material. The only ferromagnetic elements at room temperature are iron (along with some *steels*, which are alloys containing iron), nickel and cobalt. Other ferromagnetic materials include chromium dioxide and alnico (<u>al</u>uminum, <u>ni</u>ckel, <u>co</u>balt).

2.) Where does the Earth's magnetic field come from?

There is no definitive answer to this question; it remains today an area of research. It is thought that the origin of the field is in the circulation of liquid parts of the Earth's core, which results in a large electrical current.

3.) In an episode ("Nightmares") of the television series *MacGyver* (Paramount Pictures, 1985-1992), MacGyver was poisoned. Toward the end of the show, the antidote, in an iron canister, fell into a storm drain. To extract it, MacGyver picked up an iron rod and struck a fire hydrant with it several times. In doing so, the rod became magnetized and he was able to pick up the canister. What's going on here?

This is an example of the <u>magnetostrictive</u> effect, in which an object will deform if it becomes magnetized. Actually, MacGyver was using the reverse of this effect. He was deforming the rod by striking it on a hard surface, resulting in a net magnetization of the iron rod. An explanation of the magnetostrictive effect is beyond the scope of this book, but can be found in books on magnetism.

𝕸𝖆𝖌𝖎𝖈:

An Attraction for Dollars

The attraction is between the magnet and the ink on the bill. The ink contains small amounts of iron compounds, which are attracted to the magnet.

The Amazing Balancing Act

When the razor blades and/or paper clips are placed on the magnet, they become magnetized, since they are made of ferromagnetic material (hopefully, you didn't use plastic paper clips!). Because they are relatively light, the repulsion between the upper pole on the magnet and the like pole created on the razor blade or paper clip is strong enough to lift this end of the item off the magnet. Be careful not to interpret this as "the razor blade lines up with the magnetic field lines". The actual direction of the blade or clip results from a combination of the magnetic repulsion and the gravitational attraction (the weight).

Being "Click-ish"

A loudspeaker uses the principle that a magnet exerts a force on a current-carrying wire. When you touch the wires in this demonstration to the battery, a current suddenly flows through the coil of the loudspeaker. Thus, there is a sudden magnetic force, which moves the speaker cone suddenly. What you hear is the resulting pulse of sound that leaves the speaker.

Being "Foiled"

This demonstration shows clearly the fact that a magnet exerts a force on a current-carrying wire. In this case, the wire is the strip of aluminum foil. When it carries the current, it feels a force from the magnetic field of the loudspeaker magnet. Since the foil's mass is relatively low, it can respond to the force by moving.

𝔐𝔶𝔱𝔥:

1.) The General *Motors* Corporation, the Ford *Motor* Company, Chrysler *Motors*, American *Motors*, Bavarian *Motor* Works (BMW), etc.

Notice how all of these company names have the word *motor* in them. But the device that runs an automobile is not a *motor*, it is an *engine*. A motor is a device that transforms electrical energy input into kinetic energy, by means of a magnetic force. An engine transforms potential energy in a fuel into kinetic energy. While automobiles may have a variety of motors to operate windshield wipers, power windows and seats, etc., the word in the company name stems from a misunderstanding between motors and engines. This confusion also exists in laser printers for computers - the drive mechanism for the paper is described as an *engine*, when, in reality, it is a *motor*! Finally, if the auto companies end up developing electrical automobiles, then their names will be correct!

In reality, the definitions of motor and engine are worthy of a new *M* - **muddled**! In defense of the auto companies, the two terms are often used interchangeably, even in science. According to the *McGraw-Hill Dictionary of Physics and Mathematics* (McGraw-Hill, New York, 1978),

Motor: *A machine that converts electric energy into mechanical energy by utilizing forces produced by magnetic fields on current-carrying conductors. Also known as electric motor.*

The word *engine* is not defined by itself in this source, but *heat engine* is:

Heat engine: *A thermodynamic system which undergoes a cyclic process during which a positive amount of work is done by the system; some heat flows into the system and a smaller amount flows out in each cycle.*

A similar definition appears in *A New Dictionary of Physics* (Longman, London, 1975):

Motor: *Electric motor. A machine that converts electrical energy into mechanical energy .*

Heat engine: *A device that transforms heat into mechanical energy.*

These resources make a clear distinction between the two devices, consistent with the discussion above. Other resources are not so clear cut, however. In the *Concise Dictionary of Physics* (Oxford University Press, Oxford, 1985), we find a more muddled description:

Motor: *Any device for converting chemical energy or electrical energy into mechanical energy .*
(Notice that this definition combines the concepts of a motor and an engine.)

Engine: *Any device for converting some other form of energy into mechanical work.*
(Notice that this definition is very broad - a nuclear power plant includes an engine, for example, according to this definition.)

In *McGraw-Hill Encyclopedia of Science and Technology* (McGraw-Hill, New York, 1982), we find a distinct definition of a motor but a muddled definition of an engine:

Motor: *An electric rotating machine which converts electrical energy into mechanical energy.*

Engine: *A machine designed for the conversion of energy into useful mechanical motion.*

A look at non-scientific dictionaries is not much help. The Random House Dictionary of the English Language (J. Stein (Ed.), Random House, New York, 1987) lists the first definition for an engine as "a machine for converting thermal energy into mechanical energy...", which is consistent with our approach. The dictionary then muddles the situation by following with the fourth definition, "any mechanical contrivance" (following this is the sixth definition, "an instrument of torture, esp. the rack"!). The word motor is defined in this dictionary as "a comparatively small and powerful engine". The concept of converting electrical energy does not occur until the fourth definition, where this usage is indicated with the Restrictive Label *Electricity*.

In summary, then, let us not take this Myth too seriously - if even the experts cannot agree on what is an engine and what is a motor, then how can we fault the automobile companies?

2.) Compasses point toward the North magnetic pole.

It is common experience that the North pole of a compass needle points toward the north. If one thinks a bit more about this, however, in light of the fact that like poles repel and unlike poles attract, it should be clear that this requires a South magnetic pole to located near the North geographic pole, which is indeed the case.

3.) Compasses point toward the North.

The location of the South magnetic pole (See Myth #2) of the effective bar magnet that we can imagine running through the Earth does not coincide with the geographic North pole of the Earth. At present, the magnetic pole is located at Hudson Bay, Canada, which is about 1300 km away from the geographic North pole. Since the geographic and magnetic poles do not coincide, compasses will not point in the direction of geographic North. The deviation of compass headings from true North is called the *declination angle*, and it depends on the location of the compass. In the middle of the United States, the declination angle is close to 0°, that is, compasses actually do point North. As one moves to the coasts, however, the declination angle rises to between 12° and 15°.

Furthermore, the line connecting points of equal declination angle is not smooth, as the magnetic field also depends on variations in the composition of the Earth at various locations.

What's more, the magnetic North pole does not stay fixed. For reasons that are not understood, the pole moves. For example, the position of the pole in 1904 was 770 km from its current position. In the 240 years preceding 1820, there was a 35° shift in the declination angle in the city of London!

4.) Newton's Third Law - if A exerts a force on B, then B exerts an equal force on A in the opposite direction.

This is the usual form in which Newton's Law is heard, but actually it is not quite true in some cases. To see that this is the case, consider two positively charged particles, *A* and *B*, moving toward a point in space with their velocity vectors oriented at right angles to each other, as shown in the diagram below.

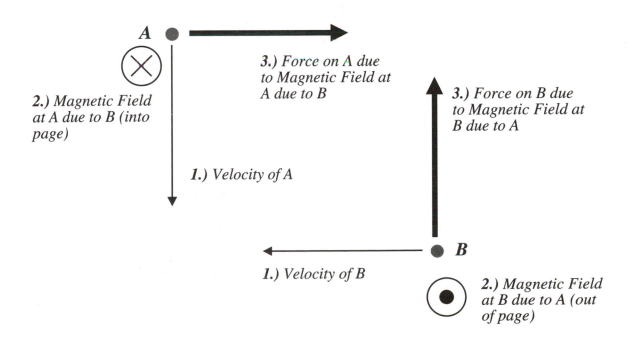

A

3.) Force on A due to Magnetic Field at A due to B

3.) Force on B due to Magnetic Field at B due to A

2.) Magnetic Field at A due to B (into page)

1.) Velocity of A

1.) Velocity of B

B

2.) Magnetic Field at B due to A (out of page)

If we go by the numbers, consider the labels marked with *1.)* first. These are the velocity vectors of particles *A* and *B*. Because of these velocities, each particle establishes a magnetic field at the location of the other, indicated by *2.)* and determined by a right hand rule (Point your right thumb in the direction of the velocity of a positive particle and your fingers curl in the direction of the circular magnetic field lines around the velocity vector.). Now, because of these magnetic fields, each particle feels a force indicated by *3.)* and the heavy arrows and determined by a right hand rule (there are various right hand rules here - one is to point the fingers of your right hand in the direction of the velocity vector, with the magnetic field vector coming up out of your palm. Your thumb then points in the direction of the magnetic force.). Notice that the two force vectors are not in opposite directions.

This is a disturbing result. Its full-scale solution is even more disturbing to the reader looking for an easy resolution and can only be fully satisfied by grinding through some hard-nosed vector calculus in textbooks on electrodynamics. We will address the solution to this problem only briefly and indirectly here, by appealing to the concept of momentum. Newton's Third Law and momentum conservation are intimately related. So if we can accept that momentum is conserved here, then perhaps we will feel better about Newton's Third Law. The important concept to realize is that the electromagnetic fields themselves will carry momentum. In the diagram, there seems to be a net force which would cause a change in momentum of the center of mass of the particles toward the upper right. If the charges were to respond to the forces shown, the electromagnetic fields would carry momentum in the net direction toward the lower left to exactly cancel the change in the particle momentum. As a result, the momentum of the center of mass of the charges would not change, as required by momentum conservation.

For a more detailed discussion of this idea, see textbooks on electrodynamics, such as D. J. Griffiths, *Introduction to Electrodynamics*, Prentice Hall, Englewood Cliffs, N.J., 1981.

Chapter 23
Electromagnetic Induction

𝕸ysteries:

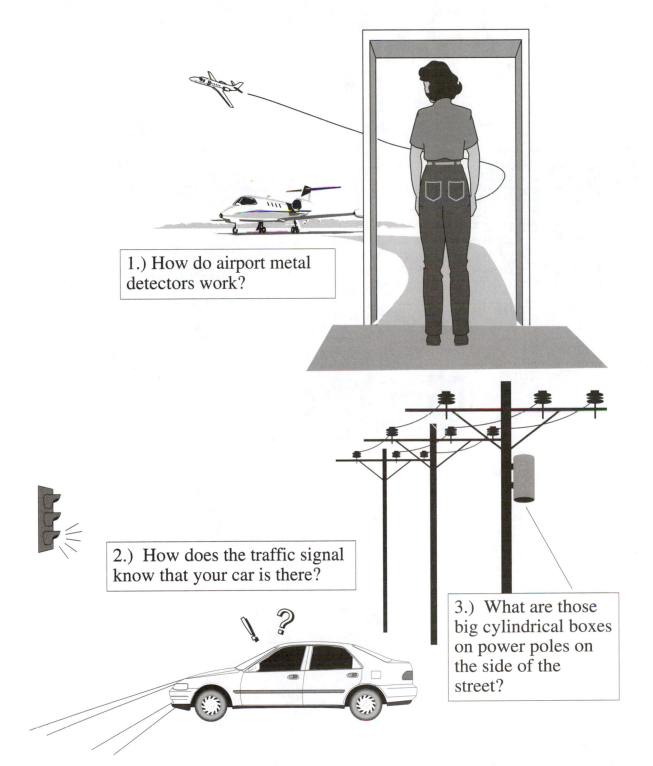

1.) How do airport metal detectors work?

2.) How does the traffic signal know that your car is there?

3.) What are those big cylindrical boxes on power poles on the side of the street?

Mysteries:

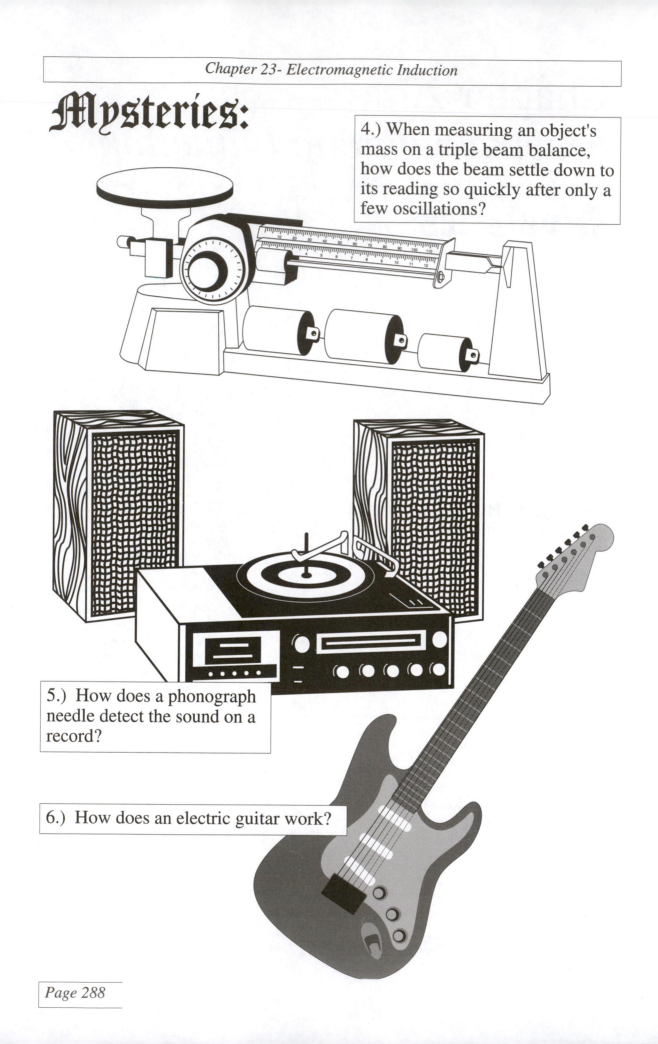

4.) When measuring an object's mass on a triple beam balance, how does the beam settle down to its reading so quickly after only a few oscillations?

5.) How does a phonograph needle detect the sound on a record?

6.) How does an electric guitar work?

Mysteries:

7.) If the rotating part of a motorized device, such as a blender or mixer, is prevented from turning, the motor overheats quickly. Why?

Magic:

Another Use for the Neodymium Magnet

In a Magic trick in Chapter 22 (An Attraction for Dollars), we used a neodymium magnet. Drop the same magnet into a vertical metal tube whose diameter is slightly larger than that of the magnet. Notice the time required for the magnet to fall through the tube.

Myth:

1.) To use an ammeter, you must break the circuit and insert the ammeter.

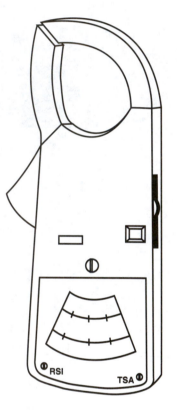

Concepts of Electromagnetic Induction

The discovery by Faraday of electromagnetic induction brought symmetry to the fledgling field of electromagnetism in the nineteenth century and has provided us with an extremely useful technological tool. The ability of changing magnetic fields to induce currents, combined with the creation of magnetic fields by moving charges leads to much deeper understanding of the unification of electricity and magnetism.

> *Electromagnetic Induction was the final step needed to unify electricity and magnetism.*

We can understand electromagnetic induction simply by realizing that a moving charge in a magnetic field experiences a force, as discussed in Chapter 22. Imagine moving a wire upward in the magnetic field below:

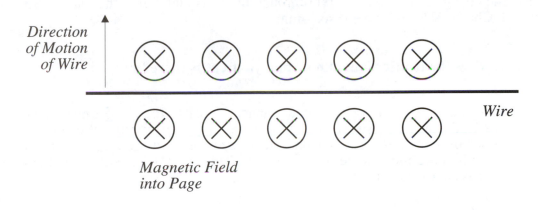

The wire contains free charges, which will feel a magnetic force due to this motion. By using a "right hand rule"*, we can see that a current toward the left is induced in the wire above. Since this is <u>along</u> the wire, the charges can move and a current in the wire is established.

> *Induction is based simply on the magnetic forces on charges.*

We can generalize from this situation to the statement that any magnetic change will induce a current if there are charges in the vicinity. The details of the induced current are summarized in two laws:

Faraday's Law, which predicts the magnitude of an electromotive force driving the current from the geometry of the situation and the rate of change in the magnetic flux**.

Lenz' Law, which predicts that the direction of an induced current is such as to create a magnetic field that tries to oppose the original change. This law is a consequence of conservation of energy.

---

*This is the same right hand rule as that mentioned in Myth #4, Chapter 22 - point the fingers of your right hand in the direction of the velocity vector, with the magnetic field vector coming out of your palm. Your thumb then points in the direction of the magnetic force and, consequently, the induced current.

**$\mathscr{E} = -N\Delta\phi/\Delta t$, where $\mathscr{E}$ is the induced emf, ϕ is the magnetic flux and N is the number of turns in the coil of wire. The negative sign is the mathematical way to represent Lenz' Law.

Discussions; Chapter 23 - Electromagnetic Induction

𝔐𝔶𝔰𝔱𝔢𝔯𝔦𝔢𝔰:

1.) How do airport metal detectors work?

The airport metal detector is a large rectangle that passengers walk through before proceeding to the gate. Around the rectangle are several coils of wire that are connected to an electrical circuit. There is a magnetic field in the detector due to current in these coils of wire. As long as there is no change in the magnetic field, the system is in a stable condition. When metal passes through the detector, it alters the magnetic field, resulting in a self-induced current in the coils of wire. The electronics of the device sense the change in the total current through the coils and signal the alarm. See Mystery #3, Chapter 24 for further discussion.

2.) How does the traffic signal know that your car is there?

The traffic signal detectors operate in a manner similar to the airport metal detector in Mystery #1. Buried in the street are coils of wire carrying a current. When the metal of a car passes over the coils, the magnetic field is altered and the electronics associated with the coils sense the change in the current. This signals the traffic light controller that a car has approached the light.

3.) What are those big cylindrical boxes on power poles on the side of the street?

These are <u>transformers</u>, which use electromagnetic induction to change the voltage, either making it larger (a <u>step-up</u> transformer) or smaller (a <u>step-down</u> transformer). The ability of an efficient device like the transformer to change voltages is the reason that our electrical system is based on AC (<u>A</u>lternating <u>C</u>urrent - see Chapter 24) rather than DC (<u>D</u>irect <u>C</u>urrent). The cylindrical boxes on poles in some neighborhoods are likely to be transformers that lower the voltage from the high value used for long-distance and intra-city transmission (see Mystery #3, Chapter 20) to the 220 volts used in households.

4.) When measuring an object's mass on a triple beam balance, how does the beam settle down to its reading so quickly after only a few oscillations?

At the end of a triple beam balance, there is an aluminum plate which passes through the plates of a permanent magnet. When the balance is in use and the plate oscillates in the magnetic field, currents (*eddy* currents) are induced in the plate. According to Lenz' Law, these currents establish a magnetic field which tends to oppose the original change. Thus, the effect is <u>magnetic friction</u> or <u>magnetic braking</u>, and the oscillations of the balance die out rapidly.

5.) How does a phonograph needle detect the sound on a record?

There are two major types of phonograph needles. A ceramic needle has the stylus imbedded in a crystal. The oscillations of the stylus are transformed into voltages across the crystal by the piezoelectric effect. This is not the type of needle we want to look at in this Mystery. A magnetic needle has the stylus attached to a small coil of wire, between the poles of a magnet. When the needle oscillates in response to the music recorded on the record, an electromotive force is induced in the coil of wire, according to Faraday's Law. This varying voltage signal is passed on to the amplifier where it is amplified, and eventually reaches the loudspeakers.

6.) How does an electric guitar work?

Underneath each string of an electric guitar is a small permanent magnet, around which is wrapped a coil of wire. When the metal guitar string vibrates, the magnetic field of the magnet is periodically altered and an electromotive force is induced in the coil of wire. This voltage is passed on to the amplifier where it is amplified, and eventually reaches the loudspeakers.

7.) If the rotating part of a motorized device, such as a blender or mixer, is prevented from turning, the motor overheats quickly. Why?

When a motor is first turned on, a large current surges through its coils. In this condition, it needs to draw a relatively large amount of electrical power to provide the work to accelerate the armature of the motor. This is why the lights of an air-conditioned house dim when the air conditioner turns on. As the motor is starting, it draws a very large amount of current. This reduces the terminal voltage for the supply to the house (See Mystery #4, Chapter 20 for a similar effect in your car), causing the lights to dim.

As the motor armature starts to rotate, we have a moving coil in a magnetic field. Thus, an electromotive force is induced in the coil. According to Lenz' Law, this emf is in a direction which tries to oppose the original change. Thus, it is called a *back emf*. As the motor increases its speed, the back emf rises and less current is drawn. When the motor has reached its final speed, the net voltage across the motor terminals is $V - \mathscr{E}$, where V is the applied voltage from the battery or other power source and $\mathscr{E}$ is the back emf. If the motor is turning freely, these two voltages are very similar in size, so that the net voltage is relatively small. When the motor is operating in this mode, it only needs to draw enough electrical power to overcome dissipative effects, such as friction in the bearings and "Joule heating" in the wires.

When the motor is doing work (turning a fan against viscous forces from the air, mixing cement, running a compressor, etc.), the difference between the applied voltage and the back emf is larger, to account for this extra energy transfer. This difference results in a larger current through the motor than in the freely rotating condition.

Now, what about the situation in which the armature of the motor is prevented from turning, as in the blender or mixer in this Mystery? If the armature of the motor is not rotating, <u>there is no back emf</u>. Thus, a very large current is <u>continuously</u> drawn by the device. Although the coils of wire in the motor of the device can easily withstand a momentary surge in current at the beginning of normal operation, they are not designed to handle a continuous large flow of current. The wires will become very hot by "Joule heating".

𝔐agic:

Another Use for the Neodymium Magnet

As the magnet falls through the copper tube, it will induce currents in the walls of the tube. According to Lenz' Law, these currents will establish magnetic fields which will tend to oppose the original change, which is due to the falling of the magnet. A neodymium magnet is quite strong for its weight. Thus, it can induce strong opposition fields which are very effective at countering the relatively small gravitational force on the magnet. As a result, the magnet takes a surprisingly long time to pass through the tube.

𝔐yth:

1.) To use an ammeter, you must break the circuit and insert the ammeter.

While this is true for normal ammeters and always true for DC circuits, we can measure the current in an AC circuit without breaking the circuit. We use an <u>induction ammeter</u>. In this device, a set of metal jaws is designed to surround a current-carrying wire. Around part of the jaws is wrapped a coil of wire, which is connected to a calibrated ammeter. When the jaws surround a current-carrying wire (AC, remember), part of the wire's magnetic field exists in the metal jaws, and, therefore, passes through the coil of wire of the ammeter. This induces a current in the coil, which is measured by the ammeter. With proper calibration, the ammeter can be adjusted so as to read the current in the wire that is passing through the jaws.

Chapter 24
AC Circuits

𝔐ysteries:

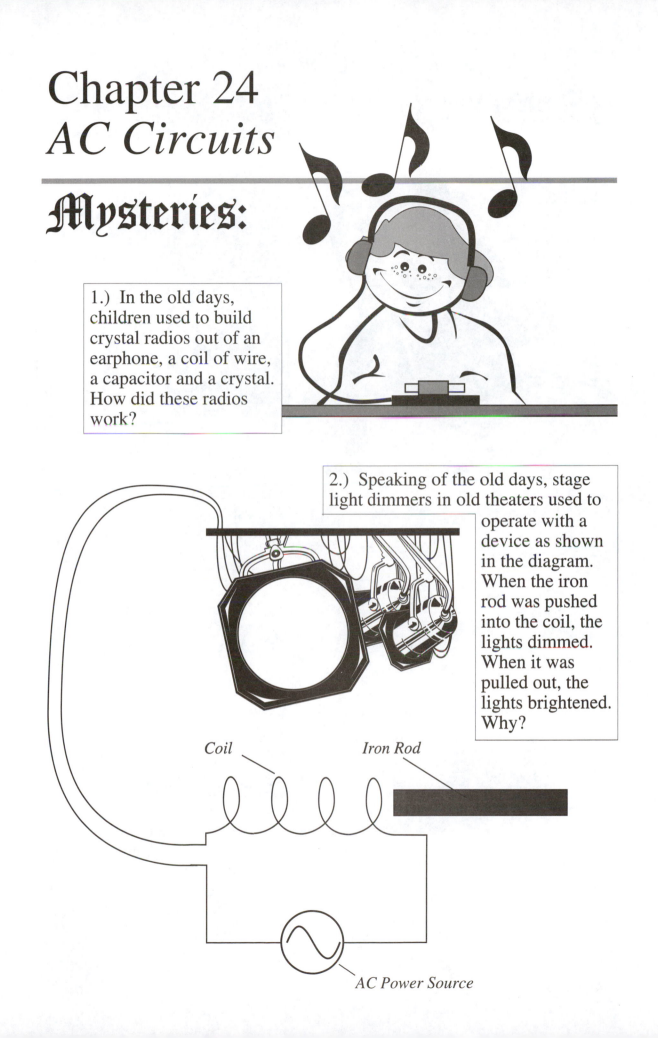

1.) In the old days, children used to build crystal radios out of an earphone, a coil of wire, a capacitor and a crystal. How did these radios work?

2.) Speaking of the old days, stage light dimmers in old theaters used to operate with a device as shown in the diagram. When the iron rod was pushed into the coil, the lights dimmed. When it was pulled out, the lights brightened. Why?

Coil

Iron Rod

AC Power Source

Mysteries:

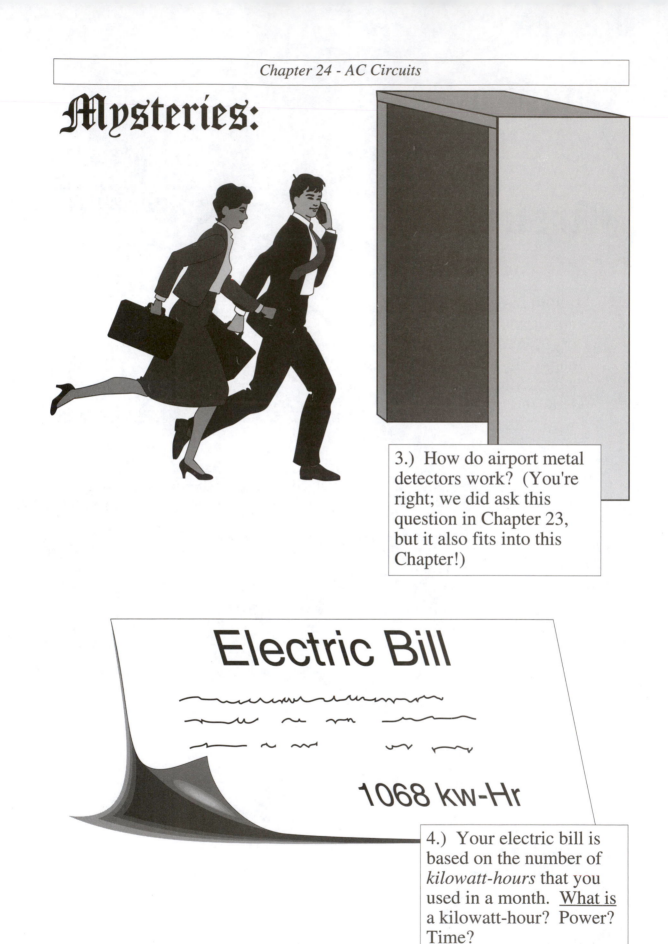

3.) How do airport metal detectors work? (You're right; we did ask this question in Chapter 23, but it also fits into this Chapter!)

Electric Bill

1068 kw-Hr

4.) Your electric bill is based on the number of *kilowatt-hours* that you used in a month. <u>What is</u> a kilowatt-hour? Power? Time?

Magic:

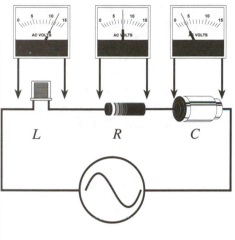

Conservation of Energy?

Set up a series circuit of a resistor, an inductor and a capacitor. Apply power to it with an alternating current source and measure the voltages across each of the three circuit elements as well as that across the output terminals of the power source. The three element voltages do not add up to the source voltage. Since voltage is energy per charge, is this a violation of conservation of energy?

Circle of Light

Use an oscillator or function generator to apply power to a light emitting diode by means of a cable about 1 to 1.5 m long. Set the frequency of the oscillator to about 100 Hz. The light from the diode will appear to be steady and continuous. Now, twirl the LED on the cable in a circle. You will be able to see the individual on and off cycles of the diode.

Fluorescent Stroboscope

Find a wheel that has spokes or some form of radial markings. Spin the wheel on its axle while viewing it in the light of a fluorescent light bulb. You will see the radial markings or spokes move forward and backward at varying speeds. Try this again under the light of an incandescent bulb and no such effect is seen.

Magic:

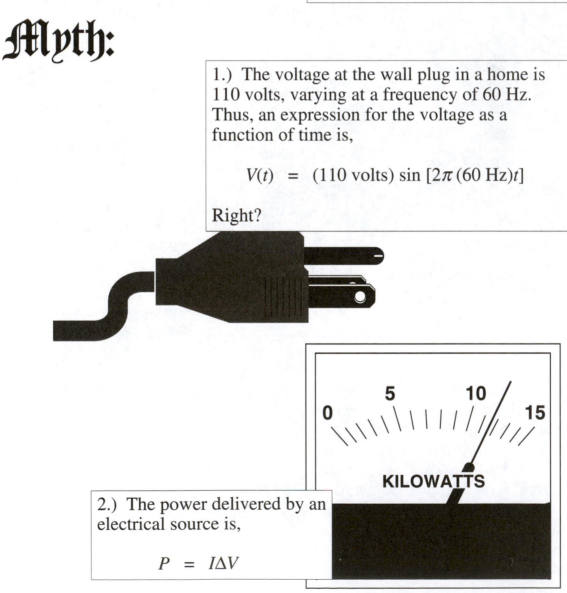

L

R

dB - 40 . 20 . . 5 . 0 . 10 +

Hi-Fi | TIMER 🔘🔘 PROG START STOP CH CATV HRC IRC

STEREO | PLAY 🔲

SAP | REC SP LP EP

1:45:42 SIMUL ⟳

AM ⊙⊷

PM VCR

Constant Display?

While looking at a time display on a video recorder or microwave oven, shake your head from side to side. What does the display look like?

Myth:

1.) The voltage at the wall plug in a home is 110 volts, varying at a frequency of 60 Hz. Thus, an expression for the voltage as a function of time is,

$$V(t) \; = \; (110 \text{ volts}) \sin [2\pi \, (60 \text{ Hz})t]$$

Right?

2.) The power delivered by an electrical source is,

$$P \; = \; I \Delta V$$

5   **10**

0   **15**

KILOWATTS

Concepts of AC Circuits

In Chapter 21, we investigated electrical circuits, many of which were *DC circuits*, in which the power source was a battery or a power supply. These types of sources provide a current flowing in one direction only. In an *AC circuit*, the current oscillates back and forth at some frequency. The typical source of AC voltage is an electrical generator, which uses the principle of electromagnetic induction (Chapter 23). In a generator, a coil of wire is rotated in a magnetic field. As a result, an oscillating current is induced, oscillating at the frequency of rotation of the coil.

> *The typical source of an AC voltage is a <u>generator</u>.*

Typically, three types of circuit elements are discussed in terms of AC circuits - resistors, capacitors and inductors. The voltage across a resistance, *R*, in an AC circuit is simply given by what is called "Ohm's Law" by many people (see Myth #2, Chapter 20):

$$\Delta V_R \ = \ IR$$

We find that the voltage and current are <u>in phase</u> in the resistor.

> *Resistor:*
> *Voltage is in phase with current*

For a capacitor, the voltage depends on the <u>charge</u>, *q*, on the capacitor:

$$\Delta V_C \ = \ \frac{q}{C}$$

> *Capacitor:*
> *Voltage lags the current*

where *C* is the capacitance. The charge (and, therefore, the voltage difference) maximizes only when the current has <u>stopped</u> flowing in one direction (and is just ready to begin flowing in the opposite direction). Thus, the voltage and current are <u>out of phase</u>. The voltage lags the current by one fourth of a cycle.

For an inductor, the voltage depends on the rate of change of the current, *I*, through the inductor:

$$\Delta V_L \ = \ L\frac{\Delta I}{\Delta t}$$

> *Inductor:*
> *Current lags the voltage*

where *L* is the inductance. When the current reaches a maximum (or a minimum), it is momentarily not changing, so that the voltage is zero. Thus, the voltage and current are out of phase in an inductor, also. In this case, the current lags the voltage by one fourth of a cycle.

For all three elements in a series connection, the circuit exhibits electrical *resonance* (see Chapter 16 for mechanical resonance). The circuit will draw the most power from the source when the source frequency is

> *AC circuits exhibit <u>resonance</u>, showing the same type of behavior as driven, damped mechanical systems.*

$$\omega \ = \ 2\pi f \ = \ \frac{1}{\sqrt{LC}}$$

which is the *natural frequency* of the circuit.

Discussions; Chapter 24 - AC Circuits

𝕸𝖞𝖘𝖙𝖊𝖗𝖎𝖊𝖘:

1.) In the old days, children used to build crystal radios out of an earphone, a coil of wire, a capacitor and a crystal. How did these radios work?

In Mystery #1 of Chapter 17, we discussed the operation of radio reception. There are two major steps: 1.) detection of the signal and selection of a station; and 2.) demodulation of the radio signal to extract out the audio information. In the normal radio that we listen to, a third step is the amplification of the sound signal for use by loudspeakers. In the crystal radio, this step is not present. There is no power source in a crystal radio, so there can be no amplification. The sound from a crystal radio can only be heard with earphones.

The circuit diagram of a crystal radio is shown below:

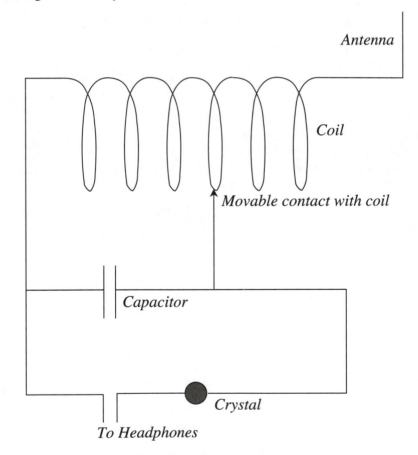

In the upper half of the circuit diagram, we see an *RLC* circuit. This is composed of the inductance of the coil, the capacitance of the capacitor and the resistance of the wires making up the coil and the rest of the circuit. This circuit is driven by the AC voltage of the radio signal from the antenna. By resonance, it will respond with a large power absorption for a given station frequency and reject others. The station frequency is selected by adjusting the movable contact on the coil. This changes the inductance in the circuit and, therefore, the resonance frequency, as discussed in the Concepts section. Thus, the upper half of the circuit acts as the detector and station selector.

The bottom half acts as the demodulator. If the output of the RLC circuit were simply supplied to the earphones, there would be no response. Since the oscillations of the signal are symmetric around zero, there would be no net power delivered. We must first remove the symmetry. This is done by the crystal, which is a <u>diode</u>. It only allows current to pass through in one direction. Thus, only half of the signal passes through to the earphones. Now, the earphone diaphragm is too massive to be able to respond to the very high frequency radio waves. But it can respond to the lower frequency modulation of the radio waves, which is the actual sound signal. Thus, the demodulation is performed by the earphones!

2.) Speaking of the old days, stage light dimmers in old theaters used to operate with a device as shown in the diagram. When the iron rod was pushed into the coil, the lights dimmed. When it was pulled out, the lights brightened. Why?

The lights on the stage were part of an AC circuit, being in series with the coil in the diagram. When the iron rod was pushed into the coil, the effect of the iron was to increase the inductance of the coil. As a result, the voltage drop across the coil increased. This happened at the expense of voltage somewhere else in the circuit, which was that across the lights. Thus, the lights dimmed. When the iron core was pulled out, the voltage across the inductor decreased, the voltage across the lights increased and the lights brightened.

We can also explain this effect from an energy approach. When the iron rod is pushed into the coil, the increased inductance of the coil results in increased inductive reactance to the applied AC voltage. As a result, the circuit of the coil and the lights offers more opposition to current flow, and less current, and, therefore, less power is drawn from the source. This results in dimmer lights.

3.) How do airport metal detectors work? (You're right; we did ask this question in Chapter 23, but it also fits into this Chapter!)

In Chapter 23, we discussed the fact that the metal caused a change in the inductance of the coil in the metal detector gate. In this chapter, we can add the fact that the coil is part of an AC circuit that is designed to be in resonance when no metal is present in the gate. When there is metal, and the coil inductance changes, the circuit goes off-resonance, resulting in a large change in the current. This change in current is detected easily and triggers the alarm.

4.) Your electric bill is based on the number of *kilowatt-hours* that you used in a month. <u>What is</u> a kilowatt-hour? Power? Time?

This unit sounds similar to *light-year,* which is sometimes heard being used as a unit of time (see Myth #4, Chapter 2). The kilowatt-hour is not a unit of time, however. It is the product of power and time, which is <u>energy transfer</u>. Thus, it measures how much energy was transferred from the electrical company to your home by electrical transmission. A kilowatt-hour is equivalent to 3,600,000 joules.

Magic:

Conservation of Energy?

The three voltage readings on the voltmeter will generally add up to more than the source voltage. This is because the voltmeter is reading a time-averaged voltage and is not sensitive to the <u>phase differences</u> among the three circuit elements. For example, the voltages across the inductor and capacitor are always one half cycle out of phase, so that they tend to subtract from each other at all times. Thus, the measured voltage across the series combination of the inductor and capacitor will be the <u>difference</u> between the voltages across the two elements. A particularly striking result for these two members of the circuit occurs when the circuit is in resonance, for which the potential differences across the inductor and the capacitor have the same magnitude. In this case, the measured voltage across this combination would be zero, even though the voltages across each of the elements may be large!

If we were able to measure the voltages for each of the three elements in this circuit at a given instant of time, rather than time averages, then the measurements that we make would add up (vectorially) to the correct value.

Circle of Light

The eye is not able to resolve the individual pulses of light from the AC-powered LED in <u>time</u>. By twirling the LED in a circle, however, the pulsations in time are converted into variations over <u>distance</u>, which the eye can resolve. For more demonstrations involving light emitting diodes, see J. Jewett, "Get the LED Out - Physics Demonstrations Using Light Emitting Diodes", *The Physics Teacher*, **29**, 530 (1991) and G. C. Lisensky, R. Penn, M. J. Geselbracht and A. B. Ellis, "Periodic Properties in a Family of Common Semiconductors; Experiments with Light Emitting Diodes", *Journal of Chemical Education*, **69**, 151 (1992).

For a related demonstration of the variations in light output for an AC-powered street light, see C. F. Bohren, What *Light Through Yonder Window Breaks,* Wiley, New York, 1991. On page 56, Bohren shows photographs of an incandescent light, for which the variations are not visible and a mercury vapor lamp, for which they are. In our demonstration with the LED, we rotated the light source. In Bohren's photographs, he rotated the camera.

Fluorescent Stroboscope

This effect is strongly related to the wagon wheel effect of Myth #1 in Chapter 10. In that case, the stroboscopic source was the motion picture, in which 24 pictures were taken each second. In the case of the wheel spinning in the fluorescent lights, the stroboscopic light source is the fluorescent lights themselves, flashing at 120 times per second. Your eye cannot resolve these flashes in <u>time</u>, as in Circle of Light in this Chapter. But when the flashing light is illuminating a moving object that has a similar frequency, you see the stroboscopic effect as a phenomenon over <u>space</u>.

The effect is not visible under incandescent lighting. With this type of light bulb, the light is <u>thermal</u> in origin, rather than arising from discrete atomic transitions in a fluorescent coating. As the AC current oscillates, the period of an oscillation is much smaller than the time it takes for the light bulb filament to cool and cease the emission of light. The temperature of the filament cannot "keep up" with the rapidly varying voltage. Thus, the light from an incandescent bulb is essentially continuous.

Constant Display?

This is similar to Circle of Light. The display is powered by an alternating voltage. In Circle of Light, we moved the light source so that we could resolve the variations in light intensity. It's hard to shake a microwave oven back and forth, so in this demonstration, we kept the source stationary and moved our eyes by shaking our head. The result is similar - we will see a series of flashing displays repeated over a region of space.

A related demonstration is to wiggle your eyes back and forth while watching television in a darkened room. In this case, the images that you see will be <u>skewed to the side</u>, that is, the normally rectangular shape of the stationary image appears as a parallelogram shape. This is due to the fact that the image on a television screen takes time to be displayed - it is formed by a line-by-line sweep of the electron beam. Thus, the entire picture is not available at a given instant of time. As your eye sweeps to the side, each new line on the screen is displayed when your eye is looking slightly farther to the side of the screen. As a result, the entire image appears to be tilted.

𝕸𝖞𝖙𝖍:

1.) **The voltage at the wall plug in a home is 110 volts, varying at a frequency of 60 Hz. Thus, an expression for the voltage as a function of time is,**

$$V(t) \ = \ (110 \text{ volts}) \sin [2\pi (60 \text{ Hz})t]$$

Right?

The problem with this statement is that the 110 volts is an *rms* (root-mean-square) *average* voltage, not the peak voltage. The rms average of a sinusoidally varying function, such as the voltage at the wall socket, is related to the peak value of the function as follows:

$$V_{rms} \ = \ \frac{V_{peak}}{\sqrt{2}}$$

The correct voltage expression would be as follows, with the peak voltage calculated from the rms voltage of 110 volts:

$$V(t) = V_{peak} \sin [2\pi (60 \text{ Hz})t]$$

$$= \sqrt{2} \ (110 \text{ volts}) \sin \ [2\pi (60 \text{ Hz})t] = (156 \text{ volts}) \sin \ [2\pi (60 \text{ Hz})t]$$

2.) The power delivered by an electrical source is,

$$P = I\Delta V$$

While it is still true that power delivered to an AC circuit is the product of the voltage and the current, we need to remember that the voltage and current are varying sinusoidally and will not generally be in phase. It is useful to express the power in terms of readings from a meter, which will be *rms* (root-mean-square) *averages*. If we do this, we find that the power is given by

$$P = I_{rms} \Delta V_{rms} \cos \phi$$

where ϕ is the phase angle between the current and the voltage (A full cycle phase difference is associated with a phase angle of 360°. Smaller phase differences are represented by proportionally smaller angles.). Thus, the usual expression for the power is modified by the presence of the *power factor*, $\cos \phi$. The phase angle is given by,

$$\tan \phi = \frac{\omega L - \dfrac{1}{\omega C}}{R}$$

where R, L and C are the resistance, inductance and capacitance, respectively, in the circuit and $\omega \ (= 2\pi f)$ is the angular frequency of the source.

Chapter 25
Electromagnetic Waves

𝔐ysteries:

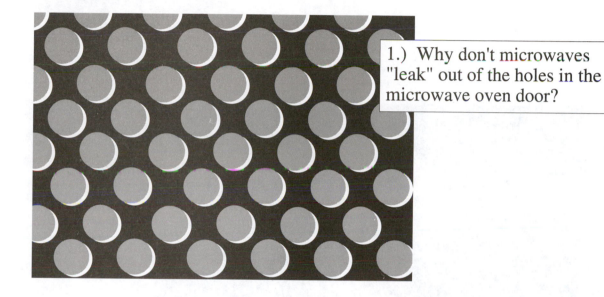

1.) Why don't microwaves "leak" out of the holes in the microwave oven door?

2.) Why is there such a difference in the range of distances over which you can hear AM and FM radio stations?

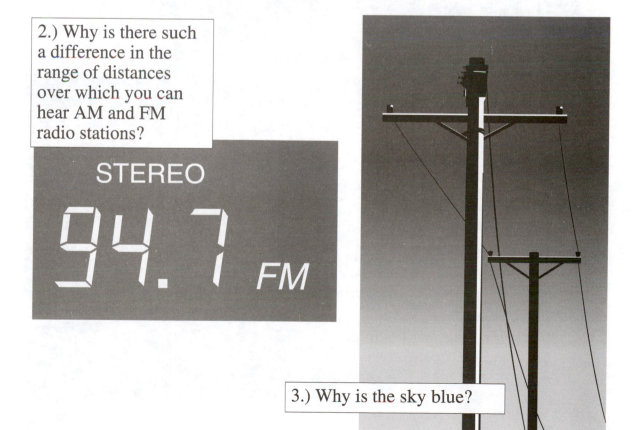

3.) Why is the sky blue?

Mysteries:

4.) Why is the ocean blue?

5.) Why is it easier to see a dim star if you move your eyes slightly to the side, rather than looking directly at it?

Look Here

6.) Where did the phrase "once in a blue moon" come from?

Mysteries:

7.) Many drinks, such as root beer or apple juice, when poured in a glass, result in a "head" of foam. Why is the foam <u>white</u>, or, at the least, <u>light-colored</u>, when the liquid is dark?

Wet Sand

Dry Sand

8.) If you are at the beach, you can tell the wet sand from the dry sand by the color. Why is wet sand darker than dry sand?

9.) If you look inside a slide projector, you may find a flat glass slab. It is clear that this is not a lens. What optical purpose does it serve?

Mysteries:

1.) Red
2.) Blue
3.) Green
4.) ?????

10.) Many textbooks today are published with illustrations using "four-color printing". Why are *four* colors necessary? After all, televisions do just fine with three!

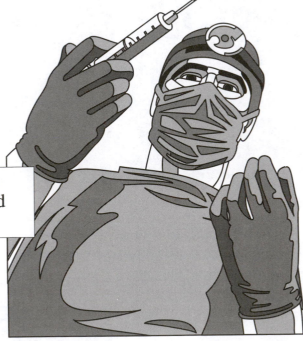

11.) Why are operating room scrub suits most often colored blue or green?

Mysteries:

12.) If you look at a distant scene through "non-reflective" glass, you cannot see through it. Yet when this glass is incorporated in a picture frame, it is perfectly clear. What's going on here?

13.) How do microwaves cook food in a microwave oven?

𝕸𝖆𝖌𝖎𝖈:

Adding Colors

On a white wall or screen, direct the light of three floodlamps of colors red, blue and green. The area of the screen where all three colors overlap will be close to white. The actual color will depend on the purity of the floodlamp colors. This demonstration becomes much more effective if the floodlamps are equipped with light dimmers so that their intensities can be individually adjusted to give the best white.

In addition, place a meter stick in the light from the floodlamps. You will see three shadows of the meter stick, each of a different color!

Subtracting Colors

Prepare three cups of water colored with food coloring - red, blue and green. Now, mix the contents of the cups together in a larger glass container. The same three colors have been combined as in the previous demonstration, but now the result is black, not white!

𝕸𝖆𝖌𝖎𝖈:

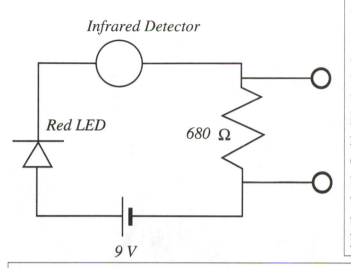

Infrared Detector

Red LED

680 Ω

9 V

A Remote Possibility

Set up the circuit shown in the diagram, using an infrared detector available from an electronics supply store. Attach a loudspeaker across the 680 Ω resistor. Now, point a remote control from an electronic device (TV, VCR, stereo, etc.) at the detector and push the buttons. What do you see and hear?

Communicating with Light

Use the same circuit as in A Remote Possibility. Attach an infrared or red light emitting diode to an oscillator or function generator. Direct the LED toward the IR detector and listen for the sound on the speaker.

The Vanishing Light Beam

Make up a <u>very</u> diluted mixture of water and blue toilet bowl cleaner, such as Vanish (Drackett Products Co., Cincinnati, Ohio). The approximate concentration can be made in the kitchen by measuring one cup of water and then adding a <u>very</u> small amount of Vanish. If a teaspoon is held <u>vertically</u> and just dipped into the Vanish and then into the water, 2 or 3 such transfers will provide about the right mixture. You should have a very light blue mixture after this process. Now, shine two red light sources through the mixture onto a sheet of white paper - a helium-neon laser and a red LED with a clear package (or a solid state laser pointer). Do both red light beams make it through the liquid?

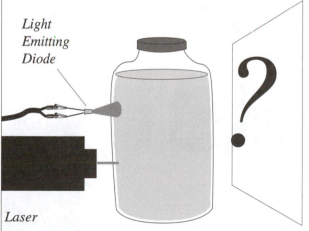

Light Emitting Diode

Laser

Dilute Toilet Bowl Cleaner

Magic:

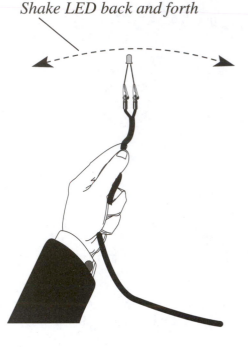

Shake LED back and forth

The Floating Light

Activate a red LED with a power supply or battery. In a dimly lit room, with the LED glowing softly, move the LED back and forth over a few inches. You may notice that the light from the LED seems to "lag behind" the motion of your hand. It may be easier to see if you view the LED peripherally rather than straight on.

Myth:

1.) The "Greenhouse" Effect.

CO_2

H_2O

2.) Smudgepots are used in an orchard during cold weather to warm up the air in the orchard.

Myth:

RED RED RED RED RED RED RED RED RED RED RED RED RED RED RED RED RED

BLUE BLUE BLUE BLUE BLUE BLUE BLUE BLUE BLUE BLUE BLUE BLUE BLUE BLUE

3.) The color of light is determined by its wavelength.

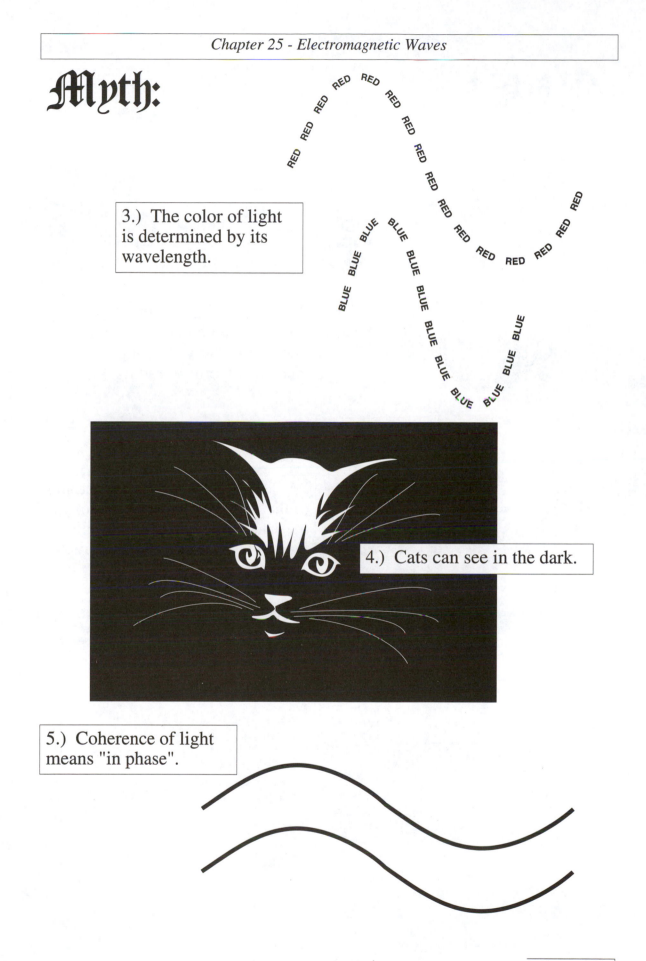

4.) Cats can see in the dark.

5.) Coherence of light means "in phase".

Myth:

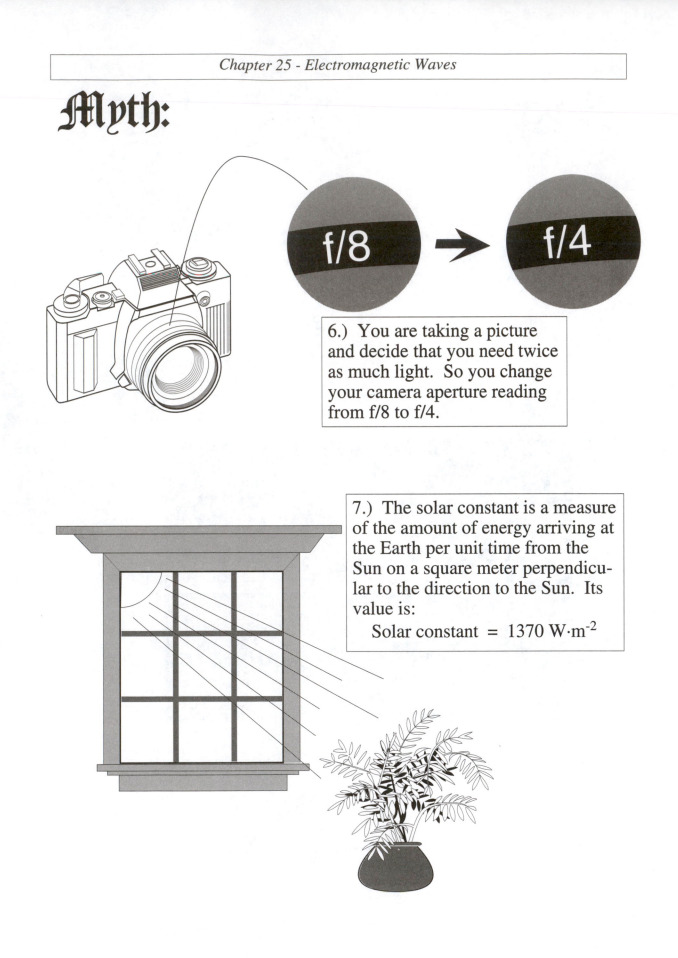

6.) You are taking a picture and decide that you need twice as much light. So you change your camera aperture reading from f/8 to f/4.

7.) The solar constant is a measure of the amount of energy arriving at the Earth per unit time from the Sun on a square meter perpendicular to the direction to the Sun. Its value is:

Solar constant = 1370 W·m^{-2}

Concepts of Electromagnetic Waves

A fundamental principle of electromagnetism is that a changing magnetic field induces an electric field and that a changing electric field induces a magnetic field. Thus, the changing fields reinforce one another which leads to the possibility of electromagnetic waves. These waves are not mechanical in nature, like sound waves, but, rather, consist of oscillating electric and magnetic fields. As non-mechanical waves, they do not require a medium and can, therefore, travel through empty space. The speed of propagation is determined by the properties of space, namely the permittivity of free space (ε_o), an electrical property, and the permeability of free space (μ_o), a magnetic property.

> *Electromagnetic waves are not mechanical; therefore, they do not require a medium.*

> *Human eyes are sensitive to a very small part of the electromagnetic spectrum.*

Electromagnetic waves have a frequency range varying from 0 to about 10^{23} Hz. In principle, we can have frequencies higher than this, but they have not been detected. Human eyes are only sensitive to a very narrow range of these frequencies, from about 430 teraHz (tera = 10^{12}), for red light, to about 750 teraHz for violet. This range of frequencies is called *visible light*.

Other frequency ranges have been given specific names, based either on historical factors or on their usage in our society. None of the boundaries between these ranges is definite. These ranges are roughly defined as follows:

Approximate Frequency	Approximate Wavelength	Name	Typical Origin	Comments
10^{23} Hz - 10^{19} Hz	10^{-15} m - 10^{-11} m	Gamma Rays	Nuclear Transitions	*Also can be called "Hard X-rays"*
10^{19} Hz - 10^{16} Hz	10^{-11} m - 10^{-8} m	X-rays	Atomic Transitions Braking of Electrons	*Penetrating Radiation*
10^{16} Hz - 7.5×10^{14} Hz	10^{-8} m - 4×10^{-7} m	Ultraviolet	Atomic Transitions Thermal Vibrations	*Responsible for sunburns*
7.5×10^{14} Hz - 4.3×10^{14} Hz	4×10^{-7} m - 7×10^{-7} m	Visible	Atomic Transitions Thermal Vibrations	*We see these!*
4.3×10^{14} Hz - 10^{11} Hz	7×10^{-7} m - 10^{-3} m	Infrared	Atomic Transitions Thermal Vibrations	*Incorrectly called "Heat radiation"*
10^{11} Hz - 10^{9} Hz	10^{-3} m - 10^{-1} m	Microwaves	Charges accelerated over macroscopic distances	*Ovens, TV Transmission*
10^{9} Hz - 10^{4} Hz	10^{-1} m - 10^{4} m	Radio	Charges accelerated over macroscopic distances	*Also includes TV*

Discussions; Chapter 25 - Electromagnetic Waves

𝕸𝖞𝖘𝖙𝖊𝖗𝖎𝖊𝖘:

1.) Why don't microwaves "leak" out of the holes in the microwave oven door?

It is impossible to use waves to "see" anything smaller than the wavelength of the radiation that you are using. Thus, we will never directly <u>see</u> atoms using visible light, since the atoms are much smaller than light wavelengths. If you look at radio telescope antennas, you may wonder why they are <u>meshlike</u> rather than solid. The holes in the mesh are smaller than the wavelength of the radio waves, so the waves don't "see" the holes. The radio dish "looks" solid to the radio waves. In the same way, the holes in a microwave oven door are smaller than the wavelength of the microwaves, which is about 12 cm. Thus, the waves don't "see" the holes and the radiation does not leak out. On the other hand, we would like to see into the oven to monitor the cooking progress - this is what the holes are for! Although microwaves cannot pass through the holes, visible light, of <u>much</u> smaller wavelength can easily pass through, allowing us to see inside. For more information on microwave cooking, see Mystery #13 and A. Steyn-Ross and A. Riddell, "Standing Waves in a Microwave Oven", *The Physics Teacher*, **28**, 474 (1990).

2.) Why is there such a difference in the range of distances over which you can hear AM and FM radio stations?

AM radio stations can be heard at surprising long distances. This is due primarily to reflection of the radio waves off the *ionosphere*, a layer of ions in the atmosphere starting about 35 miles from the Earth's surface (during the daytime). At night, the lower region of the ionosphere disappears, due to the absence of the Sun's radiation, so that the reflecting surface is effectively higher from the ground. This results in even further transmissions of radio signals. Some stations are even required to cut power during the night to avoid interference with other stations elsewhere in the world.

FM signals, on the other hand, are of much higher frequency than AM. The ionosphere is essentially transparent to these signals, so that the ionosphere cannot be used to help transmit FM over long distances. Thus, FM stations (as well as television stations) depend almost entirely on direct, line of sight communication to local antennas.

3.) Why is the sky blue?

The blueness of the sky is due to a process called <u>Rayleigh scattering</u>. This is the type of scattering that occurs when light is incident on a collection of particles that are much smaller than the wavelength of the light. As Rayleigh found in the latter half of the nineteenth century, the amount of scattering is proportional to the fourth power of the light frequency. Thus, light at the violet end of the spectrum is much more effectively scattered than at the red end.

This scattering occurs as light from the Sun is incident on molecules of air. The molecules are much smaller than the wavelength of the light, so much radiation from the violet end of the spectrum is scattered. Now, our eyes are not very sensitive to violet, so we tend to see the blue that is also scattered rather than violet. If we look at the Sun at sunrise or sunset, the light has traveled through a large amount of atmosphere. Thus, much of the violet and blue has been scattered away, leaving relatively large amounts of red light, leading to red sunrises and sunsets.

For more information on light scattering, see Chapter 5 in R. Greenler, *Rainbows, Halos and Glories*, Cambridge University Press, Cambridge, 1980.

4.) Why is the ocean blue?

We can imagine three possible answers to this question: 1.) The blue of the sky reflects off the water; 2.) Water absorbs more red light than blue light; and 3.) Water molecules scatter more blue light than red light. In reality, all three of these effects combine in varying amounts to give the blue color of the ocean. For a clear discussion of these effects, see Chapter 20 in C. F. Bohren, *Clouds in A Glass of Beer*, Wiley, New York, 1987.

5.) Why is it easier to see a dim star if you move your eyes slightly to the side, rather than looking directly at it?

If you look directly at an object, the image appears on the *fovea*, which is an area of the retina that has only cones (light-sensitive cells which detect colors and are used mostly for bright light) and no rods (light-sensitive cells used to detect black and white and low levels of illumination). Thus, when the dim image of the star is on the fovea, it is very difficult to see. By looking to the side, the image is placed on an area of the retina that has many rods, and is thus able to be seen more clearly.

6.) Where did the phrase "once in a blue moon" come from?

If white light is scattered by particles smaller than the wavelength of light, blue is scattered much more than red. This is the origin of the blue sky (Mystery #3), since light is scattered by air molecules. If light is scattered by particles larger than the wavelength of light, all colors are scattered about equally. This is the origin of the white color of clouds, since sunlight is scattered by large (relative to light wavelengths) water droplets. When the scatterers are on the same order of size as the wavelength of light, however, we find that, for certain sizes of the scatterers, <u>red light is scattered a bit more preferentially than blue</u>. Thus, light coming from the moon through these particles has the red light scattered and the transmitted light is bluish - hence, a "blue moon".

Now, what kind of particles will participate in this scattering? Generally, these particles come from volcanoes or large fires. For example, a blue moon was observed in London in 1950 and was attributed to particles traveling over the Atlantic Ocean from forest fires in Alberta, Canada. Since the creation of particles of this sort, with just the right size, occurs rarely, "once in a blue moon" is synonymous with "not very often"!

For more information on the blue moon, see W. M. Porch, "Blue Moons and Large Fires", *Applied Optics*, **28**, 1778 (1989).

7.) Many drinks, such as root beer or apple juice, when poured in a glass, result in a "head" of foam. Why is the foam <u>white</u>, or, at the least, <u>light-colored</u>, when the liquid is dark?

The color of the liquid is due to absorption of a particular mixture of frequencies of the light as it passes through the liquid. The foam formed on top of the liquid consists of many bubbles. The bubbles are very thin films of the liquid material. As light passes through the thin films, there are not enough molecules to effect significant absorption, so that the white light incident on the foam is almost completely reflected or transmitted and thus appears white or light-colored.

8.) If you are at the beach, you can tell the wet sand from the dry sand by the color. Why is wet sand darker than dry sand?

The light coming to you from the surface of sand is light that entered the sand, was scattered several times by the sand particles, and then exited toward your eye. While light is generally scattered in all directions by particles, the distribution of scattered light is a function of the size of the particles. For large particles like sand, we have the general rule that the larger the sand particle, the more light tends to be scattered in the forward direction (that is, small-angle scattering) compared to that which is scattered in a direction perpendicular to the propagation of the light. Now, what do we mean by <u>large</u>? We mean large compared to the wavelength of the light. For dry sand, this refers to the wavelength of light in air. For wet sand, this refers to the wavelength of light traveling in the water that surrounds the sand. The wavelength of light in water is less than that in air. Thus, the sand particles appear larger to the light in water than they do to the light in air. As a result, the distribution of scattered light is even more in the forward direction for wet sand particles than dry sand particles. Now, with a more forward distribution of scattering in wet sand, light from the sky must undergo more scattering events on the average to "turn around" and exit the wet sand than the dry sand. This increases the chance that the light will be <u>absorbed</u> before exiting the sand and, thus, the wet sand emits less light and appears darker.

Craig Bohren (C. F. Bohren, *Clouds in a Glass of Beer*, Wiley, New York, 1987, Chapter 15) has done a nice study of wet and dry sand.

9.) If you look inside a slide projector, you may find a flat glass slab. It is clear that this is not a lens. What optical purpose does it serve?

We only need the <u>visible</u> radiation from the projector light bulb to see the slides. But the light bulb also puts out a significant amount of infrared radiation. The glass slab is an infrared filter which reduces the amount of energy transmitted to the slides, thus reducing the chance of delivering so much energy to them that they are damaged.

10.) Many textbooks today are published with illustrations using "four-color printing". Why are *four* colors necessary? After all, televisions do just fine with three!

The colors that are combined in printing are <u>yellow</u>, <u>magenta</u> and <u>cyan</u>. In principle, these three colors should be sufficient to display all of the psychological colors that we would need in the illustration. Let us suppose that we want a black area in the illustration, so that these three colors are printed together and all three overlap. The desired result is that no light should be reflected. The inks for these colors are not spectrally pure, however - each transmits some regions of the spectrum in addition to the desired range. As a result, they do not completely absorb the incident light when the three colors are printed together. The region of the illustration that should be black, then, is a dark brown. To solve this problem, a black ink overlay is added to the illustration. This is the fourth "color" - <u>black</u>.

11.) Why are operating room scrub suits most often colored blue or green?

When the eye looks for a long time at a strong color, the cones in the retina for that color become bleached and fatigued. If the eye then looks at a white surface, there is a strong afterimage of the complementary color. You may be familiar with an activity related to this effect, in which you stare at an American flag or other familiar image, which is presented in the complementary colors to the actual colors. When you then move your eyes to a white surface, the image with the correct colors appears.

Surgeons stare intently for long periods of time at scenes that appear bright red, due to blood, under bright lights. Thus, the red photoreceptors in the surgeon's eyes become fatigued. If he or she were then to look at an assistant dressed in a white scrub suit, there would be a disturbing cyan afterimage. This effect is counteracted by making the scrub suits blue or green (cyan is blue-green!) so that the afterimage is almost unnoticeable.

12.) If you look at a distant scene through "non-reflective" glass, you cannot see through it. Yet when this glass is incorporated in a picture frame, it is perfectly clear. What's going on here?

The non-reflective glass has a surface which is rough, so that light from the environment is diffusely reflected. Thus, as you look at a picture in a picture frame with non-reflecting glass, you will not see any harsh reflections of light sources such as windows, lamps or television screens.

Now, let us look at light which is coming <u>through</u> the glass, which is "diffusely transmitted"! The diagram on the next page shows the path of one possible light ray as it leaves an object which is far from the glass and arrives at the eye of an observer. The original path of a light ray from a point on the object is toward the upper right, indicated by the solid line and extended by the short dashed line after the light passes through the glass. One possible path of the light ray is that which is deflected downward (by "diffuse transmission") by an angle θ, as indicated in the diagram. Now, light will be scattered in all directions, but we imagine θ to be some representative angle, such as the angle at which the scattered intensity is some fraction of that scattered in the original direction. Now, if we extend the line of sight of the observer, based on the light ray coming toward him or her, we see that the point on the object can be described as being diffused into an area (with a circular cross section) of half-angle β. Thus, light leaving a <u>point</u> on the object results in a <u>circle</u> of light when viewed through the glass.

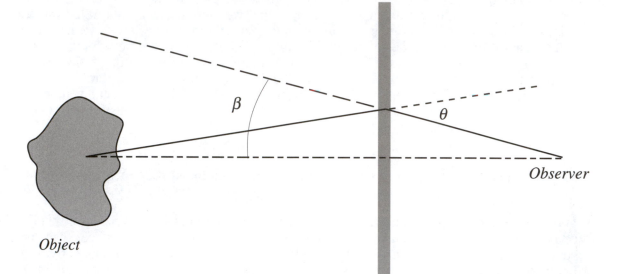

Now, let us see what happens if we move the object closer to the diffusing glass. This is shown in the next diagram:

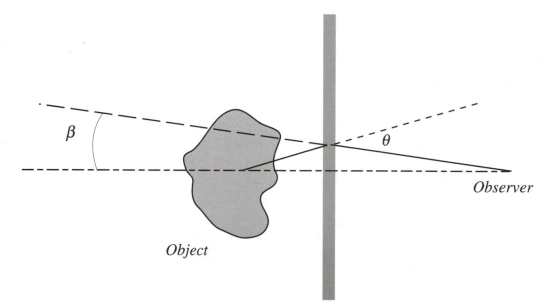

If we maintain our representative angle θ, while we move the object closer, we see that the angle β has <u>decreased</u>. Thus, the light from a point on the object now subtends a smaller angle than before. If we take the limit as the object approaches the screen, we can imagine that the angle β will approach zero. Thus, for objects right next to the glass, the glass is <u>clear</u>. But distant objects appear blurry. This is the origin of the use of frosted glass in bathroom windows and shower doors.

Miller and Benedek (*Intraocular Light Scattering*, Charles C. Thomas, Publisher, Springfield, Illinois, 1973) discuss the misunderstandings that this phenomenon may cause for the novice clinical ophthalmologist. Looking into a patient's eyes with a cloudy cornea, or other cause of a scattering surface of the eye, the ophthalmologist may be able to see the retina quite clearly. This seems to belie the patient's claims that the eye has become opaque. We can understand this from the discussion above - the retina is close to the diffusing surface, so that the image is clear for the ophthalmologist. The patient, however, is looking at objects which are far from the diffusing surface and are therefore not clearly seen, if at all.

A useful activity to study this effect can be performed with Scotch (3M Co., St. Paul, Minnesota) mending tape (the dull type, not the clear cellophane type). This type of tape is a diffuse transmitter. When the tape is placed over printed words or pictures on the paper, they show through clearly. But if the tape is held a few millimeters from the paper, the images cannot be seen. It is instructive to watch the clearing up of the image as the tape is brought closer and closer to the paper.

13.) How do microwaves cook food in a microwave oven?

In Myth #1, Chapter 4, we made the simple statement that the flipping back and forth of water molecules resulted in friction, causing the food to become hotter. Let us give more detail now, armed with our knowledge of electric fields and electromagnetic radiation. The water (H_2O) molecule, which is prevalent in foods, has an electric dipole moment, due to the non-symmetric nature of the molecule, as shown below:

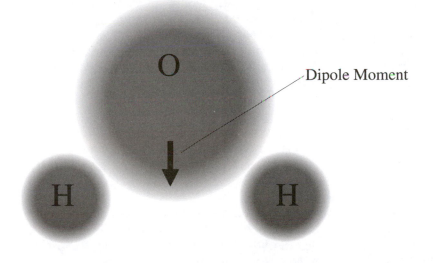

The electrons shared by the hydrogen and oxygen molecules are closer to the oxygen on the average than the hydrogen atoms, so that there is an average negative charge near the oxygen and an average positive charge between the hydrogen atoms, leading to the dipole moment indicated in the diagram.

Due to this separation of charge, it is possible for a positive "end" of one molecule to be attracted to the negative "end" of another molecule and form a bond. Typically, 2 to 5 molecules will bond together to form microscopic crystals. The diagram below shows three such molecules bonded together:

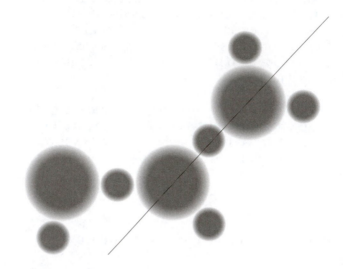

Now imagine we allow microwaves to strike this collection of bonded molecules. Since the electric field vector of the microwaves is oscillating, the dipole moment will be subject to an alternating force. The collection of bonded molecules will attempt to rotate in response to this force. Thus, <u>if the left-hand molecule in the diagram were absent</u>, the bonded pair represented by the other two molecules could rotate fairly easily around the axis represented in the diagram by the light line (The dipole moment of the upper right molecule in the diagram is parallel to this line, but the middle molecule has a component perpendicular to the axis.). This would <u>not</u> cause the water to absorb energy. But with the left-hand molecule present, <u>there is a much larger moment of inertia</u> (Chapter 11) around the rotation axis. Thus, the collection of bonded molecules offers much more resistance to rotation. In this case, it is likely that the bond joining the left-hand molecule to the middle molecule will be <u>broken</u> as the bonded molecules attempt to rotate in response to the electric field of the microwave radiation.

The result of this situation is that the <u>microwaves have delivered energy</u> to the molecules - the energy necessary to break the bond. When the bonds re-form, this energy will be transferred to the entire collection of water molecules and, in turn, to the food, as internal energy. Thus, the temperature of the water, and the food in which it resides, increases.

It is sometimes stated that the microwave frequency is in <u>resonance</u> with the rotation frequency of the water molecule. This is not true. There is indeed a resonance frequency for the water molecule, which is the inverse of the *relaxation time*. This time is that required for a given fraction of the polar molecules to return to their random orientations after the electric field of the microwaves goes to zero. A home microwave oven operates at a frequency of 2.45 GHz, which is <u>significantly below</u> the resonance frequency of water, which is about 9 GHz at $T = 0°C$. Why do we not use the resonance frequency, at which energy would seem to be absorbed most rapidly?

The answer to this question is related to the interdependence of the microwave frequency and the *penetration depth*, which is a measure of how deeply the microwaves penetrate into the volume of the food sample. The penetration depth decreases quite rapidly with increasing frequency (a graph of this dependence is available in D. A. Copson, *Microwave Heating*, AVI, Westport, CT, 1975). The penetration depth at 2.45 GHz is 15 to 20 times larger than at 9 GHz. Thus, if we were to raise the microwave frequency to that of the resonance for the water, we would have very rapid heating, but only in a shallow region near the surface of the food sample, with very little heating in the interior. To prevent burned surfaces and cold centers, we lower the frequency, accepting a slower rate of heating for the benefit of more uniform cooking.

One final note on microwave cooking will bring this discussion to a close. We have discussed dipole rotation as the primary absorption mechanism for microwaves. A second absorption method is *ionic conduction*. If the food contains ions, these ions will move in response to the microwave electric field, resulting in a flow of current. The food sample will provide resistance to this current, resulting in Joule heating (see Mystery #3, Chapter 20) within the food material, resulting in an increased temperature.

The difference between dipole rotation and ionic conduction can be clearly demonstrated by comparing the boiling times in a microwave oven for pure water and water with salt added. The salted water will boil in a significantly shorter time, due to the added absorption due to ionic conduction of the sodium and chlorine ions. Compare this to Myth #6, Chapter 14, in which adding salt to water cooked on a stovetop increases the boiling time!

𝔐𝔞𝔤𝔦𝔠:

Adding Colors

By combining lights of three colors, you are performing *additive color mixing*, so that the result of all three colors is white. With the meter stick, or any other object, in the path of the light, a shadow will be formed from each of the three light sources. Since there must necessarily be some lateral distance between the light sources, the three shadows will be separated. The shadow from the blue light, for example, will have no blue light on the wall behind the meter stick, but will be receiving red and green. The result will be the combination of these two, a color that we call *yellow*. The other shadows will also exhibit combinations - *cyan* (blue and green) and *magenta* (blue and red).

Subtracting Colors

In the previous demonstration, the light was described as having a given color because it "contained" that color, in terms of its frequency. For colored water, the water is of a given color because it absorbs all colors from white light, except for the given color. Thus, this is a subtractive process. The colored water subtracts colors from white light. Thus, red colored water can be imagined to subtract out blue and green from the incident light. Now, if we combine red, blue and green colored water, each color subtracts the other two from the incident light. As a result, all of the colors of the white light are subtracted out by the mixture and there is no light leaving the mixture - it is black.

A Remote Possibility

The infrared detector is a <u>photoresistor</u> - its resistance depends on the amount of (infrared) radiation incident upon it. Thus, when it is irradiated, its resistance changes and, as a result, the current in the circuit changes. Your remote control unit emits encoded infrared signals. When you use your remote control on the detector, the varying current causes the loudspeaker to emit sound. Different remote controls will give you different sounds. You will also see the red LED lighting up in response to the remote control signal. If you replace the loudspeaker with an oscilloscope across the 680 Ω resistor, you will be able to study the waveform of the signal.

Communicating with Light

This demonstration depends on the same concepts as the previous one. In this case, however, <u>you</u> are doing the encoding by modulating the light beam with the signal from the oscillator. This is very similar to radio transmission (Mystery #1, Chapter 17), except that you are modulating light waves instead of radio waves.

The Vanishing Light Beam

The chemical in Vanish that makes it blue (copper phthalocyanine) has a strong and narrow absorption peak at about 630 nm, which is in the red range of the visible spectrum. A helium-neon laser emits light at 633 nm, which is almost totally absorbed by the mixture, so that very little light passes through. The red LED (or solid state laser) emits light which looks very similar to that of the helium-neon laser, but is at about 670 nm. Because the absorption peak of the Vanish is so narrow, this is far enough off the peak that a significant amount of the light from this source passes through.

For a graphical absorption spectrum of Vanish, see K. Brecher, "Do Atoms Really 'Emit' Absorption Lines?", *The Physics Teacher*, **29**, 454 (1991).

The Floating Light

This is an example of a phenomenon called "Fluttering Hearts", which was reported to the British Association for the Advancement of Science in 1844. The name for the demonstration comes from the usual shape of concentric red and blue areas that were in juxtaposition on a sheet of paper. When the red and blue hearts were observed in dim light and vibrated, there was a "fluttering" effect, due to the apparent optical "inertia" of the red areas. In 1936, the artist Marcel Duchamp presented a very early example of "Op Art", called, "Fluttering Hearts", which was - you guessed it - a combination of red and blue hearts.

The origin of this illusion has been debated in the literature. An early theory due to Helmholtz (*Handbuch der Physiologischen Optik*, 1867), claimed that the visual system processed different colors at different speeds, so that one of the colors' image "stayed" on the retina longer. Another theory proposed by von Kries ("Über die Wirkung kurzdauernder Lichtreize auf das Sehorgan", *Z. Psychol. Physiol. Sinnesorg.*, **12**, 81 (1896)) and supported by McDougall ("The Illusion of the 'Fluttering Heart' and the Visual Functions of the Rods in the Retina", *Br. J. Psychol.*, **1**, 428 (1904)) declared that the rods are excited primarily by the blue light and that the red light primarily excites cones. According to this theory, the two types of cells have different speeds of processing information, which leads to the effect.

In the mid-70's, Michael W. von Grünau published a definitive series of three articles on research he had performed on the Fluttering Heart phenomenon (M. W. von Grünau, "The 'Fluttering Heart' and Spatio-Temporal Characteristics of Color Processing - I; Reversibility and the Influence of Luminance", *Vision Research*, **15**, 431 (1975); M. W. von Grünau, "The 'Fluttering Heart' and Spatio-Temporal Characteristics of Color Processing - II; Lateral Interactions Across the Chromatic Border", *Vision Research*, **15**, 437 (1975); M. W. von Grünau, "The 'Fluttering Heart' and Spatio-Temporal Characteristics of Color Processing - III; Interactions Between the Systems of the Rods and the Long-Wavelength Cones", *Vision Research*, **16**, 397 (1976)). In these articles, he discounts both of the theories described above and presents convincing experimental evidence that the effect is due to an inhibitory interaction between the rods and the cones. In agreement with von Kries, the theory claims that the blue light excites primarily the rods and the red light, the cones. Because of interactions between the two systems, however, the excitation of one system, say, the rods, inhibits the other system, the cones. This results in a time delay in the processing of the information by the inhibited system. This delay leads to the lagging effect of the second color. It is not clear what the mechanism of the cross-system interaction is, but its existence seems clear.

The articles by von Grünau also indicate that the effect is seen with a combination of red and grey. This is related to the effect that we see in this Magic demonstration. In the dim light, the bright red LED excites the red cones in the eye and is seen against the dimly lit hand, which excites the rods. As the LED is moved, then, the "Fluttering Heart" phenomenon is seen. For more information, see J. W. Jewett, "LED's and the 'Fluttering Heart' Phenomenon", *The Physics Teacher*, **31**, 180 (1993).

𝔐𝔶𝔱𝔥:

1.) The "Greenhouse" Effect.

The Greenhouse Effect refers to the trapping of radiation by pollutants ("greenhouse gases", such as carbon dioxide) in the atmosphere. These pollutants tend to be transparent to the range of electromagnetic frequencies radiated in the highest amounts from the Sun, such as visible light. This radiation is absorbed by the Earth and then reradiated in the long wavelength infrared range. The pollutant gases are opaque to this radiation, so that the radiation is absorbed by the atmosphere. This causes a larger net absorption of radiation than is the case in the absence of greenhouse gases, thus causing the temperature of the earth to rise.

This sounds very much like the way that a greenhouse works, but there is actually a controversy about the theory of greenhouses. Some feel that the greenhouse glass passes visible radiation but is opaque to the infrared radiation emitted by the warmed ground inside the greenhouse. If this is the case, then the Greenhouse Effect is aptly named.

But others maintain that a greenhouse merely provides shelter from convective energy losses, so that air warmed by the Sun cannot mix with cooler air. Indeed, the literature even shows data in which the air in a greenhouse was <u>cooler</u> than the outside air, since the warmer outside air could not mix with the cool air inside (K. Hanson, "The Radiative Effectiveness of Plastic Films for Greenhouses", *Journal of Applied Meteorology*, **2**, 793 (1963)). You might also think about your car - if you leave it in the Sun with the windows open, does it get as hot inside as it does with the windows closed? If this is the explanation of the greenhouse, then the Greenhouse Effect is misnamed.

As with most controversies such as this, there is probably some truth to both claims. In any case, the Greenhouse Effect is well-established in our language, so, whatever the final outcome, it will most likely have no effect on popular usage of the term. For some calculations to estimate the inside temperature of your car, see R. H. Garstang, "How Hot Does Your Parked Car Become?", *The Physics Teacher*, **29**, 589 (1991). For general information on greenhouse gases, see G. J. Aubrecht, "Trace Gases, CO_2, Climate and the Greenhouse Effect", *The Physics Teacher*, **26**, 145 (1988).

2.) Smudgepots are used in an orchard during cold weather to warm up the air in the orchard.

The number of pots used in an orchard is far too few to perform any significant warming. The operational principle here is radiation absorption. After a day of sunlight, the ground is warm. At night, there is no input of radiation and the ground will cool off by radiating into space (this radiation occurs during the day also, but is dominated by the radiation input from the Sun). The role of the smudgepot is to create a <u>cloud</u> (from combustion products) close to the ground. This cloud will absorb the radiation from the ground and reradiate it back to the ground. This results in a longer time for the ground to cool, enabling it to survive the night without freezing until additional energy input is available from the Sun the next morning.

3.) The color of light is determined by its wavelength.

This statement can be found in encyclopedias everywhere, as well as other reference sources. It is incorrect, however, as can be easily seen by taking a red object into a swimming pool and looking at it underwater. Light traveling in water has a different wavelength than it had in the air. Yet the red object still looks red underwater.

One point which should be made for the alert reader is as follows. Even though the light from the red object underwater in the previous discussion is different from that in air, the wavelength of the light <u>passing through the vitreous humor of the eye and striking the retina</u> is unrelated to whether the light entered the eye from air or water.

In reality, we should modify the statement to say that the color of light (actually, this is incorrect also - light has no color; the color is a psychological sensation in our own brain) is determined by its *frequency*. Frequency is the wave parameter that does not change when light enters a new medium, explaining our red object under the water. What's more, color is sensed by three types of cone cells in the eye that respond to different ranges of frequencies. The absorption of photons by these cells will depend on the <u>energy</u> of the photon, which is proportional to the frequency, not the wavelength.

4.) Cats can see in the dark.

This is clearly not true, since seeing requires a light source to illuminate objects. Cats can, however, see much better than humans in dim light. The eye of the cat has evolved for nocturnal hunting and other activities. We will discuss two contributions to this ability in the cat eye.

First, the pupil of the cat can open much wider (relative to the focal length) than that of a human, allowing more light to enter the eye. To quantify this, we use the familiar notion of an f-stop from photography (see Myth #6). The f-number of a lens is the ratio of the focal length of the lens to the diameter. Thus, <u>larger</u> diameters correspond to <u>smaller</u> f-numbers. A fast camera lens has an f-number of 1.4. The human eye has an f-number of about 2.4. An owl has an f-number of 1.3, while <u>the cat's f-number is 0.09</u>!

Cats' eyes (as well as whales', dolphins', horses', fishes', crocodiles' and some other creatures' eyes) have a reflecting layer of cells just behind the retina, called the *tapetum*. This is designed to reflect light within the eye so that there is an increased probability of photons that miss a photoreceptive cell upon arriving at the retina striking another cell after reflection. The tapetum is the origin of the "glowing" of some animals' eyes when seen at night by means of a flashlight or automobile headlight.

5.) Coherence of light means "in phase".

In explaining the operation of lasers, many teachers will use the term *coherent* to mean that all of the light waves are exactly in phase with each other, which they are. This is too strict, however. The term simply means that <u>light waves have a fixed phase relationship to each other, but the phase angle does not have to be zero.</u>

6.) You are taking a picture and decide that you need twice as much light. So you change your camera aperture reading from f/8 to f/4.

While the numbers given in the statement are indeed in the ratio of 2:1, we need to remember that the f-number is a ratio of the focal length to the <u>diameter</u> of the lens, as discussed in Myth #4. The amount of light entering the camera, however, is proportional to the <u>area</u> of the opening, not the diameter. Since the area is proportional to the square of the diameter, we would need to change the f-stop by the <u>square root of 2</u> to double the light. Thus, if we started at f/8, we would need to reduce the f-stop to f/5.6.

Now, an extra Mystery related to this Myth. Why do f-stops on cameras have the particular numbers as follows:

$$1.4, \quad 2, \quad 2.8, \quad 4, \quad 5.6, \quad 8, \quad 11, \quad 16 \ ?$$

We can understand this in terms of the above discussion and the realization that this series of numbers can be expressed as,

$$f = (2)^{\frac{n}{2}}$$

where *n* starts at *n* = 1 and increases by 1 for each new f-stop. Thus, the f-numbers are designed so that <u>each change from one number to the next represents a doubling (or halving) of the amount of light entering the lens</u>.

7.) The solar constant is a measure of the amount of energy arriving at the Earth per unit time from the Sun on a square meter perpendicular to the direction to the Sun. Its value is:

$$\text{Solar constant} = 1370 \text{ W·m}^{-2}$$

The problem with this statement is the word *constant*. First of all, the solar "constant" is not a fundamental and universal constant like π, or Planck's constant or the charge on the electron. It's simply the intensity of radiation from <u>our</u> Sun on <u>our</u> Earth. But what's worse is that <u>it is not a constant</u>. It is only an average or typical value. For example, one contribution to the lack of constancy is the fact that the Earth's orbit is slightly elliptical. Thus, the variations in the distance between the Earth and the Sun result in a small but easily measurable variation in the solar "constant".

Chapter 26
Reflection and Refraction

𝕸ysteries:

1.) What is a rainbow?

2.) Speaking of rainbows, what causes a *double* rainbow?

3.) Speaking of rainbows, why is the sky darker <u>above</u> a rainbow than below, or, if the double rainbow is evident, <u>between</u> the rainbows?

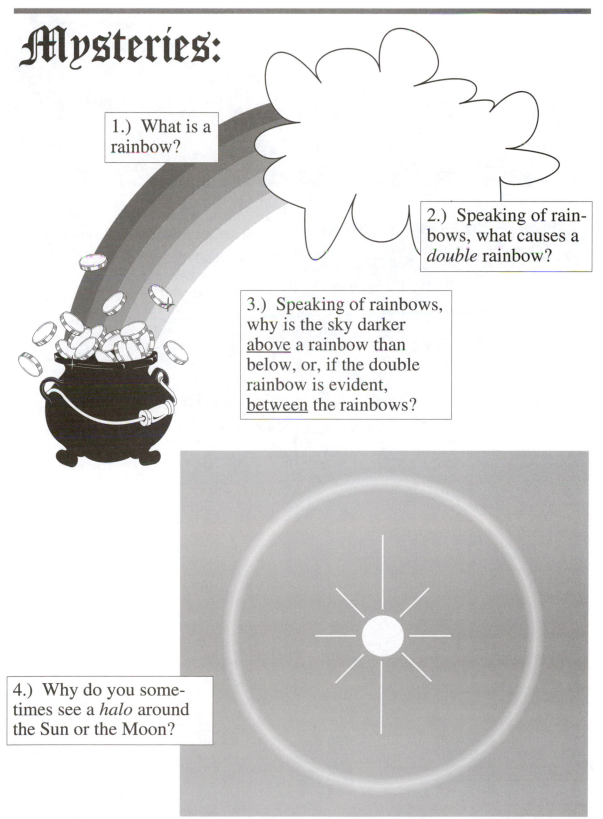

4.) Why do you sometimes see a *halo* around the Sun or the Moon?

Mysteries:

5.) When driving on a black roadway on a hot day, you may see what appears to be a puddle of water up ahead of you. But when you get there, it is dry. Why?

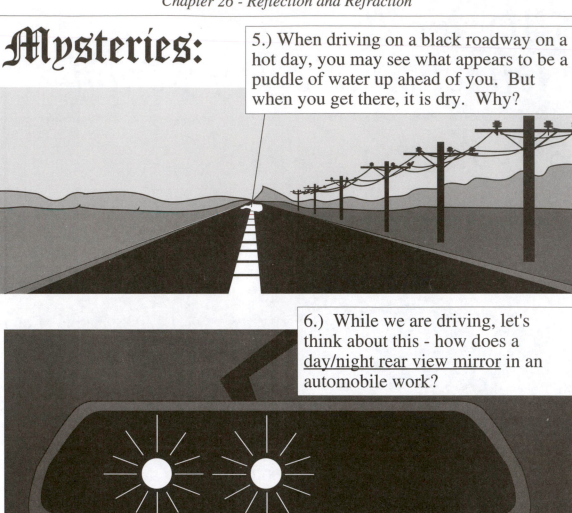

6.) While we are driving, let's think about this - how does a <u>day/night rear view mirror</u> in an automobile work?

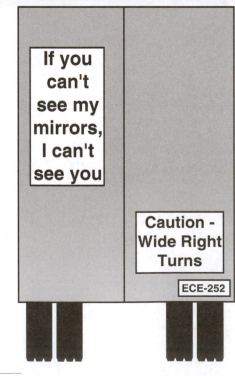

If you can't see my mirrors, I can't see you

Caution - Wide Right Turns

ECE-252

7.) Continuing with the driving theme, the following sign was observed on the back of a large truck: "If you can't see my mirrors, I can't see you". Why is this?

Mysteries:

8.) Why does a prism separate white light into colors?

Laser Gun

9.) Two hunters are aiming at the same fish in a pond - one with a laser gun and the other with a bow and arrow. They aim in <u>different directions</u>. Why?

10.) Why do you sometimes appear in photographs with red eyes?

Mysteries:

11.) When the astronauts walked on the Moon, they saw bright lights around the shadows of their heads. Why was this?

12.) When you look at the Moon over the ocean or a large lake, you see a blaze of light coming across the water toward you. Why?

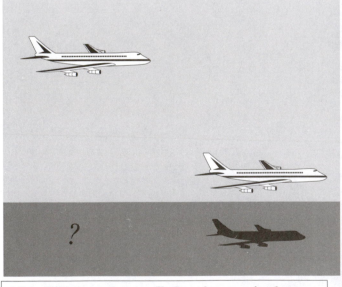

13.) When you are flying in an airplane, you can only see the shadow of the airplane when it is close to the ground. Why can't you see the shadow when the airplane is higher?

Magic:

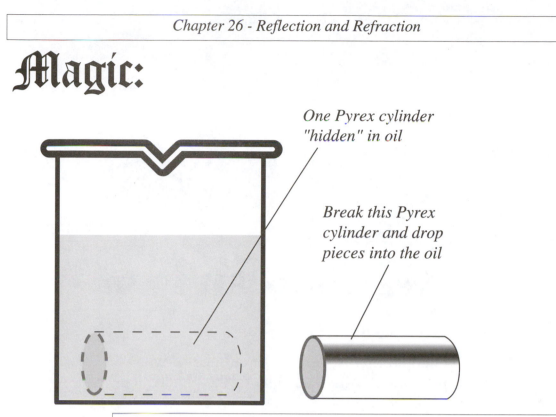

One Pyrex cylinder "hidden" in oil

Break this Pyrex cylinder and drop pieces into the oil

The Magical Glass Repairing Fluid

Secretly immerse a piece of Pyrex tubing in supermarket vegetable oil. It will disappear! Now, in sight of an unsuspecting volunteer, break another identical piece of Pyrex and dump the pieces into the vegetable oil. Reach in with tongs and pull out the first piece. The magical fluid has repaired the glass!

The Straightest Line You'll Ever See

Shine a laser beam across the room and make the beam visible with a smoke machine or by clapping blackboard erasers together.

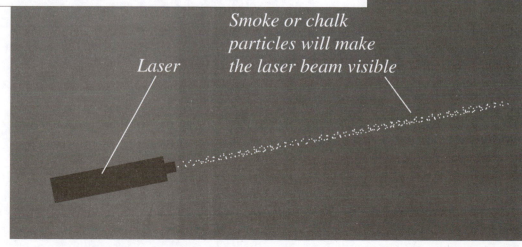

Smoke or chalk particles will make the laser beam visible

Laser

Magic:

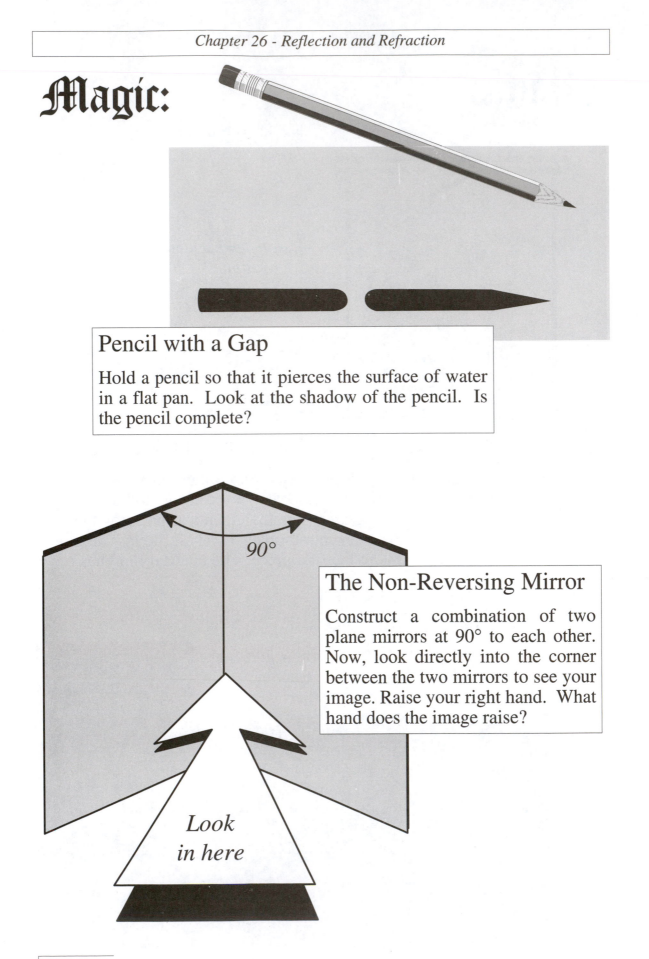

Pencil with a Gap

Hold a pencil so that it pierces the surface of water in a flat pan. Look at the shadow of the pencil. Is the pencil complete?

90°

The Non-Reversing Mirror

Construct a combination of two plane mirrors at 90° to each other. Now, look directly into the corner between the two mirrors to see your image. Raise your right hand. What hand does the image raise?

Look in here

𝔐agic:

The Polarized Rainbow

When the rainbow is present, look at it through a polarizing filter and turn the filter. How is the rainbow polarized?

The Broken Rainbow

This demonstration may require a long wait for the conditions to be right, but look at the rainbow over the ocean such that you are looking through some ocean spray. Do you notice anything interesting?

The Disappearing Shadow

Take the lamp shade off a table lamp or floor lamp and place the lamp about 30 - 60 cm from a white wall. Hold a paper clip in your fingers and place your hand close to the wall so that clear shadows of your hand and the paper clip are projected on the wall. Now, move your hand slowly toward the light bulb. The shadow of your hand becomes blurry but what happens to the shadow of the paper clip?

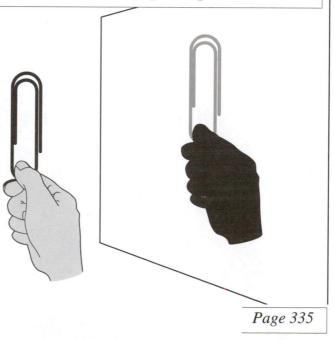

Myth:

1.) Stars twinkle.

2.) One way mirrors.

3.) Full-length mirrors.

𝔐𝔶𝔱𝔥:

4.) The Moon is larger when it is near the horizon than when it is near the zenith.

SPRING

5.) On the days of the vernal and autumnal equinoxes, the times for the rising and setting Sun are the same (one AM and one PM, of course), since the day and the night are each 12 hours long.

FALL

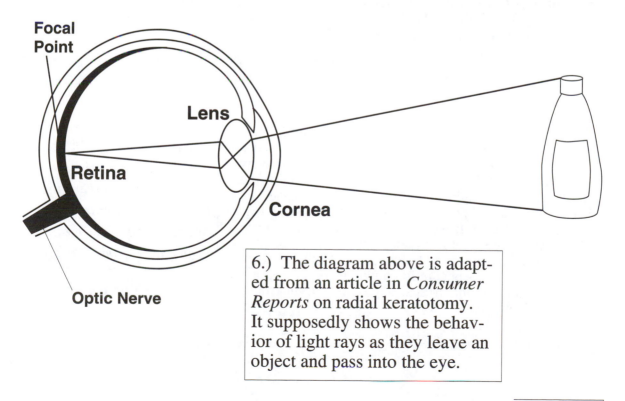

Focal Point

Lens

Retina

Cornea

Optic Nerve

6.) The diagram above is adapted from an article in *Consumer Reports* on radial keratotomy. It supposedly shows the behavior of light rays as they leave an object and pass into the eye.

Concepts of Reflection and Refraction

These words, *reflection* and *refraction*, refer to our macroscopic description of the two possibilities that can occur when light is incident on a boundary between two media.

Reflection: Angle of incidence equals angle of reflection.

The *law of reflection* simply states that the angle of incidence (the angle between the incoming ray of light and the normal to the boundary) and the angle of reflection (the angle between the outgoing ray and the normal to the boundary) are the same. If the boundary is optically flat (irregularities are small compared to the wavelength of the light), then we have *specular reflection*, where the law of reflection is clearly obvious. A plane mirror in your bathroom exhibits acceptable specular reflection. This quality of specular reflection is not acceptable, however, for research spectrometers, which require highly polished optical mirrors. If a flashlight is shone on your bathroom mirror, you can clearly see the reflected ray obeying the law of reflection. If the boundary is not optically flat, we have *diffuse reflection*, such as that from a piece of paper. If we shine a flashlight on a piece of paper, we do not see a beam bouncing off at the appropriate reflection angle. Instead, we see the light scattered in all directions. If we were able to investigate the boundary microscopically, we would see the law of reflection satisfied for each ray that hits the paper. Since the paper is not optically flat, however, rays of light hit small bits of the surface that are oriented in a variety of directions, so that the incoming rays are scattered widely.

The *law of refraction* is also known as *Snell's Law*, which, symbolically, is as follows:

$$n_1 \sin \theta_1 \ = \ n_2 \sin \theta_2$$

Here, θ_1 is the angle of incidence and θ_2 is the angle of refraction (the angle between the outgoing ray, in the new medium, and the normal to the boundary). The n's are *indices of refraction*, which provide descriptions of the optical properties of the material making up the medium. The index of refraction for a material is defined as the ratio of the speed of light in a vacuum to that in the material:

$$n \ = \ \frac{c}{v} \ = \ \frac{\text{speed of light in vacuum}}{\text{speed of light in material}}$$

As light enters an optically more dense material, the angle of refraction is smaller than the angle of incidence - the light bends toward the normal. Likewise, if light passes from an optically dense medium to a less dense medium, the light bends away from the normal. In this latter case, we can imagine the situation that the light bending away from the normal leaves the material at an angle of 90°, propagating parallel to the surface. If we make the incident angle even larger, then there is no refracted ray. This is the phenomenon of *total internal reflection* and the incident angle at which this occurs is known as the <u>critical angle</u>. We can find the critical angle for total internal reflection by setting the refracted angle to 90° in Snell's Law:

Refraction: Light entering an optically more dense medium bends toward the normal. Light entering an optically less dense medium bends away from the normal.

$$n_1 \sin \theta_1 \ = \ n_2 \sin \theta_2 \ \Rightarrow \ n_1 \sin \theta_c \ = \ n_2 \sin 90° \ \Rightarrow \ \theta_c = \sin^{-1} \left(\frac{n_2}{n_1} \right)$$

Discussions; Chapter 26 - Reflection and Refraction

𝔐𝔶𝔰𝔱𝔢𝔯𝔦𝔢𝔰:

1.) What is a rainbow?

The origin of the rainbow is in water droplets in the sky, and is due to the refraction of sunlight at the surface of the droplets and reflection from the "back". Let us imagine a set of six horizontal light rays incident on an ideal spherical raindrop (very small raindrops are very close to spherical, since the surface tension effects are large compared to the air resistance effects mentioned in Myth #6, Chapter 13), as shown below:

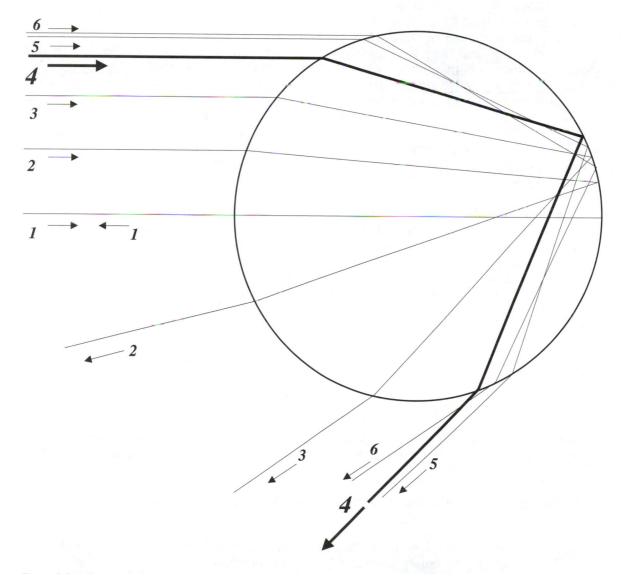

Ray **1** is directed toward the center of the raindrop and simply returns along the same line after reflection from the back surface. Higher numbered rays correspond to moving the point of incidence vertically farther away from a line going through the center, that is, farther from ray *1*. Applying the laws of refraction and reflection, we obtain the paths of 5 more rays as shown in the diagram. Ray *4* (the bold ray in the diagram) is what we call the *Descartes Ray*. It represents the ray that emerges from the raindrop with the <u>largest angle relative to the horizontal</u> (ray *1*). If we continue to move the incident rays upward on the raindrop from ray *4*, the emerging rays exit at shallower angles again, as shown by rays *5* and *6*. This behavior is shown in the graph at the right, where the exit angle is plotted against the entry point, showing that the curve maximizes for the Descartes Ray, with a resulting exit angle (for red light) of 42°.

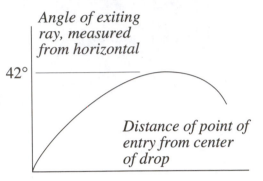

Now, since ray *4* represents a direction where the rate of change of the angle of exit (as the incoming ray is moved upward on the drop) has "slowed down" and the angle itself reverses, there will be a large number of rays that exit the raindrop in the vicinity of ray *4*. Thus, <u>more light will be refracted by the raindrop in this general direction than in any other direction</u>. This is the origin of the brightness of a rainbow. Now, the index of refraction for water is a function of frequency of the light (see Mystery #8). As a result, the angle of concentration of exiting rays will be different for different colors. We find that, at the red end of the spectrum, as mentioned above, the exiting angle (from the line through the center of the drop) for the brightest light is 42°, while it is 40° for the violet end, with the other colors spanning the difference.

Thus, raindrops that are high in the sky (42° from the <u>anti-solar point</u>, which is in the direction from your head exactly 180° from the direction to the Sun) will send red light to the observer on the ground. As we look toward lower drops, these will send the other colors toward the observer, until we reach 40°, which corresponds to drops sending violet light to the observer. This describes the separation of the colors in the rainbow, with red on the top and violet on the bottom. But why is it a <u>bow</u>? An observation angle of a given number of degrees actually describes a <u>cone</u>. Thus, as we look away from the Sun, any drop that is on this cone will participate in the rainbow. Thus, we see the bow as a part of the circle that is the "end" of the cone as we look out from the vertex of the cone. In reality, of course, there is no end of the cone, since raindrops at any distance can participate in the rainbow.

Nice color photographs of the rainbow, with supernumerary bows (see Mystery #8, Chapter 29) are available on the cover of *The Physics Teacher*, April 1989 and April 1990. A procedure for demonstrating the rainbow in the classroom is available in H. A. Daw, "A 360° Rainbow Demonstration", *American Journal of Physics*, **58**, 593 (1990).

2.) **Speaking of rainbows, what causes a *double* rainbow?**

Sometimes, a second rainbow can be seen, fainter than the ordinary rainbow and above it. What's more, the colors are reversed in this rainbow. This is due to light which makes <u>two</u> reflections from the inside of the raindrop before refracting out. Since each interaction with the surface results in some light refracting and some reflecting, the amount of light which is available after making two reflections is necessarily less than that available after making one. Thus, the second rainbow is fainter than the first. If we use the laws of reflection and refraction to trace rays making two reflections, similar to our analysis in Mystery #1, we find that the Descartes Ray for this situation is at about 51°. Thus, the second rainbow is higher in the sky than the first. We also find that the angle for violet in this case is steeper than for red, so that the colors are reversed.

If the light can reflect twice, how about three times and four? In fact, many rainbows have been seen in the laboratory (see J. Walker, "Multiple Rainbows from Single Drops of Water and Other Liquids", *American Journal of Physics*, **44**, 421 (1976)). These are not generally seen in the sky, since they are very faint and are overshadowed by background light from the sky.

3.) Speaking of rainbows, why is the sky darker <u>above</u> a rainbow than below, or, if the double rainbow is evident, <u>between</u> the rainbows?

As seen in our discussion of Mystery #1, the laws of reflection and refraction cause the light rays exiting the raindrop to "bunch up" around the Descartes Ray. But there are other rays leaving, as seen in the diagram for Mystery #1. Notice, though, that these other rays are at <u>smaller</u> angles (with respect to the horizontal) than the Descartes Ray; <u>there are none at larger angles</u>. Thus, the drops lower in the sky than those that give us the rainbow will be sending light to our eyes also. This will result in a general, whitish light coming from that region of the sky below the rainbow. By contrast, then, the region above the rainbow, where there are no drops that send light toward your eye, will appear dark. This is called *Alexander's Band*, after Alexander of Aphrodisius.

In the case of the double rainbow, the drops sending out rays far from the Descartes Ray are <u>higher</u> in the sky than the secondary rainbow. Thus, the regions <u>above</u> the secondary and <u>below</u> the primary are bright and the region between them is dark by comparison.

4.) Why do you sometimes see a *halo* around the Sun or the Moon?

In discussing the rainbow, we considered refraction by spherical water droplets. In order to explain the halo around the Sun or Moon, we need to consider refraction by <u>ice crystals</u>. When water freezes into ice crystals, the crystals tend to have hexagonal cross sections, as shown below (the hexagonal shape of the ice crystal structure has a great deal of open space, which is why ice expands upon freezing, as discussed in the Concepts section of Chapter 14 and Mystery #3, Chapter 14):

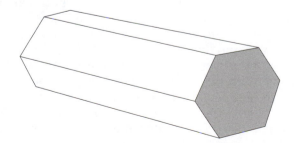

Now, let us imagine that we send light into this crystal from the side. In the top part of the diagram below, we show light entering the side <u>so that the refracted ray is parallel to the top edge of the crystal</u>. If we apply the law of refraction to this path, we find that the total deviation from the original direction is 22°.

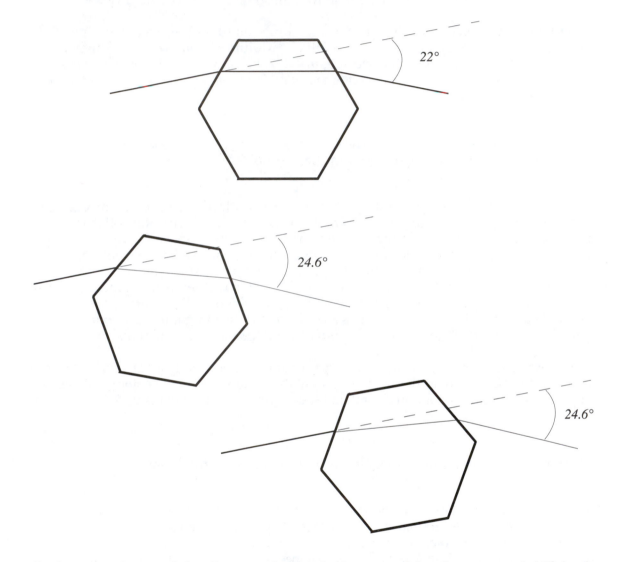

In the second part of the diagram above, the ice crystal has been rotated 10° in the clockwise direction. The light ray now strikes the first surface at a larger angle than in the top diagram, but it strikes the second surface at a smaller angle. The net result is that the deviation of the light is larger than in the top diagram, an angle of 24.6°. In the bottom picture, the ice crystal has been rotated 10° in the <u>counterclockwise</u> direction from the original orientation. It now strikes the first surface at a smaller angle but the second at a larger angle. We find that there is a 24.6° deviation of the light ray again. If we continue looking at other angles of rotation of the crystal, we find that <u>the minimum angle of deviation occurs when the crystal is in the orientation at the top of the diagram</u>. If we now imagine starting the ice crystal from a large rotation in the counterclockwise direction and turning it clockwise, we would see the angle of the deviated beam decrease, reach a minimum at 22° and then start increasing again, as shown in the graph to the right, which is analogous to the graph in Mystery #1 for light entering a raindrop. Thus, <u>this is very similar to the case of the maximum angle reached by rays exiting the raindrop in Mystery #1</u>. Just as we had in that case, <u>we will have a clustering of light rays leaving the ice crystal at an angle of 22°</u>. As a result, the ice crystals will tend to emit light preferentially at an angle of 22° from the incident light.

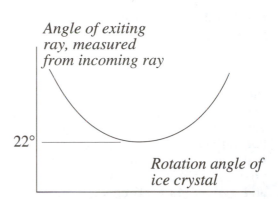

Now, let us imagine that these ice crystals are in the atmosphere and we move our gaze 22° away from the direction toward the Sun. In this location, there will be ice crystals sending a significant amount of light toward us. Of course, the 22° angle can be in <u>any</u> direction away from the Sun, so that there is a <u>circle</u> of light of angle 22° around the Sun that will have ice crystals sending light to our eyes. Thus, we see a <u>halo</u> around the Sun. If the light source happens to be the Moon, we see a halo around the Moon.

This discussion is only the beginning of refraction effects from ice crystals in the sky. If the hexagonal crystal in the diagram above is very short, so that it is essentially a hexagonal plate, then, as it falls through the air, it tends to fall in a horizontal position. Thus, the 22° refraction will send light to your eyes only for light which is in a horizontal plane with the Sun. This results in *sun dogs*, or *mock suns*, which are bright images of the Sun on either side of the Sun.

There are a variety of other halos, sun dogs, arcs, etc. that can arise from these ice crystals. A diagram of a number of them is available on page 126 in J. Walker, *The Flying Circus of Physics with Answers*, Wiley, New York, 1977. A detailed but clear description of these possibilities is available in R. Greenler, *Rainbows, Halos and Glories*, Cambridge University Press, Cambridge, 1980.

5.) When driving on a black roadway on a hot day, you may see what appears to be a puddle of water up ahead of you. But when you get there, it is dry. Why?

This is a common example of one type of *mirage*. It occurs because the air over a roadway is heated, since it is in contact with a dark roadway, whose temperature has been raised by absorption of radiation from the Sun. Near the roadway, there is a significant temperature gradient in the air. The index of refraction of air is a function of the density of the air, which is related to the temperature. The hot air nearest the ground has the lowest density and the lowest index of refraction. Thus, as light from a point off the ground and moving in a direction toward the ground enters the region of decreasing index of refraction, it is continuously refracted "away from the normal". For light making a large angle (with respect to the normal) incidence with the air near the ground, this refraction is sufficient to cause the light to "miss" the ground and be directed upward again. Thus, someone in a position to receive this light will see light appearing to come from the ground that originated at a point off the ground. If the origin of the light was the sky, the appearance of skylight against the dark roadway gives the impression of a puddle of water. If this mirage is investigated closely, one can see (inverted) images of trees, cars or any other object which is in the right position to send grazing light rays in the observer's direction.

6.) While we are driving, let's think abut this - how does a <u>day/night rear view mirror</u> in an automobile work?

The day/night mirror in an automobile is a back-silvered mirror in the form of a <u>wedge</u>. Before we explain how it works, let us first consider a more familiar situation. When you look into your bathroom mirror, you actually see two reflections - one from the silvered back surface of the mirror, the other, much fainter, from the front surface. The two reflections are easiest to see if you put your face close to the mirror and make sure that there is a dark surface in back of your head. Now, close your left eye and look at your right cheek - you will see a faint second image. You can also close your right eye and look at your left cheek - same thing. Normally, this second image is not even noticed and is not a problem for a variety of reasons - the background may not be dark, you are far away from the mirror, so that the images are very close in size, the images from both eyes are fused, which tends to wash out the effect, etc.

Now, the two images in the bathroom mirror almost coincide because the two surfaces of the mirror are parallel. They differ slightly in size, since the distances from your face to the reflecting surfaces differ by the thickness of the glass. By using a <u>wedge</u> in the day/night mirror, the images from the two surfaces end up traveling in different directions. We can follow three possible rays in the diagram below.

The dark, solid ray is the "normal" ray that will be observed in the daytime. It refracts into the mirror, reflects off the back and refracts back out again, toward the eye of the driver. It contains about 90% of the energy of the incoming ray. Now, there is also some light reflected off the front surface of the mirror, as we discussed above for the

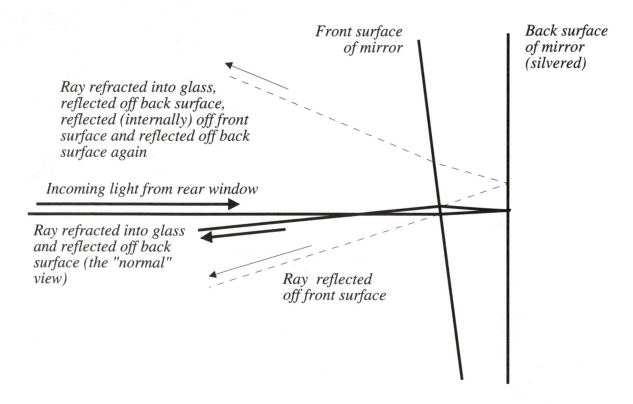

bathroom mirror. This is the lower dashed ray in the diagram. Because the surfaces of the mirror are not parallel, however, this ray comes off the mirror at a larger angle than the normal ray. It contains about 5% of the incident energy. At night, the mirror is rotated clockwise in the diagram, so that this ray enters the eye. Thus, the bright lights of cars behind the driver are reduced in intensity. It should be noted that there is a third ray, which is not used in the normal design of the mirror, but which has about the same intensity as the ray used in the night setting. This is the upper dashed ray in the diagram. This ray makes three reflections and two refractions on its way back toward the driver and exits the mirror at a steeper angle than the normal ray. If you sit in your car and move your head up and down while looking in the mirror, you will see a fainter image <u>both above and below</u> the normal image, exhibiting the existence of both of the dashed rays in the diagram (There are theoretically more images caused by further reflections off the inner surface, but these are very faint and at very steep angles.).

7.) Continuing with the driving theme, the following sign was observed on the back of a large truck: "If you can't see my mirrors, I can't see you". Why is this?

This exploits the symmetric nature of light ray tracing. For a truck driver to see your car in back of the truck, the light from your car must have a clear path to travel to the mirror and reflect into the eyes of the driver. Now, let us reverse the ray going from your car to the mirror. In order for you to see the mirror, light from the mirror must have a clear path to travel directly to your eye. Thus, if you can't see the mirror, there is no direct ray going from the mirror to you that we could reverse in order to provide an image of your car for the truck driver.

8.) Why does a prism separate white light into colors?

This is due to the phenomenon of *dispersion*. A dispersive medium is one whose index of refraction is a function of the frequency of the light. As a result, different colors are refracted by different angles. Thus, in the raindrop in Mystery #1, different colors of sunlight followed slightly different paths, so that we see the separation of colors in the rainbow. Similarly, when white light passes through a piece of material, different colors are refracted by different amounts and the result is the separation of light into colors. If the sides of the piece of material are parallel, the colors are recombined and the dispersion can only be observed by the faint red and blue edges of the exiting white light beam. If the sides are not parallel, further separation of the colors occurs at the second interface and the projection of a spectrum on a screen is possible.

9.) Two hunters are aiming at the same fish in a pond - one with a laser gun and the other with a bow and arrow. They aim in <u>different directions</u>. Why?

The hunter who is shooting the arrow has the tougher job and must understand the refraction of light in order to aim the arrow correctly. Light from the fish will bend away from the normal as it exits the water and travels to his or her eye. Thus, the fish will <u>appear</u> to be above a line drawn straight from the eye to the fish. Therefore, this hunter must aim <u>below</u> the apparent position of the fish. Using the reversible property of light rays discussed in Mystery #7, however, we realize that the light from the laser gun will simply follow the same path back to the fish that the light from the fish followed to the laser hunter. Thus, the laser hunter simply aims at where he or she sees the fish.

10.) Why do you sometimes appear in photographs with red eyes?

If you are looking right into the camera as the flash goes off, and the flash is located close to the lens of the camera, light from the flash can reflect off your retina and travel back into the camera. Since your retina is red, the reflected light is red (or is this statement redundant?). Normally, the pupil of the eye is an imperfect model for a blackbody - a cavity with a hole whose size is small compared to the size of the cavity. If the camera (and flash unit), pupil and back of the retina are all lined up, however, it is highly likely for light to enter the eye and reflect right back out again into the camera. The human retina is not as reflecting as the cat retina (see Myth #4 in Chapter 25), but will reflect enough radiation to give the "red-eye" effect in photographs.

This effect can be eliminated by making sure that there is a larger lateral distance between the camera lens and the flash so that the retroreflected light from the flash does not enter the camera lens.

11.) When the astronauts walked on the Moon, they saw bright lights around the shadows of their heads. Why was this?

This is an example of *heiligenschein*, which is easily observable on Earth. On a sunny morning when the grass is covered with dew, or anytime after you have sprinkled water on some grass, look at your shadow on the wet grass. You will see a bright white glowing area around your head. This is the *heiligenschein*. It is caused by light from the Sun which enters the water droplets and is <u>retroreflected</u> - that is, it is reflected back along the path from which it came. This is the same effect as in Mystery # 10, where the light from the camera flash is retroreflected from the eye. It is also the same principle involved in reflective highway signs or license plates. These surfaces are covered with small glass beads, which will retroreflect the light from the driver's headlights back to the driver.

Now, what was on the Moon's surface to cause the retroreflection? It certainly wasn't dew! Mixed in with the lunar soil were small glassy beads which played the role of the dewdrops and provided the heiligenschein for the astronauts.

12.) When you look at the Moon over the ocean or a large lake, you see a blaze of light coming across the water toward you. Why?

This is sometimes called a "glitter path". If the surface of the water were perfectly calm and flat, you would simply see a reflection of the moon. But the presence of waves is what gives us the bright blaze of light. Between you and the horizon, there is a very large number of waves. If you imagine the convex crests of these waves, somewhere on each wave will be a small surface area that is oriented correctly to reflect the moonlight toward your eyes. Thus, every wave reflects some light toward your eyes, which gives you the long lighted glitter path.

13.) When you are flying in an airplane, you can only see the shadow of the airplane when it is close to the ground. Why can't you see the shadow when the airplane is higher?

This is due to the angular diameter of the Sun. If the Sun were a point source, we would be able to see the shadow of the airplane from any altitude. But the Sun subtends an angle of about 0.5° from the Earth. Thus, if the airplane is high enough so that rays coming from the edges of the Sun and past edges of the airplane converge to a point above the surface of the Earth, then no shadow will be cast. The diagram below (not to scale!) shows that situation where the shadow of the length of the airplane just "lifts off" the surface of the Earth as the plane climbs.

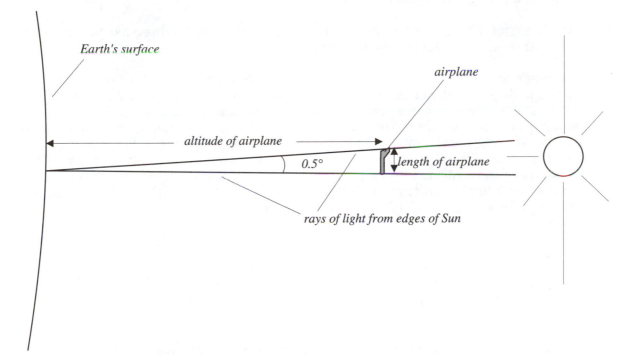

The shadow region in the diagram is called an *umbra*. From the diagram, we can estimate the altitude of the airplane at which the umbra lifts off the Earth. Since the angle is small, we can approximate the triangle formed by the length of the airplane and the two rays of light just passing its tip and tail as a sector of a circle. Then, by means of the definition of an angle in radians,

$$\theta = \frac{\text{length of airplane}}{\text{altitude of airplane}} \quad \Rightarrow \quad \text{altitude of airplane} = \frac{\text{length of airplane}}{\theta}$$

Now, the length of a Boeing 747 is 231 feet. Using this value in the above expression, we obtain the altitude of the airplane at which the umbra just lifts off the ground:

$$\text{altitude of airplane} = \frac{231 \text{ feet}}{(0.5°)(\frac{\pi \text{ radians}}{180°})} = 26{,}500 \text{ feet}$$

For anyone familiar with this phenomenon, this seems awfully high, and <u>it is</u>. <u>We have used an incorrect procedure.</u> We calculated this number using the largest dimension on the airplane, its length. But at the altitude of 26,500 feet, there is plenty of room for light from the edges of the Sun to "leak" around the fuselage and wings of the airplane to wash out the shadow. Thus, we should not use the largest dimension, we should use some smaller dimension that represents the airplane in some way. In addition, we should choose a length that will allow a shadow of enough small parts (wings, fuselage, tail) of the airplane to be cast that we recognize it as an airplane. As an estimate, let us take about 10% of the length - 23 feet, as a representative dimension of wings, etc. This will give us a much smaller altitude, about 2650 feet, at which the shadow disappears, a much more reasonable number.

The discussion above can be made clear by performing an Earth-bound version - see The Disappearing Shadow in this Chapter.

A similar situation exists for solar eclipses. Since the moon is round, its umbra is much simpler to deal with than that of the airplane. It is a cosmic coincidence that the length of the moon's umbra is similar to the distance between the Moon and the Earth's surface. During a total eclipse of the Sun, the umbra sweeps across the Earth, and anyone in its narrow path will see the total eclipse. But, if the Moon happens to be particularly far from the Earth in its elliptical orbit, the umbra may not touch the Earth's surface, just as the umbra of the airplane lifted off the Earth's surface. In this case, observers on Earth will not see a total eclipse, but rather an *annular eclipse*, in which the Sun is blocked off except for a thin ring of light around the edge of the Moon.

𝔐𝔞𝔤𝔦𝔠:

The Magical Glass Repairing Fluid

This demonstration exploits the fact that the indices of refraction for Pyrex glass and vegetable oil are very close in value. Thus, when Pyrex glass and vegetable oil are in contact, there may be a <u>physical</u> boundary between the two, but there is (practically) no <u>optical</u> boundary. We see objects that do not emit their own visible radiation only by reflection and, in order to have a reflection, there must be an optical boundary between media with two different indices of refraction. In the case of Pyrex glass and vegetable oil, this boundary does not exist, so that we cannot see the Pyrex. An extension of this demonstration is to immerse a Pyrex measuring cup into vegetable oil. While the cup itself will be almost invisible, the lettering on the side of the cup will be in plain view.

The Straightest Line You'll Ever See

This demonstration exhibits the necessity of scatterers in a beam of light in order to see the light. In a laser beam, all of the light is traveling in the same direction - that of the beam. If you are observing the beam from the side, no light is traveling toward your eye and you should see nothing. The beam is made visible by adding scatterers in the beam that will scatter the light in various directions, including toward your eye. Similarly, a searchlight is made visible by particles in the air that scatter the light toward your eye. Now, why are laser beams in science fiction movies visible if they are traveling through empty space? Is this more cinematic license, as we discussed in Myth #1 in Chapter 18? (Yes, it is!)

Pencil with a Gap

The pencil shadow will appear as two pieces, separated by a curved gap. Where the pencil enters the water, the surface tension causes a meniscus. Light entering the meniscus just to the side of the pencil will be refracted toward the normal upon entering the water. This is in the direction toward the pencil, since the water surface curves upward as we move closer to the pencil. Thus, light is refracted into the shadow region for this part of the pencil and washes out the shadow. This results in the apparently separated effect.

The Non-Reversing Mirror

A single mirror will apparently reverse right and left, so that raising your right hand will result in the image raising the left hand. With the combination of the mirrors described in this demonstration, the light exiting the combination has experienced two reflections, so that the left-right reversal is canceled. Thus, when you raise your right hand, the image raises its right hand also.

The Polarized Rainbow

The rainbow is *tangentially polarized*, meaning that the polarizing filter axis must lie tangentially along the bow for the light to be transmitted. This is due to the fact that the reflection angle inside the raindrop is close to Brewster's angle, so that the light reflected from the back of the drop (see Mystery #1) is about 96% polarized in the direction perpendicular to the plane of the refracted and reflected rays in the diagram in Mystery #1 (which is the plane of the page).

As a result, it is impossible to make a complete rainbow disappear with a polarizing filter. If only a portion of the rainbow is present, you may be able to wipe it out by placing the polarization axis perpendicular to the rainbow segment.

The Broken Rainbow

What you will see in this situation is a "broken" rainbow, as indicated in the illustration. The part of the rainbow that is seen in the ocean spray will be displaced inward from the part that is seen in the raindrops. This is due to the fact that salt water has a different index of refraction than fresh water. This difference causes the rainbow in salt water droplets to be at an angle about 1° smaller than that from fresh water droplets. If you are impatient and cannot wait for the right conditions, look on page 54 of G. P. Konnen, *Polarized Light in Nature*, Cambridge University Press, Cambridge, 1980, for an excellent picture of this effect.

The Disappearing Shadow

As mentioned in Mystery #13, this is an Earth-bound version of the disappearing shadow of an airplane. While the shadow of your hand becomes blurry, it is still there, while the shadow of the paper clip <u>disappears</u> as the clip is brought close to the light bulb. The light bulb is not a point source. As the clip is brought close to the bulb, light can "leak" around the small diameter, as described in the discussion to Mystery #13 and the umbra lifts off the wall.

𝕸𝖞𝖙𝖍:

1.) Stars twinkle.

Upon looking at the night sky, it seems that stars twinkle and planets do not. This is due only to the fact that the stars act essentially as point sources. The twinkling that is evident is due to turbulence in the Earth's atmosphere. This causes random variations in the effective index of refraction of the air, which results in variations in the direction from which light comes to enter your eye. The result is the twinkling effect. A similar effect occurs when you look at a scene over the hot hood of your car or other source of hot air. The scene appears to "shimmy" because of the turbulent air and random variations in index of refraction.

2.) One way mirrors.

There is no such device as a simple mirror that will pass light through in one direction and not the other. So called "one-way" mirrors are actually "half-silvered" mirrors that reflect and transmit about equal amounts of incident light. In use, the scene that is to be observed must be in bright light while the observers who do not wish to be seen are in dim light. Thus, from the observers' side, light from the observed side passes through but not much light is available to be reflected from the observers' dimly lit room. Thus, it is easy to see into the other room without any problems with reflected light in the mirror from inside the observers' room. From the observed side, light from the observed room reflects from the mirror and there is very little light from the observers' side to transmit through. Thus, the mirror looks as if it were a regular mirror. If the light in the observers' room is turned on, the effect disappears and observers in both sides can see into the other room.

This effect can be seen somewhat with ordinary window glass. During the day, it is easy to see outside, since both inside and outside are illuminated. At night, the interior of the house is illuminated while the outside is dark. Thus, an observer on the inside cannot see very clearly outside, but sees primarily the interior light reflected from the inside surface of the glass. An observer on the outside, however, sees easily into the house because of the bright lighting inside.

3.) Full-length mirrors.

Most full-length mirrors that are sold are not actually full-length, in the sense of being as long as a person is tall. But, on the other hand, they are usually longer than they have to be. In the diagram to the right, the dotted lines represent light rays originating at the woman's feet and top of her head. The white region on the mirror surrounded by dotted lines shows the only part of the mirror which is contributing to her reflection. Thus, we can show that one only needs a mirror that is <u>half as tall</u> as the person.

4.) The Moon is larger when it is near the horizon than when it is near the zenith.

This is the famous "Moon Illusion" and, despite your protestations to the contrary, <u>the Moon is not larger on the horizon than it is in the zenith</u>. This has been verified by photography. In fact, since you are observing the Moon from a round Earth, the Moon is closer to you (by about the radius of the Earth) at the zenith than at the horizon and thus appears slightly larger when it is overhead, just the opposite of the Moon Illusion.

The apparent size difference has been recorded since the 7th century BC and appears in the writings of the early Greeks and Chinese cultures. There have been attempts to make the Moon Illusion real by invoking explanations about refraction of light going through "more atmosphere" or "vapors" in the atmosphere when the Moon is on the horizon. Ptolemy developed a full refraction theory of the illusion and additional publications on refraction as the explanation continued into the 17th century. Despite its rejection by scientists since that time, the refraction theory is often offered by laypersons today as the reason for the effect. It should also be pointed out that the actual effect of atmospheric refraction is to "squash" the Moon (or the Sun) slightly along its vertical axis when near the horizon. Thus, it actually has a <u>smaller</u> cross-sectional area near the horizon!

Despite attempts at a physical explanation, <u>the Moon Illusion is not a physical phenomenon, it is a psychological phenomenon</u>. A satisfying theory upon which all can agree is not yet available, however. Much debate has occurred in the literature and has included such luminaries as Aristotle, Ptolemy, Roger Bacon, Leonardo da Vinci, Johannes Kepler, René Descartes, Christiaan Huygens, Jean Biot, Thomas Young, Karl Gauss, David Brewster and Hermann Helmholtz. Debate still continues in the literature and an entire book has been published on this effect (M. Hershenson (Ed.), *The Moon Illusion*, Erlbaum, Hillsdale, N.J., 1989).

Theories that have been suggested include those involving comparisons with buildings and other objects on the horizon (although the Moon Illusion also occurs over open water), differences between apparent size when looking horizontally and looking vertically, apparent size effects caused by apparent distance estimations, etc. According to Hershenson, there are at least eight major explanations proposed for the Moon Illusion.

5.) On the days of the vernal and autumnal equinoxes, the times for the rising and setting Sun are the same (one AM and one PM, of course), since the day and the night are each 12 hours long.

The falsehood of this statement is based on the difference between the <u>astronomical</u> day-night equality and the <u>optical</u> equality. The equinox dates do represent the days when a geometrical line from the Sun tangent to the Earth would take exactly 12 hours to pass from sunrise to sunset. But the effect of <u>refraction</u> in the atmosphere causes the "optical day" to be a bit longer than the "astronomical day". Indeed, times of the rising and setting Sun published in the newspaper on the dates of the equinoxes will not be 12 hours apart. As light enters the atmosphere from the rising or setting Sun, it passes through air of increasing density and, therefore, increasing index of refraction. Thus, the light is continuously bent toward the normal. As a result, in the evening, the Sun may have already set (astronomically), yet we still see it for a few minutes after this time. Similarly, in the morning, we see the Sun rise a few minutes before the astronomical sunrise. As a result, the day is actually longer than it "should be".

Thus, as we approach the vernal equinox in March, and the days get longer, the extra length of the day due to refraction causes the 12-hour separation between sunrise and sunset to occur a few days <u>earlier</u> than the equinox. As we approach the autumnal equinox in September, the atmospheric refraction keeps the days longer than they should be and the 12-hour separation occurs a few days <u>later</u> than the equinox.

The sunrise and sunset times will be affected for all times of the year, including at the solstices. Even if the refraction effect is eliminated, however, the sunrise and sunset times at the solstices are not symmetric due to contributions from mechanical considerations. See Myth #2, Chapter 11.

6.) The diagram above is adapted from an article in *Consumer Reports* on radial keratotomy. It supposedly shows the behavior of light rays as they leave an object and pass into the eye.

This is the diagram of a normal eye and is accompanied in the article (Consumers Union, "Goodbye Glasses?", *Consumer Reports*, **53**(**1**), 52 (1988)) by two other diagrams, one of a nearsighted eye and the other of a post-surgery eye. There is very little to find correct in this diagram. Two light rays leaving the bottle come from <u>different</u> points and are then focused at the same point on the retina, giving an image of zero size. Despite the discussion in the article about reshaping the cornea to correct the refraction of the light entering the eye, there is no refraction shown at the cornea. The major refraction occurs in the diagram in the lens, which is incorrect. Most of the refraction in reality occurs at the air-cornea interface. The light rays <u>cross</u>, for some reason, in the lens and then suffer another large refraction in going from the lens material to the fluid within the eye. The label "Focal Point" would only be at the retina if the object were infinitely far away, which is not the case in the diagram.

Chapter 27
Image Formation

𝕸ysteries:

1.) Why is your image in a spoon up-side-down if you look into the bowl, but right-side-up if you look into the back?

2.) How does a face mask for diving allow you to see clearly under water? After all, it's not a lens.

3.) If you look closely at the light projected onto a screen by an overhead projector, you can see faint circles. Why?

Mysteries:

4.) Why do right hand mirrors on automobiles have the inscription, "Objects in this mirror are closer than they appear to be"?

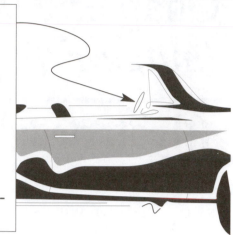

5.) Why are there a large number of circularly-shaped areas of light on the ground under a shade tree? (These are sometimes called "Sunballs".)

6.) Why do slides need to be put into a slide projector *upside-down*?

Mysteries:

7.) Contact lenses or lenses for glasses are often measured in *diopters*. What *is* a diopter?

8.) What's the difference between *binoculars* and *opera glasses*?

9.) Binoculars are described with a pair of numbers, such as 7 x 50. What do these numbers mean?

10.) Photographers know that the *depth of field* (the range of distances over which objects are in focus) increases as the camera lens aperture decreases. Why is this?

Joshua at the President's Speech

𝕸agic:

Real Image

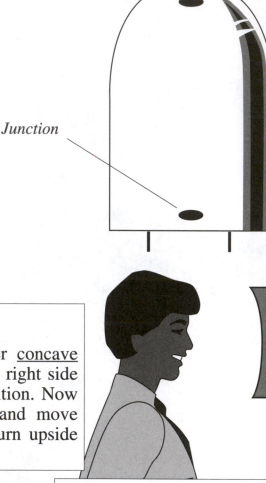

Junction

A Real Image Formed Without a Screen

Apply power with a battery or power supply to a light emitting diode having a clear plastic package and a bullet-type shape. (Do not use the type with a metal can or an odd shape.) You will see the light-emitting junction at the bottom of the package. You will also see an apparent light source at the top of the package.

Flip Your Lid

In a shaving mirror, or other concave mirror, notice that you appear right side up in the normal shaving position. Now keep looking in the mirror and move your head away. You will turn upside down.

Pictures in Space

Project a picture, such as that of the cheerleader at the left, from a slide projector onto a screen. Now, adjust the focus control so that the lens moves outward from the projector body and the image is out of focus. Wave a light colored stick up and down quickly in the light from the projector in front of the screen. With a little practice, you will find a place to wave the stick where a picture will appear on the stick and it will be in focus!

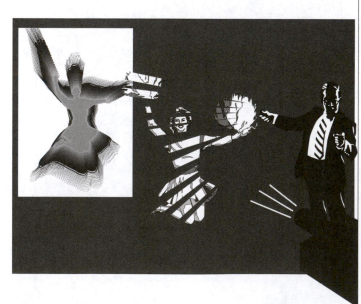

Myth:

1.) The first series of postage stamps issued by the British Commonwealth State of St. Kitts-Nevis in 1903 portrayed a picture of Columbus looking for land through a telescope.

2.) All lenses have curved surfaces.

3.) In *The Lord of the Flies*, Piggy's glasses are used to focus the sun's rays and start a fire. Later, his glasses are broken and he has trouble getting around because he is nearsighted.

4.) Professional telescopes have large diameters so that they can have large magnifications.

𝕸𝖞𝖙𝖍:

Image is here !??!

f

5.) The focal point of a lens is where the image is focused.

f   f

6.) Only two rays come from an object and only from the top!

7.) Galileo invented the telescope.

8.) The Great Wall of China is the only manmade object on the Earth visible from outer space.

There it is!

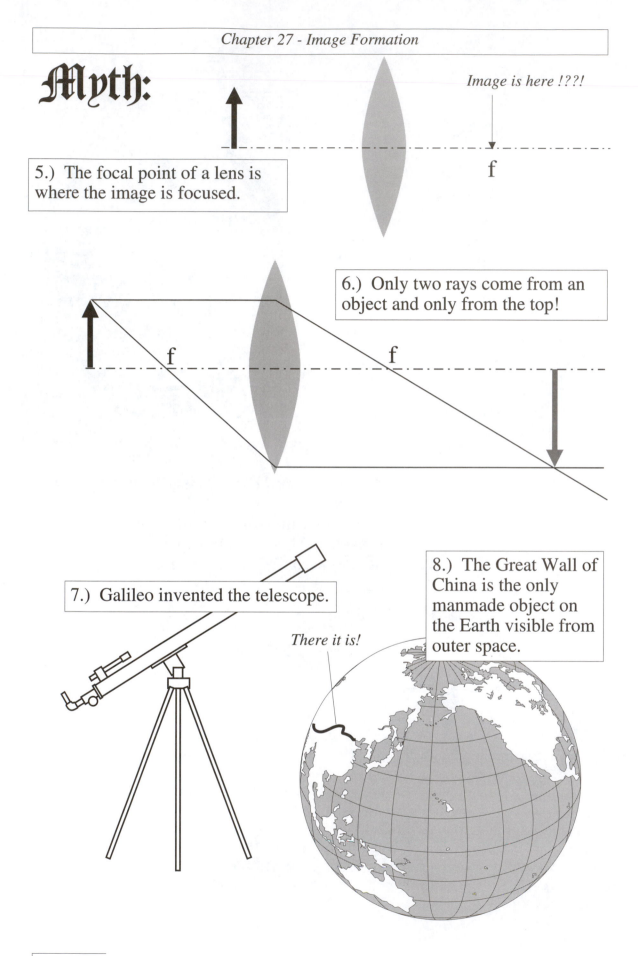

Concepts of Image Formation

The formation of images is an important application of the laws of reflection and refraction in geometrical optics. A common factor among all mirrors and lenses is the existence of a *focal point* (two points for lenses).

This is the point at which, for an idealized lens or mirror, all light impinging on the lens or mirror parallel to the axis will converge. One source of light rays moving parallel to the axis would be that from a source on the axis and infinitely far away. The focal point is located at a distance called the *focal length* from the lens or mirror. This length is representative of the optical properties of the lens or mirror and can be used in calculations to find the location of an image for a given object position.

Mirror - focal length depends only on geometry.

A mirror forms an image by the laws of reflection, while a lens forms an image by means of refraction. Thus, the focal length of a mirror depends only on geometrical considerations; the focal length is simply half the radius of curvature, *R*, of the reflecting surface:

$$f_{mirror} = \frac{R}{2}$$

Since light <u>reflects</u> from a mirror surface, the properties of the material do not effect the focal length. For the lens, however, the image-forming properties depend on refraction through the lens material; thus, the focal length will depend both on geometric factors and the index of refraction of the material of the lens. The focal length is given by the "lensmaker's formula":

Lens - focal length depends on geometry and lens material.

$$\frac{1}{f_{lens}} = (n - 1)(\frac{1}{R_1} - \frac{1}{R_2})$$

where *n* is the index of refraction of the lens material and R_1 and R_2 are the radii of curvature for the curved surfaces of the lens. The sign convention for the radii is that *R* is positive if the surface <u>encountered by the light</u> is convex, negative if it is concave.

For both lenses and mirrors, the image and object distances are related by the same equation:

$$\frac{1}{\text{object distance}} + \frac{1}{\text{image distance}} = \frac{1}{\text{focal length}}$$

There are a variety of sign conventions associated with image calculations using this equation:

Focal length: + if lens or mirror is converging, - if diverging

Object distance: + in most situations, - for a *virtual object*, possibly formed in combinations of lenses or mirrors where the image of the first lens or mirror is "beyond" the second.

Image distance: + if the image is in the "normal" place (on the same side as the source for a mirror, on the opposite side from the source for a lens), - if it is in the "abnormal" place.

Discussions; Chapter 27 - Image Formation

𝔐𝔶𝔰𝔱𝔢𝔯𝔦𝔢𝔰:

1.) Why is your image in a spoon upside-down if you look into the bowl, but right-side-up if you look into the back?

The bowl of the spoon acts as a concave mirror. The curvature of the bowl is quite sharp, so that the focal length is small. Thus, by looking into the bowl from a "normal" distance, you set yourself up as an object farther from the mirror than the focal point. As a consequence, the resulting image is upside down. If you now look into the back of a spoon, you are looking into a convex, diverging mirror and the image is right side up.

Let us go back to the bowl for a minute. For a concave mirror, you should be able to see a right side up image if you place the object closer than the focal point. Since the focal length of the spoon is so small, this is difficult to do. You may succeed, however, depending on the spoon, in placing your eye right up next to the bowl and seeing a right side up image of your eye.

2.) How does a face mask for diving allow you to see clearly under water? After all, it's not a lens.

Most of the focusing power of your eye comes from the refraction at the interface between the air and the cornea, not from the lens. The lens of the eye performs only some fine tuning to accommodate for objects at various distances. Your eye is designed to provide clear sight when the cornea is in contact with air. If the cornea is in contact with water, as it is when you open your eyes while swimming, then the refraction at the water-cornea interface is different than the refraction at the normal air-cornea interface. As a result, your vision is blurry under water. It is possible to design lenses that one could wear to account for this effect, but it is easier and more inexpensive to simply use a face mask. This provides a layer of air next to the cornea so that the normal air-cornea refraction is restored and your vision is clear.

3.) If you look closely at the light projected onto a screen by an overhead projector, you can see faint circles. Why?

One of the optical elements in many overhead projectors is a *Fresnel lens*. The design of this lens exploits the fact that the image forming ability of a lens is due to refraction at the surfaces of the lens only. There is no purpose to all of the glass (or whatever other material from which the lens is made) <u>between</u> the surfaces. Thus, for a Fresnel lens, we can imagine dividing a circular, plano-convex lens up into many annular rings, as shown in the side view diagram on the next page. The vertical lines represent our arbitrarily chosen dividing lines. At each division, a horizontal line is drawn. This forms a set of rectangular areas in the diagram, but we must remember that this is only a cross section of the actual lens. If we imagine rotating the above diagram about a vertical ax-

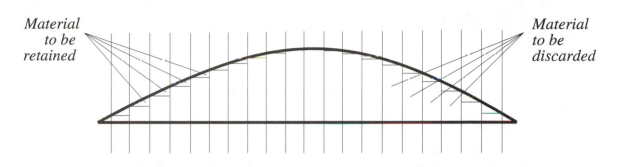

Material to be retained

Material to be discarded

is (this axis is in the plane of the paper in the diagram above), the rectangular areas become annular rings of rectangular cross section. These are the pieces of lens material that don't participate in the focusing process and that we don't need. Thus, we imagine cutting these pieces of material out and then dropping the remaining annular pieces with "triangular" cross section in the diagram down onto the horizontal surface at the bottom of the diagram. The result is a flat lens with circular ridges that reproduces the curvature of the lens surface in pieces without wasting all of the lens material between the surfaces.

Now, the circles that you see on the screen illuminated by the overhead projector are the faint shadows of the many boundaries between the ridges in the Fresnel lens.

4.) Why do right hand mirrors on automobiles have the inscription, "Objects in this mirror are closer than they appear to be"?

Some right hand rear view mirrors on automobiles are designed as <u>convex</u> mirrors, which will give a wider field of view of the area behind the car than a flat mirror. As a consequence, images of objects seen in the mirror will be smaller than what you expect to see in a mirror. After years of experience, your eye-brain system has developed a sense of how far away something is by how large it appears (as well as other clues). Thus, the smaller-than-normal looking cars in your rear view mirror may fool your eye-brain system into thinking that the cars are farther away than they really are.

5.) Why are there a large number of circularly-shaped areas of light on the ground under a shade tree? (These are sometimes called "Sunballs".)

This is actually a "pinhole camera" effect and the circularly-shaped areas of light are images of the Sun. The pinholes are the small openings between the leaves of the tree.

Observation of these images is especially effective (and safe!) during an eclipse, as the ground under the tree is covered with many images of the partially covered Sun.

6.) Why do slides need to be put into a slide projector *upside-down*?

The slide projector lens forms a real image of the slide on the screen. As for all real images formed by a single lens, the image will be inverted from the object. Thus, the object (the slide) must be inverted in the slide projector in order for the image to appear right side up to the audience.

7.) Contact lenses or lenses for glasses are often measured in *diopters*. What *is* a diopter?

The diopter is a method of measuring the strength, or *power* (this is a common usage of this word among those dealing with lenses, but is unrelated to energy transfer per unit time as discussed in Chapter 7), of a lens where we are interpreting the strength as the ability to focus an image close to the lens. The focal length is certainly a valid way of measuring the strength, also, but, as the focal length increases, the strength decreases. The strength of a lens (measured in diopters) is the inverse of its focal length (measured in meters). This gives us a unit which increases as the strength increases, which many people, such as optometrists, find more comforting.

Another advantage to the diopter approach is seen when lenses are combined. For two lenses in contact, as shown below, the effective focal length of the combination of lens-

es is found by adding the inverses of the focal lengths:

$$\frac{1}{f_{combination}} = \frac{1}{f_{lens\ 1}} + \frac{1}{f_{lens\ 2}}$$

Since the strength of a lens expressed in diopters *is* the inverse of the focal length, we see that the combined strength of two lenses in contact is simply the sum of the strengths in diopters of each of the two lenses.

8.) What's the difference between *binoculars* and *opera glasses*?

Both of these devices are a pair of refracting telescopes used for binocular viewing. A normal telescope, however, provides an inverted image, which is not a problem for astronomers but is a problem for those using devices which are designed for viewing scenes on the Earth. Thus, the image must be turned right-side-up again before viewing. In binoculars, this is performed with a pair of reflecting prisms. These prisms require that the light entering the binocular through the objective lenses be reflected sideways and vertically before being passed on to the eyepiece lenses, so that each objective lens is offset from the eyepiece lens. In opera glasses, the image is reinverted by using a diverging lens for the eyepiece and locating it closer to the objective lens than the would-be image of the objective lens. As a result, opera glasses are straight tubes - there is no offset such as that which betrays the existence of prisms in a pair of binoculars.

9.) Binoculars are described with a pair of numbers, such as 7 x 50. What do these numbers mean?

The first number is the angular magnification of the optical system in the binocular. The second number is the diameter of the objective lens in millimeters.

10.) Photographers know that the *depth of field* (the range of distances over which objects are in focus) increases as the camera lens aperture decreases. Why is this?

In order for a picture to be "in focus", <u>points</u> on the object being photographed must be represented as <u>circular images</u> on the film, with the radius of the circle small enough so that the observer does not describe the image as blurry. The ideal situation, of course, is that the radius be zero, but there is a radius, of what is called the "noticeable blur circle", below which the focus is said to be acceptable. The depth of field is defined as the range of object distances over which the images of points result in circles on the film whose radii are smaller than that of the noticeable blur circle. Of course, only one object distance will give a "perfect" focus, but a range of object distances will give "acceptable" focus.

We can understand the dependence of depth of field on aperture by imagining a noticeable blur circle on the film, as in the diagram below. From the edges of the noticeable blur circle, we extend two light rays back to the object axis of the lens. These are actually the reverse of the light rays leaving the object and hitting the film, but that's okay. The first "leaves" (we're going backward, remember) the lower edge of the noticeable blur circle, passes through the image axis (at point *A'*) and refracts through the lens so as to <u>just miss</u> the inner edge of the aperture. This is the solid ray shown in the diagram. This ray determines the farthest object distance in the depth of field, at point *A*.

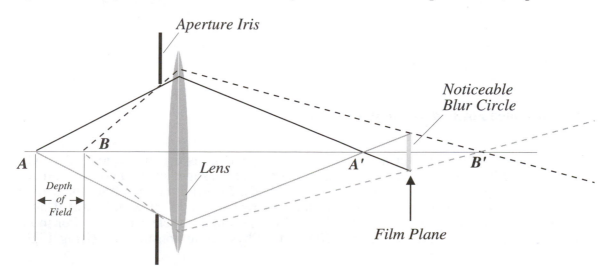

Now, do the same thing for a light ray which "leaves" the upper edge of the noticeable blur circle, <u>does not</u> cross the image axis, and refracts through the lens so as to just miss the inner edge of the aperture. This is the dotted line in the diagram (which we have also extended the other way, so that it crosses the axis at **B'**). This defines the near point in the depth of field, point **B**. Each of the above rays has been repeated in a lighter tint on the lower side of the axis, so that the object and image points are clear at their intersections with the axis. The range of distances from **A** to **B** is the depth of field.

Now, let us repeat this drawing with the aperture reduced, as shown below. We start drawing from the edge of the noticeable blur circle again. For the ray represented by the solid line, we must come in toward the lens at a shallower angle to just miss the iris. This is equivalent to moving point **A'**, the image position, <u>closer</u> to the lens. Thus, the corresponding object position **A** must move <u>farther</u> from the lens. If we consider the dotted line, this ray must also come in toward the lens at a shallower angle. But, since the ray has not crossed the axis between the lens and the blur circle, this is equivalent to moving the image point **B'** <u>farther</u> from the lens, as shown by the extension of the dotted line. Thus, the object position **B** is <u>closer</u> to the lens. As a result, the depth of field has lengthened in both directions, due to the closing down of the aperture.

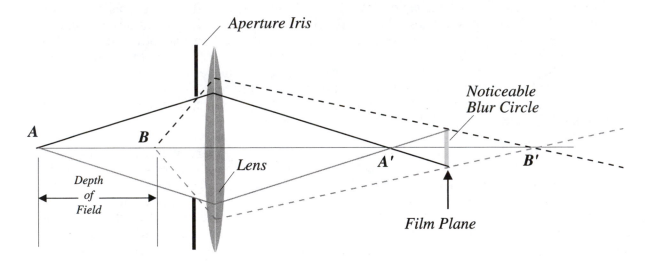

𝕸𝖆𝖌𝖎𝖈:

A Real Image Formed Without a Screen

The apparent light source at the top of the LED package is a <u>real image</u> of the light source at the bottom. The inside surface of the plastic package acts as a concave mirror (most of the light refracts out, but some is reflected), focusing the rays from the source and forming the real image. This is an unusual real image, as it is not formed on a screen, as is the usual situation, but is formed in the middle of clear plastic. For more information, see J. Jewett, "Get the LED Out - Physics Demonstrations Using Light

Emitting Diodes", *The Physics Teacher*, **29**, 530 (1991).

A virtual image is also possible with this type of LED. If the curved end of the plastic package is held very close (1 - 2 cm) to the eye, the refraction at the surface will result in a magnified image of the diode junction at the bottom of the package.

Flip Your Lid

This is a demonstration of the difference between being closer than the focal point to a concave mirror and farther than the focal point. Shaving mirrors are designed so that your face is generally closer than the focal point and your image is right side up, which is desirable for shaving. Moving your face farther away takes it beyond the focal point and the image inverts.

It is interesting to note, in this situation of the inverted image, that the image is <u>real</u> (unlike the <u>virtual</u> image formed when your face is close to the mirror) and formed in the region of space <u>between</u> the mirror and the eye. This causes some psychological difficulty in focusing on the image, since we are used to focusing our eyes on a region <u>behind</u> the mirror!

Pictures in Space

Adjusting the focus control so that the lens moves away from the projector body increases the distance between the object and the lens. As a result, the distance between the image and the lens decreases. Thus, the focused image is now somewhere in front of the screen. If you quickly wave a light-colored stick in the light of the projector, the eye's persistence of vision will allow a full image of the picture to be formed on the stick. As you move the waving stick toward and away from the projector, you should be able to find a point where the image on the stick is in clear focus.

Myth:

1.) The first series of postage stamps issued by the British Commonwealth State of St. Kitts-Nevis in 1903 portrayed a picture of Columbus looking for land through a telescope.

If we recall our elementary school history, Columbus' voyages took place at the end of the fifteenth century. But the telescope was not invented until the early part of the seventeenth century.

2.) All lenses have curved surfaces.

While this is indeed true for most lenses that we encounter, it is not always true. We could claim that a Fresnel lens (see discussion of Mystery #3) is flat, although this is cheating a bit, since there are ridges on the Fresnel lens that duplicate the collapsed curved surface. A better argument against curved lens is to present the concept of the *gradient index lens*. A lens focuses light by arranging for the light farther from the axis of the lens to be refracted through a larger angle. This is accomplished in a normal lens by curving the surface, so that light farther from the axis strikes the surface at an increasing angle with respect to the normal to the surface. But there is another way. Suppose we keep the lens surfaces flat and increase the index of refraction of the material as we move away from the axis? Then, we once again have achieved our goal of arranging for the light farther from the axis of the lens to be refracted through a larger angle. This is the principle behind the gradient index lens, which is used most often in photocopying machines.

While a gradient index lens is difficult to manufacture, a gradient index sugar solution can be formed from water and corn syrup. Experiments and demonstrations involving this system can be found in W. M. Strouse, "Bouncing Light Beams", *American Journal of Physics*, **40**, 913 (1972) and K. C. Mamola, W. F. Mueller and B. J. Regittko, "Light Rays in Gradient Index Media: A Laboratory Exercise", *American Journal of Physics*, **60**, 527 (1992).

3.) In *The Lord of the Flies*, Piggy's glasses are used to focus the sun's rays and start a fire. Later, his glasses are broken and he has trouble getting around because he is nearsighted.

On pages 40-41 of one particular edition of this novel (W. Golding, *Lord of the Flies*, Perigee Books, New York, 1954), Ralph and Jack snatch Piggy's glasses and use them to start a fire in a pile of rotten wood. Without his glasses, Piggy says, "Jus' blurs, that's all. Hardly see my hand -". Later on, (page 169), Piggy's glasses have been broken and he is described as "expressionless behind the luminous wall of his myopia". Well, now, wait a minute. If Piggy is myopic (nearsighted), then he needs diverging lenses to correct his vision. But diverging lenses could not be used back on pages 40-41 to converge the rays of the Sun on the wood to start the fire. This problem has been noted by a number of physicists, including a 12-year-old student in introductory physics at the University of British Columbia (S. Prytulak, "Interdisciplinary Application", *The Physics Teacher*, **29**, 135 (1991)). A trick to turn a diverging lens into a converging lens is described by J. K. Huhn, "You Can Be Myopic and Still Survive on a Desert Island!", *The Physics Teacher*, **29**, 577 (1991). This feat is performed by simply holding the lens horizontally, with the concave side upward, and filling the hollow of the lens with water!

4.) Professional telescopes have large diameters so that they can have large magnifications.

While it is true that professional telescopes have large diameters, the goal is not increased magnification. In fact, professional astronomical photographs are often taken at relatively low magnification. The goal with large diameters is to increase the amount of light that enters the telescope. Astronomical photographs require long periods of exposure time to capture the relatively weak light that hits the film. The more light-gathering power the telescope can provide, the shorter the exposure time can be. In addition, the larger diameter telescopes have fewer problems with diffraction (Chapter 29), so that closely spaced stars, for example, can be resolved more effectively with a larger telescope than with a small one.

5.) The focal point of a lens is where the image is focused.

This is strictly true only if the object is infinitely far away. As the object is brought toward the lens, the image moves away from the focal point.

6.) Only two rays come from an object and only from the top!

The drawing of ray diagrams is often performed by using an arrow to represent the object, placing the bottom of the arrow on the lens axis and drawing two rays from the top of the arrow. One of the rays travels parallel to the lens axis and passes through the focal point on the other side. The second ray passes through the near focal point and exits the lens on the other side parallel to the axis. The intersection of these two rays then locates the image. While these two rays are certainly sufficient to locate the image, it should be remembered that there are rays leaving the top of the object (as well as all other points on the object!) in all possible directions. We simply choose two rays out of the multitude of rays for which it is easy for us to determine the refracted directions.

7.) Galileo invented the telescope.

Although Galileo certainly made great discoveries with his telescope (Saturn's rings, the moons of Jupiter, craters on the Moon), it is generally believed that the telescope was invented in 1608 by a Dutch optician named Hans Lippershey.

8.) The Great Wall of China is the only manmade object on the Earth visible from outer space.

This myth may be based on the fact that the Great Wall is such a large structure. But, despite the great <u>length</u> of the wall, it will not be visible unless its <u>width</u> subtends an angle larger than the minimum necessary for human visual acuity. As an analog, no matter how long a human hair is, you will not see it if you cannot resolve the width of the hair from your viewing location. This is an important principle for those responsible for making humans fly in theatrical presentations, since a thin supporting wire may not be visible from the audience.

The maximum width of the Great Wall is 7 m and the minimum angle for human visual acuity is about 3×10^{-4} rad. Imagining the width of the Great Wall to be the arc length of a sector of a circle with a sector angle given by 3×10^{-4} rad, we calculate the radius of the circle to be 23.3 km. Thus, if we are farther from the Earth's surface than 23.3 km, we will not be able to see the Great Wall. This distance (about 14.4 miles) is still well within the atmosphere of the Earth and hardly qualifies as "outer space".

Chapter 28
Polarization

𝕸ysteries:

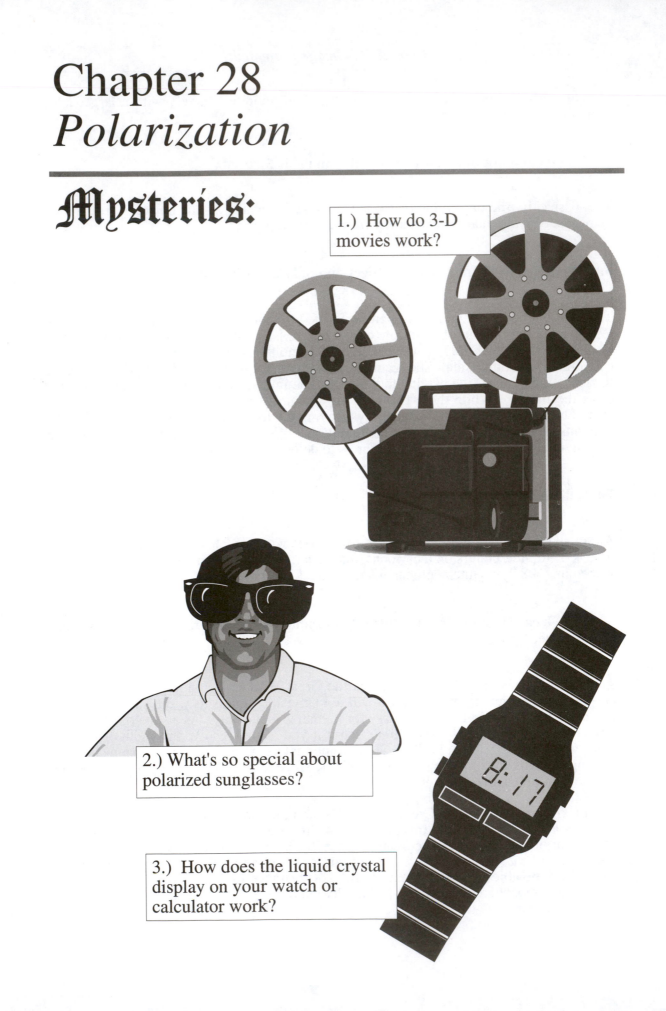

1.) How do 3-D movies work?

2.) What's so special about polarized sunglasses?

3.) How does the liquid crystal display on your watch or calculator work?

Mysteries:

4.) A polarizing filter is used in photography to improve the contrast between clouds and the sky. Why does this work?

5.) If you walk through a parking lot wearing polarized sunglasses, you may notice regular patterns of spots in the rear windows of automobiles. What causes these?

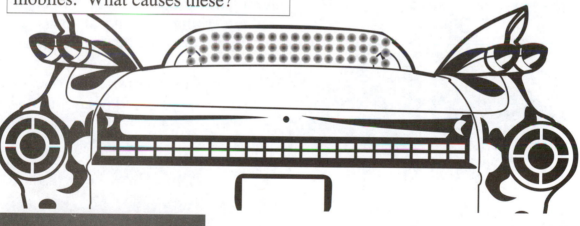

6.) Back in the days before cable television, an antiquated system was used. Television signals were broadcast by electromagnetic radiation and picked up by an antenna placed on your roof. Putting facetiousness aside, this process is still carried out today, especially in rural areas. If you go on a TV antenna observing tour in the US, you will find the elements of antennae in a <u>horizontal</u> position. But if you carry out this tour in England, you will find antennae with <u>vertical</u> elements. Why?

Magic:

The Disappearing Light Fixtures

Wearing polarized sunglasses, walk down a long hallway in a building equipped with light fixtures in the ceiling and look at the polished floor down the hallway. You will notice that some of the fixtures appear reflected in the floor but others are missing. As you walk down the hallway, the missing fixtures seem to stay a given distance in front of you. You can make the missing fixtures reappear by turning your head to the side.

So You Think You Understand Polarizers...

Try this demonstration on an unsuspecting volunteer. Beforehand, cut two pieces of polarizing filter material in squares so that the polarizing axis is *along the diagonal*. Now, show the volunteer that, if one filter is turned 90°, the correct result occurs. But now pick up the top polarizer and rotate it around an axis parallel to one of the sides. An unexpected result occurs!

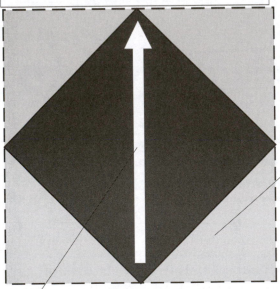

Cut off corners

Polarization axis

Light From the Sky

Looking through a polarizing filter, view the portion of the sky about 90° from the location of the sun. Turn the polarizer and notice that the sky appears to get darker and or lighter as the polarizer turns. Why?

𝔐𝔞𝔤𝔦𝔠:

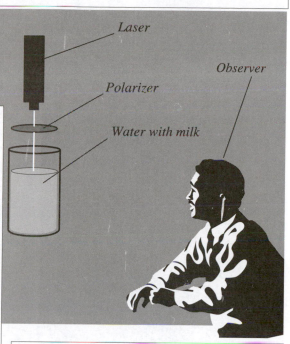

Polarization of a Light Beam

Support a laser (or a white light source, such as a slide projector, if a laser is not available) vertically, with the beam projected downward. In the beam, place a container of water with a small amount of scatterers added, such as milk. Now, place a polarizing filter between the light source and the water and rotate it while viewing the container from the side. The beam in the water will be bright for some orientations of the polarizer and dim for others. The effect may also be observed by viewing the beam through the polarizer from the side of the container and rotating the polarizer.

Polarization in Your Living Room

We can see some polarization effects with a television. If you observe the television through a polarizing filter and rotate the filter, you should see no effect - the light from the screen is not polarized <u>in the normal viewing position</u>. But now, move yourself so that you are <u>next</u> to the television, looking at the screen at grazing incidence. If you now view through the polarizer, you will see some effect of rotating the polarization.

Many televisions are located in entertainment centers with glass doors covering the stereo section. If this is the case with yours, adjust one of the doors so that you can see a reflection of your television screen in the door. Now look at the reflection of the screen through your polarizing filter. You should be able to find a position of the door at which the image on the screen will almost completely disappear with the polarizer in the right orientation.

Magic:

Polarization in the Streets

Use a polarizing filter to observe the light on a dry street below a street lamp shining at night. You should be standing about 5 - 10 m from the light. As you rotate the filter, what do you see?

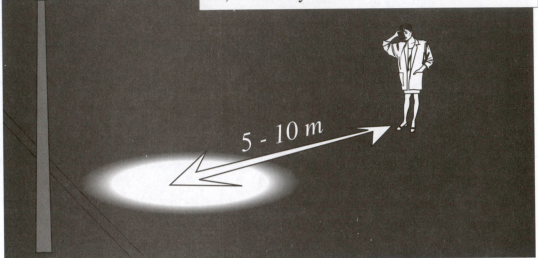

5 - 10 m

The Dark Water

This "demonstration" will require a still pond and must be performed at twilight. Facing in a direction 90° from the Sun (northward or southward), look at the reflection of the sky in the water. You will see one part of the reflection that is surprisingly darker than the sky above the water. (Note - in the illustration, the hiker will see the dark water phenomenon. From our vantage point looking at the picture, we will not see it. The dark water effect is drawn in the illustration just to demonstrate the appearance.)

Myth:

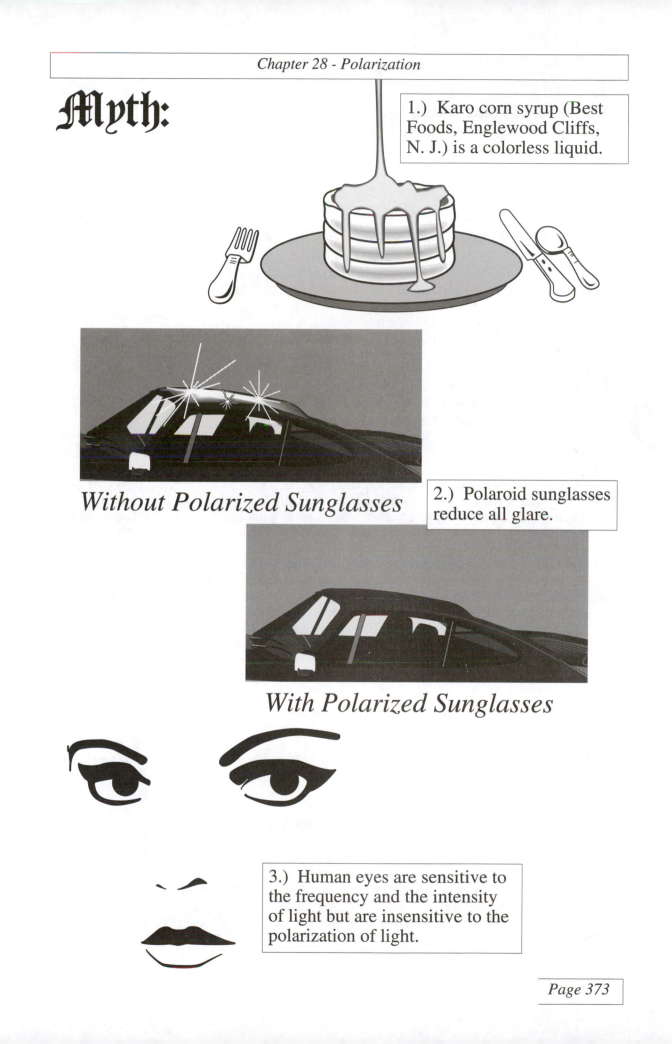

1.) Karo corn syrup (Best Foods, Englewood Cliffs, N. J.) is a colorless liquid.

Without Polarized Sunglasses

2.) Polaroid sunglasses reduce all glare.

With Polarized Sunglasses

3.) Human eyes are sensitive to the frequency and the intensity of light but are insensitive to the polarization of light.

Concepts of Polarization

The concept of polarization arises because electromagnetic radiation is a <u>transverse</u> wave. There are an infinite number of choices of a direction perpendicular to the propagation to represent the direction of the electric field vector of the wave. But since these choices all lie in a plane perpendicular to the propagation direction, we could represent any arbitrary direction by giving two components relative to some choice of two perpendicular axes. It is common and often convenient to choose vertical and horizontal axes as our reference frame.

Polarization is determined by the direction of the electric field vector.

Unpolarized light shows no preferred direction for the electric field oscillations. We can polarize this light in a variety of ways, some of which are exemplified in this Chapter - scattering, reflection and filtration. For the last method, a polarizing filter is a material with long molecules that have been aligned in the material, often by stretching. When light with polarization parallel to the molecules strikes the molecules, dipole oscillations can occur within the molecules. Thus, the incoming radiation is transformed in two ways - the radiation is absorbed and re-emitted in all directions radially outward from the molecules, and some of the energy carried in by the light is converted into internal energy within the material. Thus, the amount of light in the original direction is reduced significantly. Light polarized perpendicularly to the molecules cannot induce these oscillations within the molecules, so the light is passed through.

Polarization can be achieved by scattering, reflection and filtering, along with other methods.

When light reflects from a surface, it is often polarized. At a certain incident angle (Brewster's angle), the reflected light is <u>totally</u> polarized in the direction parallel to the surface of the material. Brewster's angle is that angle of incidence at which the reflected light makes an angle of 90° with respect to the refracted light. Using this fact, we can find an expression for Brewster's angle for light in a medium of index of refraction n_1 encountering a boundary with a medium of index n_2:

$$n_1 \sin \theta_1 = n_2 \sin \theta_2 \quad \Rightarrow \quad n_1 \sin \theta_B = n_2 \sin (90° - \theta_B) = n_2 \cos \theta_B$$

$$\Rightarrow \quad \theta_B = \tan^{-1} \frac{n_2}{n_1}$$

While Brewster received the honor of having his name associated with this angle for his quantitative work, the original semi-quantitative investigation of polarization upon reflection was performed by Malus.

While an arbitrary linear polarization can be resolved into components parallel to two reference axes, it is sometimes useful to resolve it into two <u>circular polarizations</u>, one clockwise and one counterclockwise. Circularly polarized light is that in which the tip of the electric field vector traces out a helix as the wave propagates. If the cross section of the helix is not circular, but rather elliptical in shape, then the light is called <u>elliptically polarized</u>. This is the most general type of polarization for light. Linear polarization is elliptical polarization with a minor axis of zero. Circular polarization represents an elliptical polarization with equal major and minor axes.

Discussions; Chapter 28 - Polarization

Mysteries:

1.) How do 3-D movies work?

In general, 3-D movies work by photographing the scene from two vantage points, slightly separated in the horizontal direction, and providing some means for presenting these separate views to the separate eyes of the members of the audience. The result is the illusion of enhanced "depth" in the projected scene. Early 3-D movies maintained the separation of the two images on film by projecting them in two different colors, such as red and green. Members of the audience wore glasses with a red filter for one eye and a green filter for the other, which separated the two views for the observer.

This method, however, was restricted to black and white movies - it could not be used to show movies in color, since color was used to separate the images. Modern 3-D movies can be shown in color because the separation of the images is maintained by using perpendicular polarizations of light. One scene is projected with light polarized vertically and the other horizontally. Members of the audience wear corresponding polarizing filters on each eye, which separates the images for the 3-D effect.

2.) What's so special about polarized sunglasses?

Polarized sunglasses exploit the fact that light reflected from many surfaces tends to be at least partially polarized, in the direction parallel to the surface. Thus, light reflected from horizontal surfaces has a horizontal polarization (not necessarily complete but some of the vertical polarization from the incident beam has been removed). By wearing polarizing filters with a vertical axis, this reflected light is blocked.

Windshields in many cars are at an angle such that light from the dashboard and reflected off the windshield to the driver is highly polarized. If a piece of paper is placed on the dashboard, it may be almost invisible to the driver wearing polarized sunglasses. By removing the glasses or turning the head, the paper will become visible again.

3.) How does the liquid crystal display on your watch or calculator work?

This type of display depends on the phenomenon of polarization. The diagram on the next page shows the order of items in a sandwich that makes up the liquid crystal display.

In the diagram, imagine that unpolarized light from the environment enters from the left. It is polarized vertically by the first filter. It passes through the transparent electrode and enters the liquid crystal. When the liquid crystal is not subject to an electric field, it has the effect of rotating the plane of polarization of the light through 90°. Thus, as the light passes through the second transparent electrode and strikes the second polarizer, it is in the "correct" polarization state for the second polarizer, so that it passes through. It reflects off the mirror, passes back through the polarizer, back through the liquid crystal, being rotated again, and then back through the initial polarizing filter. Thus, the appearance of the display is that of reflected light, although there is a grayish tinge due to some absorption through the various layers of materials.

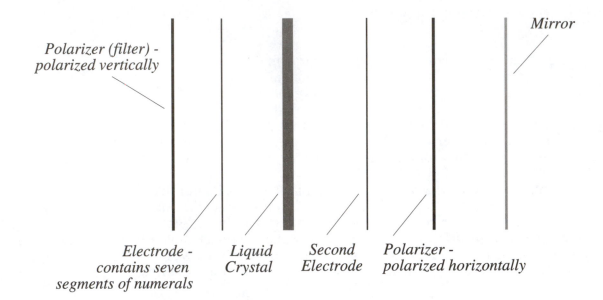

Now, let us turn on a voltage for one or more of the segments of a numeral on the electrodes. The liquid crystal in the region between the powered segments on the electrodes is now in an electric field. As a result, there is an alignment of the liquid crystal molecules such that the plane of polarization of light is no longer rotated. Now, when light passes through the first polarizer and is <u>not</u> rotated by the liquid crystal, it does not pass through the second polarizer and is not reflected out again. Thus, the segment is dark. As a result, we can turn on and off various segments to form numerals.

It is an instructive experience to view this type of display while wearing polarized sunglasses. The polarized nature of the light leaving the display is evident as you turn your head from side to side.

4.) A polarizing filter is used in photography to improve the contrast between clouds and the sky. Why does this work?

In Mysteries #3 and #6 of Chapter 25, we discussed the scattering of light by particles. If the particles are smaller than the wavelength of light, the scattering depends strongly on wavelength, hence, the blue sky. If the particles are larger than the wavelength, the scattering is uniform for all wavelengths, hence, white clouds. But there is another difference that depends on the particle size beside the dependence on wavelength. Light scattered from small particles is highly polarized, while light scattered from large particles is polarized to a much lesser extent. Thus, if we photograph a scene with clouds in the sky with the polarizing filter perpendicular to the polarization of the light from the sky, <u>the filter will darken the cloud light by about half, but darken the sky light even more</u>. The result is an increased contrast between the sky and the clouds.

5.) If you walk through a parking lot wearing polarized sunglasses, you may notice regular patterns of spots in the rear windows of automobiles. What causes these?

This question is closely related to Myth #1. If we have crossed polarizing filters, we expect no light to pass through. This is indeed the case. But, if we insert a third polarizing filter between the first two, with its axis at some arbitrary angle to those of the first two, then we see light passing. If the polarized light from the first filter has a component parallel to the axis of the second filter, that component is transmitted. The resulting polarized light similarly has a component parallel to the axis of the third filter and some light passes through it also. Sometimes the intermediate filter is described as "rotating the plane of polarization" of the light.

Now suppose we try the same trick with something other than a polarizing filter. If we stretch a piece of plastic food wrap, for example, so that the molecules are lined up, this will act as a polarizing filter which will rotate the plane of polarization. If the plastic is placed between the crossed filters, we see light passing through. We will find that different colors have their polarization planes rotated through different angles, so the display can become quite colorful. We attribute the rotation of the plane of polarization to the *birefringence* of the material. A birefringent material has two indices of refraction, one for each of two perpendicular directions of polarization. As unpolarized light enters a birefringent material, it splits into two polarization components, which pass through the material at different speeds. Upon recombining at the exit from the material, the two polarizations now have a different phase relationship, so that the exiting light is generally elliptically polarized. If the incoming light was initially linearly polarized, the recombination of the two components in a different phase relationship at the exit has the effect of rotating the direction of polarization.

Now, let us finally get to the car windows. The origin of the birefringence in these windows is in the manufacturing process. To ensure that the glass will shatter in small chunks rather than large shards upon impact, mechanical stresses are introduced by cooling the glass at regular locations. These stresses have the same effect on the glass that the stretching had on the plastic - it becomes birefringent in those areas that are stressed. Now, the light entering the birefringent areas is polarized from two sources. First, the sunlight from the sky is already partially polarized by scattering (Magic - Light from the Sky); Second, the light enters the glass and reflects off the inner surface, resulting in further polarization (Magic - The Disappearing Light Fixtures). This light then passes back through the glass, either through a stressed or a non-stressed region. The light passing through a stressed region has the plane of polarization rotated. Finally, when the light passes through your sunglasses, the different polarizations of the stressed and non-stressed regions become apparent as the array of regularly spaced spots.

6.) Back in the days before cable television, an antiquated system was used. Television signals were broadcast by electromagnetic radiation and picked up by antennas placed on your roof. Putting facetiousness aside, this process is still carried out today, especially in rural areas. If you go on a TV antenna observing tour in the US, you will find the elements of an antenna in a <u>horizontal</u> position. But if you carry out this tour in England, you will find antennae with <u>vertical</u> elements. Why?

This is a demonstration of the polarization of electromagnetic waves in the television frequency range. In the United States, television signals are transmitted with the electric field vector in a horizontal orientation. The antenna elements must be horizontal to absorb the radiation. In England, the television transmission is polarized vertically and the antenna elements must comply. The choices are arbitrary, and both systems work equally well, <u>except</u> for the following. For the horizontally-oriented elements, the long dimension of the elements must be oriented perpendicularly to the incoming waves for maximum signal reception. This results in the necessity for motorized antenna rotators, to allow for adjustment of the antenna orientation. Rotators are unnecessary in England, since vertically polarized waves from <u>all</u> directions will be absorbed by the vertically-oriented elements.

𝕸𝖆𝖌𝖎𝖈:

The Disappearing Light Fixtures

One way of polarizing light is by reflection. In particular, when the incident angle on a surface is Brewster's angle, as discussed in the Concepts section, the reflected light is polarized completely in the direction parallel to the surface (perpendicular to the plane of incidence).

When you look at reflections of light fixtures in a long hallway, the light reflected to your eye from the fixtures strikes the polished floor at a variety of angles, depending on the location of the fixture. Those near your position will strike the floor at a small incidence angle (with respect to the normal), while those far down the hall will strike at a large angle. For some fixtures, the light arriving at your eye will have struck the floor at or near the Brewster angle, so that the reflected light will be highly polarized parallel to the floor. Your sunglasses have a polarization axis perpendicular to the floor, so this light is blocked and these particular fixtures can only be seen barely or not at all.

So You Think You Understand Polarizers...

Consider the diagram below, showing two of the specially prepared polarizers. Since the optical axis is on the diagonal of the square polarizer, as indicated by the dark ar-

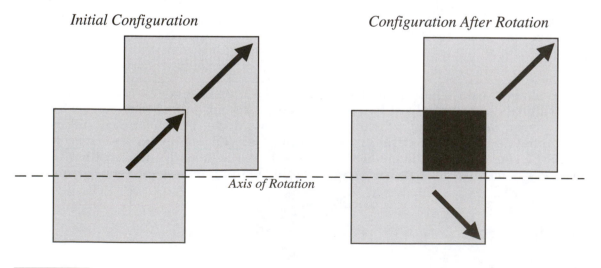

Initial Configuration   *Configuration After Rotation*

Axis of Rotation

rows, turning a polarizer on an axis parallel to one side is equivalent to a rotation of the polarization axis of 90°, as shown in the diagram. On the left, the axes are parallel and light passes through both filters. The top filter (lower left) is now rotated 180° around the axis indicated by the dashed line and parallel to the horizontal sides. The result of this rotation is shown in the right of the diagram. The two polarization axes are now perpendicular to each other and the common area of overlap of the filters is dark.

Light From the Sky

The results of this activity suggest that the light from the sky is polarized. This polarization is due to scattering from air molecules. When light is incident on a molecule, the oscillating electric field vector causes an oscillating dipole moment of the molecule. This oscillating dipole moment acts as a tiny antenna, emitting light in directions perpendicular to the oscillation. The diagram below shows unpolarized light from the Sun scattering from an air molecule. There are two possible dipole oscillation directions indicated (There are many more, but they can be resolved into components in these two directions), horizontal and vertical, relative to the ground. The vertical oscillations will propagate electromagnetic radiation (heavy arrows) in directions that will not interact with your eye. The horizontal oscillations will propagate electromagnetic radiation (heavy arrows) in directions that <u>will</u> enter your eye. As a result, the light is polarized.

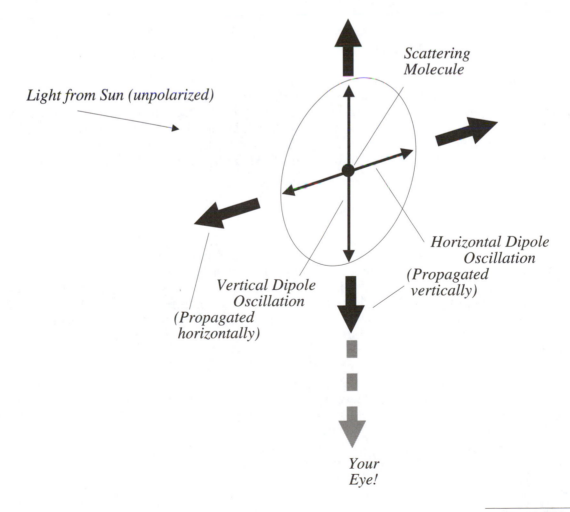

This sounds great, but now let's introduce some realism. The diagram shows the situation in which the light rays from the Sun to the molecule are perpendicular to the scattered rays from the molecule to your eye. For molecules in the location in the sky where this is the case, we see no component of the vertical dipole oscillation, since the oscillation is directed straight toward our eye. But, if we are looking at molecules located elsewhere, we <u>will</u> see a component of the vertical oscillation and, therefore, will receive some vertically polarized light. This will combine with the horizontally polarized light to give us only <u>partially</u> polarized light. This is evident with the polarizer if we notice that the region of the sky along a direction perpendicular to that to the Sun gives us the largest variation in light intensity as we rotate the polarizer.

Now, a further step toward reality - even if we look along a direction perpendicular to that to the Sun, the sky will not go completely <u>dark</u> for any orientation of the polarizer. This is because the polarization is not complete. Complete polarization would only occur for ideal spherical scatterers. But air is primarily composed of oxygen and nitrogen molecules, which are not spherically symmetric. What's more, some of the light entering our eyes will have made two or more scattering encounters with air molecules. As you can imagine, this throws the discussion about the perpendicular direction out the window, since many combinations of two scattering angles are possible. We add to this scattering by impurities in the air and light reflected from the ground (from <u>many</u> angles!) and then scattered into our eyes by the air and we can see that we have no hope for 100% polarization of skylight. The best value to be expected is about 75% - 80% in the direction perpendicular to that to the Sun.

Polarization of a Light Beam

This is a classroom demonstration of polarization by scattering, as occurs in the sky. When the polarizing axis of the filter is "aimed" at the observer, the dipole oscillations of the scatterers are also aimed at the observer (parallel to the observer's line of sight) and no light is emitted in that direction. When the filter is turned 90° from this orientation, the dipole oscillations are perpendicular to the observer and the light is observed.

Polarization in Your Living Room

The light which you see from the oblique view of the television screen has suffered refraction by passing through the glass screen. It was once part of unpolarized light from the actual picture tube, some of which reflected off the glass screen and some of which refracted to your eyes. The reflected light is partially polarized parallel to the glass surface, while the transmitted light will be partially polarized perpendicular to the glass surface. This is a small effect, but may be visible with a polarizing filter on your television screen.

If you are able to reflect the television screen off a glass door, the polarization by reflection will be easily seen. You may be able to find an angle where the light from some part of the screen strikes the door at Brewster's angle and that part of the screen will go dark upon proper orientation of the polarizer.

Polarization in the Streets

The light reflected from the street has a high degree of horizontal polarization due to reflection from the road surface. The light from the source at the top of the lamppost is coming almost straight down toward the roadway. Your view of the roadway is very close to 90° from this direction. Thus, the dipole oscillations of the molecules of the road material are either perpendicular to your viewpoint or directed along your line of sight. Just as is the case with polarization by scattering in the sky (Light From the Sky), the oscillations along your line of sight send no radiation in your direction, so that you see the light polarized horizontally. The polarization is much stronger for dark surfaces than for light-colored surfaces, a behavior called the *Umov Effect*. With light-colored surfaces, the light has a higher probability of making multiple reflections from different points on the rough surface (the surface is light-colored because there is a large probability of reflection rather than absorption - thus, there is an increased probability of multiple reflections) before propagating toward your eyes. These extra reflections tend to randomize the polarization.

The Dark Water

Due to the scattering of light by air molecules, as in Light From the Sky, light from the northern or southern sky at twilight is highly polarized in the vertical direction. If this light is allowed to reflect off water, and a particular light ray strikes the water at Brewster's angle, then this ray of light is not reflected. Thus, the reflection of this part of the sky will be extremely dark compared to the corresponding spot in the sky above it. At points farther and nearer the observer than the dark area, the incident angle will be slightly different than Brewster's angle and the light will reappear faintly. To the left or right of the dark area, the sky will not be sending perfectly vertically polarized light and the light will reappear also.

𝔐𝔶𝔱𝔥:

1.) Karo corn syrup (Best Foods, Englewood Cliffs, N. J.) is a colorless liquid.

This question is closely related to Mystery #5. Karo syrup is *optically active*, in that it rotates the plane of polarization of light.

So let us imagine setting up a Magic demonstration to see this effect. We place a bottle of syrup between two polarizers, place a light source on one side and look in the other side. The first polarizer results in linearly polarized light entering the syrup. Now, the syrup will rotate the plane of polarization. The angle of rotation per unit length of the syrup depends on the frequency of the light - different colors suffer different angles of rotation. Thus, for a given length of syrup, such as the diameter of the bottle, different colors will exit the syrup at different polarization angles. As the second, analyzing polarizer is rotated, then, it will let pass light of varying colors.

Optical activity is related to the spiral shape of the sugar molecules in the syrup. More details on the phenomenon can be found in textbooks on optics, such as F. A. Jenkins and H. E. White, *Fundamentals of Optics*, 4th ed., McGraw-Hill Book Company, New York, 1976.

2.) Polaroid sunglasses reduce all glare.

Polaroid sunglasses are indeed useful for cutting down glare. But the polarization axis of sunglasses is <u>vertical</u>. Thus, the glasses can only block <u>horizontal</u> glare, which comes from reflections from horizontal surfaces. The glasses are useless for reflections from vertical surfaces such as store windows.

3.) Human eyes are sensitive to the frequency and the intensity of light but are insensitive to the polarization of light.

Many creatures in the world, such as ants, bees, fruit flies and some species of fish are sensitive to polarized light and use it for orientation and navigation. In general, however, human beings are not able to distinguish polarized light from unpolarized light. Some individuals, however, are able to see a small polarization effect called *Haidinger's brush*. It was discovered by Haidinger in 1844. If one looks into the blue sky, in a direction perpendicular to that to the Sun, it is possible to see a small yellowish hourglass-shaped figure, with the long dimension of the hourglass perpendicular to the direction of polarization. It is also possible to see Haidinger's brush by looking through a polarizing filter at a bright light. Current theory relates Haidinger's brush to the particular way that blue light is preferentially (by polarization) absorbed at the fovea. The complementary color of yellow for the brush is related to this absorption.

For some tips on seeing Haidinger's brush in the classroom, see B. Reid, "Haidinger's Brush", *The Physics Teacher*, **28**, 598 (1990) and R. D. Edge, "The Optics of the Human Eye", *The Physics Teacher*, **27**, 392 (1989).

Chapter 29
Interference and Diffraction

𝕸ysteries:

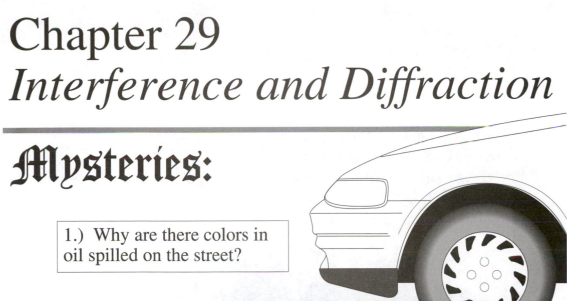

1.) Why are there colors in oil spilled on the street?

2.) Why are there colors in soap bubbles?

SOAP

BUBBLES

3.) Why are there colors in a compact disk?

Mysteries:

4.) When you are watching a marching band and the players turn and march away from you, why do the trumpets become quiet, but the clarinets stay loud?

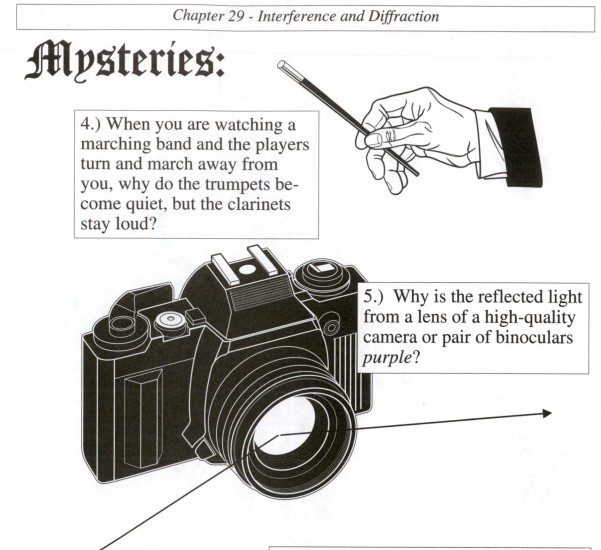

5.) Why is the reflected light from a lens of a high-quality camera or pair of binoculars *purple*?

6.) If you are traveling in an airplane over a cloud layer, you might see a circular, rainbow-like display of colors surrounding the shadow of the airplane. You may also notice that the exact center of the circle of colors corresponds to the point on the shadow of the airplane where you are seated. If you walk down the aisle, the shadow of the airplane on the clouds will move with respect to the colored rings. This phenomenon is called the *glory*. What is responsible for this phenomenon?

Mysteries:

7.) Why does a megaphone make sound louder?

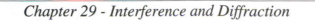

8.) If you look carefully at the primary rainbow, you may see some additional pastel-colored bows below it. These are called *supernumerary bows*. What causes these?

𝕸𝖞𝖘𝖙𝖊𝖗𝖎𝖊𝖘:

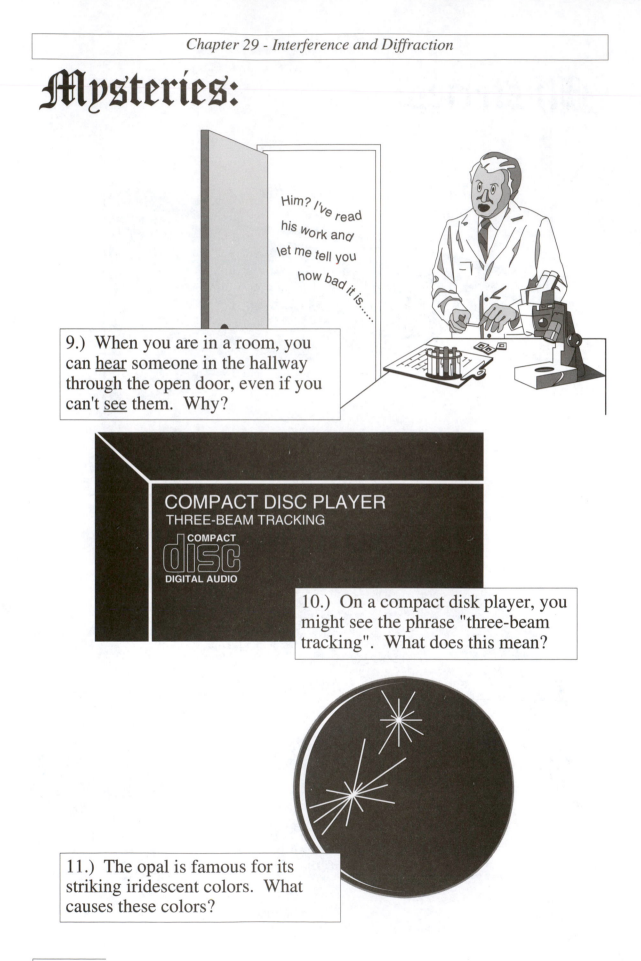

Him? I've read his work and let me tell you how bad it is......

9.) When you are in a room, you can <u>hear</u> someone in the hallway through the open door, even if you can't <u>see</u> them. Why?

COMPACT DISC PLAYER
THREE-BEAM TRACKING

COMPACT
disc
DIGITAL AUDIO

10.) On a compact disk player, you might see the phrase "three-beam tracking". What does this mean?

11.) The opal is famous for its striking iridescent colors. What causes these colors?

Magic:

Speckles on the Wall

Use a lens to spread the beam of a laser and direct the spread beam on a wall. You will see a pattern of speckles. If you are near-sighted, take off your glasses. The speckles are still in focus! Now, move your head to one side. Which way do the speckles move?

Lens to spread beam

Laser

Myth:

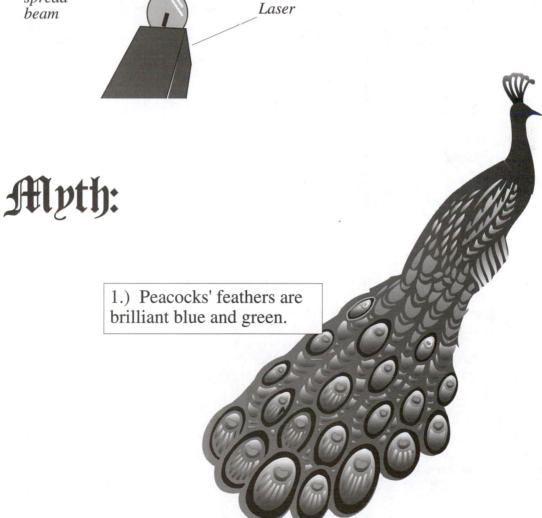

1.) Peacocks' feathers are brilliant blue and green.

Concepts of Interference and Diffraction

The phenomena of interference and diffraction cannot be explained with geometrical optics and depend on the wave nature of light for their explanation. We invoke the *principle of superposition*, which claims that the net effect of two waves coexisting in a region of space is simply the sum of the effects of each of the two waves.

Two-slit interference was investigated by Thomas Young in the very early years of the nineteenth century. It was these investigations that firmly established the wave nature of light. The light from two slits illuminated with a single source will diffract through the slits, spreading in all directions toward a screen on the other side of the slits, as shown below.

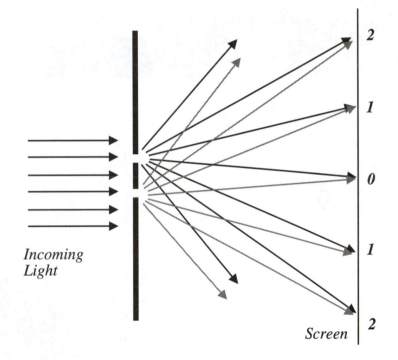

Incoming Light

Screen

We have chosen the particular light rays, as shown in the diagram, which arrive at the screen at points marked *0*, *1* and *2* (in either direction from the center point). Let us imagine that these numbers refer to the path length difference between the two rays arriving at the point from the two slits, measured in number of wavelengths. This path length difference exists because one of the rays has to travel farther from the slit to the screen than the other (except for the point marked *0*). For these particular points, then, since the path length difference is an integral number of wavelengths, there will be <u>constructive interference</u> and the area on the screen will be bright. Approximately halfway between each of these pairs of points, light will arrive from the two slits with a path length difference of half a wavelength (or one and a half, two and a half, etc.), so that we will have <u>destructive interference</u> and these

> *Interference results in areas of constructive and destructive superposition.*

regions will be dark. As a result, we will see a series of light and dark *fringes*. Similar results occur whenever two coherent rays of light interfere. These rays could come from two slits, as above, or from a variety of other physical situations, as explored in the *M's* of this chapter.

We have described interference for <u>light</u>, but it is a general wave phenomenon. In fact, a nice photograph of interfering water waves from the two flapping wings of a bee stuck in the water is available in D. R. Lapp, "Wave Interference Pattern from the Wings of a Bee", *The Physics Teacher*, **28**, 335 (1990).

We mentioned above that the light diffracts through the slits in the double slit experiment. Diffraction is a result of light encountering an aperture, an edge of an object, or a small object. It can be understood from *Huygen's Principle* - each point on the wavefront of a wave acts as a source of spherical wavelets.

Diffraction: a demonstration of Huygen's principle.

When a light wave passes through a small slit, for example, the Huygen's sources at either side of the slit are blocked and the spherical wave sources at the slit result in waves propagating in all directions. The effect of diffraction depends on the relative size of the wavelength of the light and the dimensions of the diffracting object. If the wavelength is large compared to the slit (or a diffracting object), then the diffraction is extensive. On the other hand, there is very little effect when the wavelength is small compared to the slit or the object. Diffraction <u>patterns</u> of light and dark fringes can be understood by applying the principle of superposition to a number of Huygens sources spread across the slit.

Discussions; Chapter 29 - Interference and Diffraction

Mysteries:

1.) Why are there colors in oil spilled on the street?

This is an everyday example of <u>thin film interference</u>. The two light rays which are interfering are a reflection from the upper surface and a reflection from the lower surface. Different regions of the film will be viewed at different angles and can also be of varying thickness, so that the frequency of light undergoing constructive interference will vary over the film. As a result, you see different colors in different parts of the film. The appearance of colors in oil films is a common experience due to the generally dark background of the road. Only a small percentage of the incident light is reflected at either of the two surfaces in question, so that it would be difficult to see the display against a bright background.

2.) Why are there colors in soap bubbles?

This is also an example of thin film interference, between the same type of interfering rays of light as in Mystery #1. A stationary planar soap bubble supported in a vertical frame has a much simpler variation in thickness than the oil in the street. Thus, the different colors will appear as simple horizontal bands exhibiting the spectrum in a vertical direction. The top of the bubble may be dark. This is where the bubble is too thin to provide constructive interference and we see only destructive interference between the reflection from the "front" of the bubble (which suffers a 180° phase reversal because the light strikes a medium of higher index of refraction) and the "back" of the bubble. A classic source for information on colors in soap bubbles is C. V. Boys, *Soap Bubbles - Their Colours and the Forces which Mold Them*, Dover Publications, Inc., New York, 1959. A more recent source is G. Ramme, "Colors on Soap Films - An Interference Phenomenon", *The Physics Teacher*, **28**, 479 (1990).

3.) Why are there colors in a compact disk?

On a compact disk, the audio information is encoded in binary form in a series of pits and flat areas. The pits are designed to scatter light so that the resultant low level of light detected by the detector can be differentiated from the high level reflected from the flat areas. This scattered light is what provides the colors from the disk. Light is scattered from many closely spaced lines of pits. Thus, the surface of the CD acts as a <u>reflection diffraction grating</u>. For various angles of viewing, constructive interference will occur for different frequencies of light. Thus, reflection of white light off the surface gives a spectrum of colors across the surface of the CD. More information can be found in C. Noldeke, "Compact Disc Diffraction", *The Physics Teacher*, **28**, 484 (1990). Experimental activities are discussed in H. Kruglak, "The Compact Disc as a Diffraction Grating", *Physics Education*, **25**, 255 (1990) and J. E. Kettler, "The Compact Disk as a Diffraction Grating", *American Journal of Physics*, **59**, 367 (1991).

4.) When you are watching a marching band and the players turn and march away from you, why do the trumpets become quiet, but the clarinets stay loud?

This effect can be heard when comparing a brass instrument (e.g., tuba, trombone) with a woodwind (e.g., saxophone, flute). The effect is a result of wave <u>diffraction</u> as the sound waves exit the instrument through an orifice. In a woodwind instrument, the sound exits the instrument through the tone holes along the column of the instrument. These orifices are small compared to the wavelength of the sound. As a result, there is significant diffraction of the sound upon leaving the holes. Thus, the sound is radiated effectively in all directions. In a brass instrument, the sound exits the instrument through the large bell at the end. This orifice is large compared to the tone holes, so that the effect of diffraction is less. The sound from a brass instrument is more directional, with most of the energy being radiated in the forward direction. Thus, when the trumpet players turn around and march away from you, very little sound is radiated backward to be heard by you. From the clarinets, however, the sound is radiated more uniformly and you can still hear them. This effect is particularly troublesome for French horn players in an orchestra. In this brass instrument, the sound radiates from the large bell, which is directed toward the back of the stage, if the player is facing the audience, and often impinges on a nicely absorbing curtain backdrop!

5.) Why is the reflected light from a lens of a high-quality camera or pair of binoculars *purple*?

A desirable feature of a camera is to capture as much light as possible. Light arriving at the camera lens encounters a change in the medium in going from the air into the glass of the lens. Necessarily, some of the light is reflected, working against the desirable feature just mentioned. In order to reduce this reflection, the lens is <u>coated</u> with a thin film (often formed from magnesium fluoride or calcium fluoride). This film is designed to have just the right thickness so that light reflected from the front and back surfaces will undergo destructive interference. The light reflected from the front of the lens experiences a 180° phase inversion, since it reflects from the film, which has a higher index of refraction than the air. The light reflecting from the back of the film also experiences a 180° phase inversion, since it reflects from the glass of the lens, which has a higher index of refraction than the film. Thus, to achieve destructive interference, the film must be <u>one fourth of a wavelength</u> in thickness (for normal incidence), so that the total path length (into the film and back out) is half a wavelength. This condition can only be achieved for one particular color, since the wavelength of the light in the film will vary as the color of the light from the outside changes. The condition is chosen so that light in the middle of the spectrum (yellow, green) experiences the destructive interference and enjoys very little reflection from the coated lens. On either side of the middle range of the spectrum, toward red and toward blue, the destructive interference condition is not satisfied, so that the coating tends to reflect primarily red and blue light. The combination of red and blue is interpreted by our eyes and brains as <u>purple</u>.

More details on optical coatings of this sort are discussed in P. Baumeister and G. Pincus, "Optical Interference Coatings", *Scientific American*, **223(6)**, 59 (1970).

6.) If you are traveling in an airplane over a cloud layer, you might see a circular, rainbow-like display of colors surrounding the shadow of the airplane. You may also notice that the exact center of the circle of colors corresponds to the point on the shadow of the airplane where you are seated.   If you walk down the aisle, the shadow of the airplane on the clouds will move with respect to the colored rings. This phenomenon is called the *glory*. What is responsible for this phenomenon?

The glory is a fascinating optical phenomenon. It was first described in writing by a French expedition in 1735 climbing the mountains of Peru. It was sighted by balloon-ists in the nineteenth century. Wilson was actually studying the glory in 1895 when he built his cloud chamber. He was distracted from this study when he noticed that his chamber betrayed the paths of charged particles and the rest, as they say, is history - the cloud chamber as a tool for particle research was born.

The explanation of the glory combines several concepts from optics. Despite the pessi-mism of Greenler (R. Greenler, *Rainbows, Halos and Glories*, Cambridge University Press, Cambridge, 1980), who wonders "if there is no simple model containing the physical essence of the explanation of the glory", it can be understood, but you will need to follow this explanation carefully.

Since the glory surrounds the shadow of the airplane, it arises due to light which is back-scattered by the clouds underneath the airplane near the anti-solar point - that is, it is returned by the water droplets in the clouds back along the line to the Sun to the ob-server's eyes. We will discuss the formation of glory from back-scattered light with two questions: 1.) Why does the light create colored circles? and 2.) Why is the light back-scattered in the first place?

The answer to the first question is relatively easy - it is a diffraction effect of the light coming from the edge of a circular water droplet.   Imagine that you are looking at a water droplet in the laboratory from the point directly opposite a light source of a single color, as shown in the diagram on the next page (This is not the location of the light source for the glory, but placing the source here makes the explanation of question 1 easier.).   Your eye is at the bottom of the diagram, at point *1*, looking toward the top. We will consider the light from the source that just passes the edge of the droplet and diffracts toward your eye - let's ignore any other light. As the light passes around the edge of the droplet and converges on your eye, light from all points on the edge will have traveled the same distance from the drop to your eye. Thus, the light will be in phase. Now, imagine that you start moving your head and eye toward the right. The light from the left edge of the drop will have to travel farther to your eye than that from the right edge. Eventually, you will reach a point (*2*) where the two rays are half a wavelength out of phase and the light will destructively interfere. Move further to the right and you will arrive at a point (*3*) where the two rays are one wavelength out of phase and the light will be bright again. Now, you moved to the right, but you could have moved in any direction perpendicular to the line from the light source and the re-sult would have been the same. Thus, there is a circle of points such as *3* around the drop that will exhibit a brightness of the light. If you now change the color of the light, you will find another circle of light but of a different radius, due to the change in wave-length. Thus, for white light impinging on the droplet, you should see circles of varying radii for each color, similar to the rainbow.

This argument was based on a single water droplet and displacement of the head.   If we now imagine a <u>collection</u> of illuminated water droplets that will give constructive interference for various colors at the angles appropriate to the above discussion, we see that we do not have to move our head - the various droplets will send various colors to our stationary eyes so that we will see the colored rings.

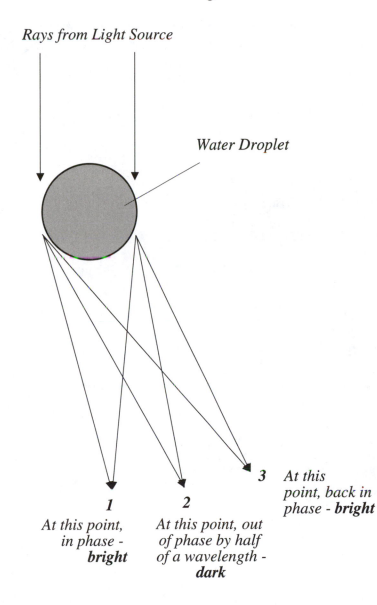

Rays from Light Source

Water Droplet

3 *At this
point, back in
phase - **bright***

1
*At this point,
in phase -
bright*

2
*At this point, out
of phase by half
of a wavelength -
dark*

Now, we address the second question as to why the light is backscattered. We consider the behavior of three rays impinging on a perfectly spherical water droplet, as shown in the diagram on the next page. The first ray, ray *1*, strikes the droplet directly in the center of the cross sectional area presented to the ray. It is reflected directly back by either the front surface or the back (internal) surface or it may escape out of the back of the raindrop. In any case, it does not deviate from its original path. This is not an edge ray, as in the discussion above, but it does play some role in the glory as we shall see soon. The second ray, ray **2**, that we consider is one that strikes the droplet anywhere between the center and the edge. This ray will either reflect off the surface in an unimportant direction or refract into the droplet where it may contribute to an ordinary rainbow. We will ignore this ray, as it plays no role in glory.

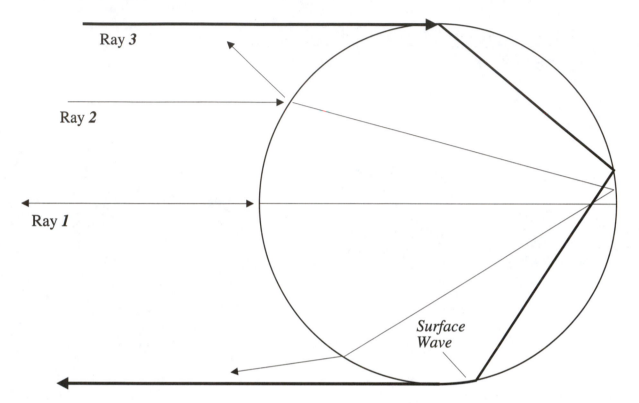

The third type of ray, ray **3**, is one that strikes at the edge of the droplet. This ray can refract into the droplet and then begin a series of reflections inside the droplet. <u>Since the incoming ray was tangent to the droplet as it entered, the refracted ray will enter the droplet at the critical angle for total internal reflection</u>. What's more, every reflection off the interior surface will occur at the critical angle (for the idealized perfect sphere). Thus, at each reflection, some of the light will refract out, parallel to a tangent to the sphere. For some number of internal reflections, a reflection will occur somewhat short of 180° around the droplet from where the light originally entered. Since the light refracted out of the sphere is parallel to the surface, it is possible that it will be carried further around the edge of the sphere as a "surface wave" before actually propagating into the air. This surface wave can make up the difference in the 180° and the result is light which is scattered directly back toward the source. It is this light directed back from the edges of the drops that participates in the diffraction phenomenon described above and gives us the glory.

One final comment is appropriate here. The intensity of the glory depends strongly on droplet size. This is due to another interference effect - that between the back-scattered rays from the droplet edges and those reflected from the center of the drop, ray **1** above. If the size of the drop is such that these rays are out of phase, the glory will be dark or absent. If the size of the drop is such that these rays are in phase, the glory will be bright.

More information about the glory, as well as photographs, can be found in H. C. Bryant and N. Jarmie, "The Glory", *Scientific American*, **231(1)**, 60 (1974), along with the Greenler text mentioned earlier. The following articles also discuss the glory, but are not for the mathematically faint-hearted: H. C. Bryant and A. J. Cox, "Mie Theory and the Glory", *Journal of the Optical Society of America*, **56**, 1529 (1966) and H. M. Nussenzveig, "Complex Angular Momentum Theory of the Rainbow and the Glory", *Journal of the Optical Society of America*, **69**, 1068 (1979).

7.) Why does a megaphone make sound louder?

The operation of the megaphone depends on two principles:

1.) The sound of a voice must leave the mouth and enter the open air. Even though the medium is air in both cases, there is a difference between the enclosed air in the mouth and the open air outside. Thus, as with any change in medium, there will be reflected sound and transmitted sound. The opposition which air provides to the passage of a sound wave can be described with an acoustic <u>impedance</u> (see Myth #6, Chapter 18 for more discussion of acoustic impedance), just as electrons moving through a wire experience electrical impedance. The acoustic impedance depends on properties of the medium. Thus, the impedances of the air in the mouth and the open air are different. The megaphone acts as an <u>impedance matcher</u>, providing a smooth transition from the mouth to the open air rather than an abrupt change. As a result, less sound is reflected and more is transmitted and the sound is louder.

2.) Sound that leaves an open mouth suffers a large amount of diffraction, since the size of the mouth opening is small compared to the wavelength of sound. Thus, there is significant spreading of the sound. When the sound leaves the end of the megaphone, it leaves a much larger aperture and thus experiences less diffraction. As a result, the sound is more concentrated in the forward direction, similar to the effect of the bell on a brass instrument in Mystery #4.

Above all, be sure <u>not</u> to say that a megaphone <u>amplifies</u> sound (Myth #2, Chapter 18), unless it is a battery powered bullhorn.

8.) If you look carefully at the primary rainbow, you may see some additional pastel-colored bows below it. These are called *supernumerary bows*. What causes these?

The supernumerary bows and, indeed, the main rainbow itself are the results of interference. In Mystery #1, Chapter 26, we drew a ray diagram for rays entering the raindrop and exiting after one internal reflection. As far as explaining the main rainbow, this discussion of geometrical optics was sufficient. But look again at the diagram. Let us concentrate on the Descartes Ray (ray *4*) and two rays striking the raindrop just above and just below the Descartes Ray. Rays *3* and *5* in the diagram are too far away. Imagine rays <u>very</u> close to the Descartes Ray. Now, imagine those rays leaving the raindrop at the bottom. Even though they were on opposite sides of the Descartes Ray upon entering, they will both leave the raindrop at an angle smaller than that of the Descartes Ray, since the Descartes Ray represents the largest possible exit angle. They will also be very close to each other, since they entered the raindrop close to each other. They will not leave the raindrop at <u>exactly</u> the same point, however, so there will be a phase difference between them as they travel toward an observer. Thus, we have the conditions for an interference pattern - two rays with a phase difference arriving at the eye and combining.

Now, we have just considered two rays. If we imagine all of the rays that enter the raindrop and bunch up near the Descartes Ray as they leave the raindrop, we can see that we have a rich supply of waves to interfere. These rays will all be at smaller angles than the Descartes Ray and will thus provide interference below the primary rainbow. We have an interference pattern which spreads the colors out, just as with a diffraction grating. The first interference maximum is near the Descartes Ray and gives us the primary rainbow. Just below the rainbow angle is a dark area where we have destructive interference. At a lower angle, the second interference maximum gives us the first supernumerary rainbow. At lower and lower angles, in principle, we can see higher order interference maxima. For more information on supernumerary rainbows, see A. B. Fraser, "Chasing Rainbows", *Weatherwise*, **36(6)**, 280 (1983).

Nice color photographs of a rainbow, with clear supernumerary bows, are available on the covers of *The Physics Teacher*, April 1989 and April 1990.

9.) When you are in a room, you can <u>hear</u> someone in the hallway through the open door, even if you can't <u>see</u> them. Why?

When someone is standing and talking in a hallway, two forms of waves leave the person, some of which propagate toward the open doorway - *sound* and *light*. For the sound, the open doorway is an aperture with a size comparable to the wavelength, so significant diffraction occurs as the sound encounters and enters the doorway. The sound is then reflected repeatedly by the walls of the room. Thus, the sound is available almost everywhere in the room. For the light, however, compared to its wavelength, the door is a <u>huge</u> aperture so that no significant diffraction takes place. Thus, the light does not spread into the room as the sound does. The only people in the room who can <u>see</u> the people in the hall are those with a line-of-sight view of them, meaning that there is a straight line from the light source, through the door, to the observer.

10.) On a compact disk player, you might see the phrase "three-beam tracking". What does this mean?

Three beam tracking is designed to keep the laser directed on the line of pits in which the audio information is encoded (see Mystery #3). The laser beam is passed through a <u>diffraction grating</u>, so that a central maximum and two secondary maxima, one on either side of the central maximum, are formed. These are the "three beams". The center beam rides along the line of pits. The two other beams ride along the flat areas to the side of the line of pits. As long as the laser is tracking correctly, the two side beams will reflect off the flat area, into their own detectors, and result in a constant signal. If the laser starts to drift to one side, one of the side beams will enter the line of pits. The side beam detector will detect the change in the reflected light and provide a feedback signal to move the laser back on track. The feedback mechanism will know which way to move the laser based on which side beam exhibited a change in reflected light.

11.)   The opal is famous for its striking iridescent colors.   What causes these colors?

These colors are caused by diffraction. An opal is formed by the evaporation of underground water containing dissolved silica (silicon dioxide). As the water evaporates, the silica molecules attach and form small spheres of radius in the range of 100 nm. As the spheres settle to the bottom of the water, they form into layers with a regular arrangement. As the water continues to evaporate, these regular arrays of spheres harden into a solid mass. As a result, the spheres act as a diffraction grating, resulting in constructive interference for different colors in different directions. Further discussion of the optics of opals is available in P. J. Darragh, et al, "Opals", *Scientific American*, **234(4)**, 84 (1976).

Magic:

Speckles on the Wall

The speckles appear because the wall is not perfectly flat. Slight variations in the surface of the wall result in light rays propagating to your eyes that are out of phase and will destructively interfere when they combine in your eye. It is important to realize that <u>the constructive and destructive interference occurs in the eye</u>, not at the wall. This is why the speckles are still visible to the near-sighted person with no glasses.

If you move your head, the speckles should move to the side. The direction of motion, however, depends on whether you are near-sighted or farsighted! For more discussion of this aspect of the speckle pattern, see J. Walker, "The Amateur Scientist - The 'Speckle' on a Surface Lit by Laser Light can be Seen with Other Types of Illumination", *Scientific American*, **246(2)**, 162 (1982) and J. L. Carlsten, "Laser Speckle", *The Physics Teacher*, **25**, 175 (1987).

Myth:

1.)   Peacocks feathers are brilliant blue and green.

The bright colors in a peacock's feathers, as well as the similarly bright colors on the throat of a hummingbird, are due to <u>interference</u>, not to absorption and reflection, as with "normal" colored objects. Structures in the feathers act as multilayer interference films that exhibit constructive interference for the colors that you see from the feathers, for example, blue and green from the peacock feather. If you have access to a peacock feather, look at it from different angles and notice how the color changes. Other creatures exhibit similar interference effects, such as the *Morpho* butterfly from South America and the beetle *Chrysochroa fulminans*, for which the interference combines with a highly glossy surface to give colors which ranges from metallic gold to green. For more information, see pp. 103 - 104 in S. J. Williamson and H. Z. Cummins, *Light and Color in Nature and Art*, Wiley, New York, 1983, or W. D. Wright, "The Rays are not Coloured", *Nature*, **198**, 1239 (1963).

Illustration Credits

While some illustrations were drawn by the author, most of the artwork in this book was obtained from computer clip art collections published by the following software distributors and used with permission. Some of the clip art illustrations were modified by the author. The names of the software packages appear under the manufacturer in the list below.

3G Graphics, Inc. (114 Second Ave. S., Suite 104, Edmonds, Washington 98020)

Business 1
Graphics & Symbols 1
People 1

T/Maker Company (1390 Villa Street, Mountain View, California 94041)

ClickArt EPS Animals & Nature
ClickArt EPS Business
ClickArt EPS Illustrations
ClickArt EPS Sports & Games
ClickArt EPS Symbols & Industry Vol. 1

Multi-Ad Services, Inc. (1720 West Detweiller Drive, Peoria, Illinois 61615)

Kwikee InHouse PAL - Professional Art Library Potpourri

Several pieces of artwork were downloaded and used with permission from America Online (Quantum Computer Services, Inc., 8619 Westwood Center Drive, Vienna, Virginia 22182):

Chapter 1, Myths #6 and #7 - Earth
Chapter 1, Myth #8 - Thermometer
Chapter 7, Mystery #4 - Candle
Chapter 9, Myth #3 - World Map
Chapter 10, Myth #2 - Compact Disk
Chapter 15, Mystery #12 - Automobile
Chapter 15, Mystery #14 - Automobile
Chapter 15, Magic (The Paper That Would Not Burn) - Candle
Chapter 19, Mysteries #1 and #2 - Lightning Bolt
Chapter 20, Mystery #8 - Light Bulb
Chapter 21, Mystery #2 - Christmas Tree
Chapter 22, Mystery #1 - Magnet
Chapter 23, Mystery #2 - Automobile
Chapter 27, Mystery #4 - Automobile
Chapter 29, Mystery #3 - Compact Disk

The following individual pieces of artwork are credited to the original source:

Chapter 3, Mystery #5 - Hourglass
Dubl-Click Software, 22521 Styles Street, Woodland Hills, California 91367

Chapter 15, Mystery #9 - Woman with Coffee Cup
Dream Maker Software, 925 West Kenyon Ave., Suite 16, Englewood,
Colorado 80110

Chapter 17, Myth #4 - Earthquake Graphic
Greg Kearney, Casper Star-Tribune, PO Box 80, Casper, Wyoming 8260

Chapter 29, Myth #1 - Peacock
Deb Skadeland, Savannah College of Art and Design, PO Box 3146,
342 Bull Street, Savannah, Georgia 31402

Bibliography

Books

Arons, A. B., *A Guide to Introductory Physics Teaching*, John Wiley & Sons, New York, 1990

Barrett, C., *Op Art*, The Viking Press, New York, 1970

Benson, H., *University Physics*, John Wiley & Sons, Inc, New York, 1991

Berg, R. E. and Stork, D. G., *The Physics of Sound*, Prentice-Hall, Inc., Englewood Cliffs, 1982

Birks, J. B. and Hart, J., *Progress in Dielectrics, Volume 3*, John Wiley & Sons, New York, 1961

Bohren, C. F., *Clouds in a Glass of Beer*, John Wiley & Sons, New York, 1987

Bohren, C. F., *What Light Through Yonder Window Breaks?*, John Wiley & Sons, New York, 1991

Bolt, B. A., *Earthquakes: A Primer*, W. H. Freeman & Co., San Francisco, 1978

Boys, C. V., *Soap Bubbles - Their Colours and the Forces which Mold Them*, Dover Publications, Inc., New York, 1959

Bullen, K. E., *An Introduction to the Theory of Seismology*, Cambridge University Press, London, 1963

Carraher, R., *Electronic Flash Photography*, Curtin & London, Inc., Somerville, 1980

Clancy, E. P., *The Tides: Pulse of the Earth*, Doubleday & Co., Garden City, New York, 1968

Copson, D. A., *Microwave Heating*, AVI, Westport, CT, 1975

Cutnell, J. D. and Johnson, K. W., *Physics*, John Wiley & Sons, New York, 1989

Daniel, V. V., *Dielectric Relaxation*, Academic Press, London, 1967

Davies, P., *Davies' Dictionary of Golfing Terms*, Simon & Schuster, New York, 1980

Davson, H., *The Eye; Comparative Physiology*, Academic Press, New York, 1974

Davson, H., *The Eye; The Photobiology of Vision*, Academic Press, New York, 1977

Discovery Toys, *Simple Science Experiments*, Hans Jurgen Press, Ravensburg, 1967

Edgerton, H. E., *Electronic Flash, Strobe*, 2nd, ed., MIT Press, Cambridge, 1979

Emery, D. and Greenberg, S., *The World Sports Record Atlas*, Facts on File Publications, New York, 1986

Feldman, D., *Imponderables*, William Morrow & Co., New York, 1987

Feldman, D., *When Do Fish Sleep?*, Harper & Row, New York, 1989

Feldman, D., *Why Do Clocks Run Clockwise?*, Harper & Row, New York, 1988

Feldman, D., *Why Do Dogs Have Wet Noses?*, Harper Perrenial, New York, 1991

Frautschi, S. C., et al, *The Mechanical Universe - Mechanics and Heat, Advanced Edition*, Cambridge University Press, Cambridge, 1986

Ganslen, R. V., *Mechanics of the Pole Vault*, John S. Swift, Inc., St. Louis, 1979

Gaskell, T. F., *The Gulf Stream*, The John Day Company, New York, 1973

Gaunt, L., *Electronic Flash Guide*, 6th ed., Focal Press, London, 1979

Godin, G., *The Analysis of Tides*, University of Toronto Press, Toronto, 1972

Golde, R. H. (Ed.), *Lightning; Volume 1 - Physics of Lightning*, Academic Press, London, 1977

Golde, R. H. (Ed.), *Lightning; Volume 2 - Lightning Protection*, Academic Press, London, 1977

Golding, W., *Lord of the Flies*, Perigee Books, New York, 1954

Greenler, R., *Rainbows, Halos and Glories*, Cambridge University Press, Cambridge, 1980

Griffing, D. F., *The Dynamics of Sports*, The Dalog Company, Oxford, Ohio, 1987

Griffiths, D. J., *Introduction to Electrodynamics*, Prentice-Hall, Inc., Englewood Cliffs, 1981

Halliday, D. and Resnick, R., *Fundamentals of Physics*, 3rd. ed., John Wiley & Sons, New York, 1988

Halliday, D., Resnick, R. and Walker, J., *Fundamentals of Physics*, 4th. ed., John Wiley & Sons, New York, 1993

Hart, W. C. and Malone, E. W., *Lightning and Lightning Protection*, Don White Consultants, Inc., Gainesville, VA, 1988

Hay, A., *The Handbook of Golf*, Salem House, Salem, New Hampshire, 1985

Helmholtz, H., *Handbuch der Physiologischen Optik* (English translation edited by J. P. C. Southall, 1924), Optical Society of America, 1867

Hershenson, M., *The Moon Illusion*, Erlbaum, Hillsdale, N.J., 1989

Hewitt, H. and Vause, A. S., *Lamps and Lighting*, American Elsevier Publishing Company, New York, 1966

Hoffman, M. S. (Ed.), *World Almanac and Book of Facts*, Pharos Books, New York, 1991

Hubel, D. H., *Eye, Brain and Vision*, W. H. Freeman & Co., New York, 1988

Iribane, J. V. and Godson, W. L., *Atmospheric Thermodynamics*, 2nd ed., D. Reidel Publishing Co., Dordrecht, Holland, 1981

Isaacs, A. (Ed.), *Concise Dictionary of Physics*, Oxford University Press, Oxford, 1985

Isenberg, C., *The Science of Soap Films and Soap Bubbles*, Tieto, Ltd., Clevedon, 1978

Jenkins, F. A. and White, H. E., *Fundamentals of Optics*, 4th ed., McGraw-Hill Book Company, New York, 1976

Johnson, O. (Ed.), *Information Please Almanac 1992*, Houghton Mifflin, Boston, 1991

Jones, E. R., *Contemporary College Physics*, Addison-Wesley Publishing Co., Reading, 1990

Kingston, H. M. and Jassie, L. B. (Eds.), *Introduction to Microwave Sample Preparation*, American Chemical Society, Washington DC, 1988

Konnen, G. P., *Polarized Light in Nature*, Cambridge University Press, Cambridge, 1980

Landsberg, H. E., *Weather and Health*, Doubleday & Co., Inc., Garden City, New York, 1969

Macauley, D., *The Way Things Work*, Houghton Mifflin, Boston, 1988

McFarlan, D. (Ed.), *Guinness Book of Records 1991*, Facts on File Publications, New York, 1990

McFarlan, D. (Ed.), *Guinness Book of World Records 1990*, Sterling Publishing Co., New York, 1989

Meine, A., *Sunsets, Twilights and Evening Skies*, Cambridge University Press, Cambridge, 1983

Miller, D. and Benedek, G., *Intraocular Light Scattering*, Charles C. Thomas, Publisher, Springfield, Ill, 1973

Neumann, G., *Ocean Currents*, Elsevier Publishing Co., Amsterdam, 1968

Ohanian, H. C., *Physics*, 2nd ed., W. W. Norton & Co., New York, 1989

Ostdiek, V. J. and Bord, D. J., *Inquiry Into Physics*, 2nd ed., West Publishing Co., St. Paul, 1991

Overheim, R. D. and Wagner, D. L., *Light and Color*, John Wiley & Sons, New York, 1982

Padgham, C. A. and Saunders, J. E., *The Perception of Light and Colour*, Academic Press, New York, 1975

Pasachoff, J. M., *Contemporary Astronomy*, W. B. Saunders Co., Philadelphia, 1977

Pedrotti, F. L. and Pedrotti, L. S., *Introduction to Optics*, Prentice-Hall, Inc., Englewood Cliffs, 1987

Püschner, H., *Heating with Microwaves*, Springer-Verlag, New York, 1966

Richter, C. F., *Elementary Seismology*, W. H. Freeman & Co., San Francisco, 1958

Robinson, J. O., *The Psychology of Visual Illusion*, Hutchinson University Library, London, 1972

Rock, I., *Perception*, Scientific American Library, New York, 1984

Ryan, F., *Pole Vault*, The Viking Press, New York, 1971

Sheppard, J. J., *Human Color Perception*, American Elsevier Publishing Co., New York, 1968

Sinclair, S., *How Animals See*, Facts on File Publications, New York, 1985

Stansfield, W. D., *The Science of Evolution*, MacMillan Publishing Co., Inc., New York, 1977

Strakosch, G. R., *Vertical Transportation: Elevators and Escalators*, John Wiley & Sons, New York, 1983

Thompson, N., *Thinking Like a Physicist*, Adam Hilger, Bristol, 1987

Van Zante, H. J., *The Microwave Oven*, Houghton Mifflin Company, Boston, 1973

Viemeister, P. E., *The Lightning Book*, Doubleday & Co., Inc., Garden City, NY, 1961

von Baeyer, H. C., *Rainbows, Snowflakes and Quarks*, McGraw-Hill Book Co., New York, 1984

Vonnegut, K., *Slapstick*, Dell Publishing Co., New York, 1976

Wade, N., *The Art and Science of Visual Illusions*, Routledge & Kegan Paul, London, 1982

Waldman, G., *Introduction to Light*, Prentice-Hall, Englewood Cliffs, 1983

Walker, J., *The Flying Circus of Physics with Answers*, John Wiley & Sons, New York, 1977

Weber, R. L., *Physics on Stamps*, A. S. Barnes & Co., Inc., San Diego, 1980

Whipple, A. B. C., *Restless Oceans*, Time-Life Books, Alexandria, VA, 1983

White, H. E. and White, D. H., *Physics and Music*, Saunders, Philadelphia, 1980

Williamson, S. J. and Cummins, H. Z., *Light and Color in Nature and Art*, John Wiley & Sons, Inc., New York, 1983

Wyatt, S. P., *Principles of Astronomy*, 3rd. ed., Allyn & Bacon, Inc., Boston, 1977

Wylie, F. E., *Tides and the Pull of the Moon*, The Stephen Greene Press, Brattleboro, Vermont, 1979

Zetterberg, J. P., *Evolution versus Creationism: The Public Education Controversy*, Oryx Press, Phoenix, 1983

Journal Articles

Adler, C., "Shadow-Sausage Effect", *American Journal of Physics*, **35**, 774, 1967

Akridge, R., "Cartoon Physics", *The Physics Teacher*, **28**, 336, 1990

Aljishi, S. and Tatarkiewicz, J., "Why does Heating Water in a Kettle Produce Sound?", *American Journal of Physics*, **59**, 628, 1991

Alpert, S. S., "A Simple Explanation of the Depth of Field Properties of an Ideal Lens", *American Journal of Physics*, **38**, 1355, 1970

Arons, A. B., "Developing the Energy Concepts in Introductory Physics", *The Physics Teacher*, **27**, 506, 1989

Aubrecht, G. J., "Trace Gases, CO_2, Climate and the Greenhouse Effect", *The Physics Teacher*, **26**, 145, 1988

Bachman, C. H., "The Equinox Displaced", *The Physics Teacher*, **28**, 536, 1990

Baierlein, R., "The Meaning of Temperature", *The Physics Teacher*, **28**, 94, 1990

Baird, J. C., et al, "A Simple But Powerful Theory of the Moon Illusion", *Journal of Experimental Psychology: Human Perception and Performance*, **16**, 675, 1990

Baker, D. J., "Demonstration of Fluid Flow in a Rotating System", *American Journal of Physics*, **34**, 647, 1966

Barnes, G., "Jackrabbit Ears and Other Physics Problems", *The Physics Teacher*, **28**, 156, 1990

Bartels, R. A., "Do Darker Objects Really Cool Faster?", *American Journal of Physics*, **58**, 244, 1990

Bauman, R. P., "Physics the Textbook Writers Usually Get Wrong - I; Work", *The Physics Teacher*, **30**, 264, 1992

Baumeister, P. and Pincus, G., "Optical Interference Coatings", *Scientific American*, **223(6)**, 59, 1970

Billah, K. Y. and Scanlan, R. H., "Resonance, Tacoma Narrows Bridge Failure, and Undergraduate Physics Textbooks", *American Journal of Physics*, **59**, 118, 1991

Bohren, C. F. and Fraser, A. B., "At What Altitude Does the Horizon Cease to Exist?", *American Journal of Physics*, **54**, 222, 1986

Boring, E. G., "The Moon Illusion", *American Journal of Physics*, **11**, 55, 1943

Brecher, K., "Do Atoms Really 'Emit' Absorption Lines?", *The Physics Teacher*, **29**, 454, 1991

Brindley, G. S., "Afterimages", *Scientific American*, **209(4)**, 84, 1963

Bryant, H. C. and Cox, A. J., "Mie Theory and the Glory", *Journal of the Optical Society of America*, **56**, 1529, 1966

Bryant, H. C. and Jarmie, N., "The Glory", *Scientific American*, **231(1)**, 60, 1974

Bundy, F. P., "Stresses in Freely Falling Chimneys and Columns", *Journal of Applied Physics*, **11**, 112, 1940

Carlsten, J. L., "Laser Speckle", *The Physics Teacher*, **25**, 175, 1987

Chandler, D., "Weightlessness and Microgravity", *The Physics Teacher*, **29**, 312, 1991

Chiaverina, C., "Reviewing the Troops", *The Physics Teacher*, **27**, 268, 1989

Consumers Union, "Goodbye Glasses?", *Consumer Reports*, **53(1)**, 52, 1988

Coren, S. and Aks, D. J., "Moon Illusion in Pictures: A Multimechanism Approach", *Journal of Experimental Psychology: Human Perception and Performance*, **16**, 365, 1990

Crane, H. R., "How Things Work - A Puzzle About Rear View Mirrors", *The Physics Teacher*, **23**, 238, 1984

Crane, H. R. , "Brrrr! The Origin of the Wind Chill Factor", *The Physics Teacher*, **27**, 59, 1989

Crane, H. R., "How Things Work - A Tornado in a Soda Bottle and Angular Momentum in the Wash Basin", *The Physics Teacher*, **25**, 516, 1987

Crane, H. R., "How Things Work - Halogen Lamps", *The Physics Teacher*, **23**, 41, 1985

Crawford, F. S., "Hot Water, Fresh Beer, and Salt", *American Journal of Physics*, **58**, 1033, 1990

Crawford, F. S., "The Hot Chocolate Effect", *American Journal of Physics*, **50**, 398, 1982

Dadourian, H. M., "The Moon Illusion", *American Journal of Physics*, **14**, 65, 1946

Darragh, P. J., et al, "Opals", *Scientific American*, **234(4)**, 84, 1976

Daw, H. A., "A 360° Rainbow Demonstration", *American Journal of Physics*, **58**, 593, 1990

DeBuvitz, W., "Christmas Tree Lights - A 'Continuing Series'?", *The Physics Teacher*, **30**, 530, 1992

di Cicco, D., "Watching for 'Pinatubo Sunsets' ", *Sky and Telescope*, **82**, 677, 1991

Edge, R. D., "The Optics of the Human Eye", *The Physics Teacher*, **27**, 392, 1989

Edmiston, M. D., "Does Skating Melt Ice?", *The Physics Teacher*, **27**, 327, 1989

Egler, R. A., "Supernova Core Bounce: A Demonstration", *The Physics Teacher*, **28**, 558, 1990

Erlichson, H., "Maximum Projectile Range with Drag and Lift, with Particular Application to Golf", *American Journal of Physics*, **51**, 357, 1983

Escobar, C., "Amusement Park Physics", *The Physics Teacher*, **28**, 446, 1990

Evans, L., "The Science of Traffic Safety", *The Physics Teacher*, **26**, 426, 1988

Faucher, G., "Ferromagnetism and the Secret Agent", *The Physics Teacher*, **26**, 30, 1988

Faucher, G., "Pushing or Pulling a Wheelbarrow ", *The Physics Teacher*, **27**, 379, 1989

Favreau, O. E. and Corballis, M. C., "Negative Aftereffects in Visual Perception", *Scientific American*, **235(6)**, 42, 1976

Few, A. A., "Thunder", *Scientific American*, **233(1)**, 80, 1975

Fraser, A. B., "Chasing Rainbows", *Weatherwise*, **36(6)**, 281, 1983

Fraser, A. B. and Mach, W. H., "Mirages", *Scientific American*, **234(1)**, 102, 1976

Fredrickson, J. E., "The Tail-less Cat in Free Fall", *The Physics Teacher*, **27**, 620, 1989

Gardner, M., "Physics Trick of the Month - The Falling Keys", *The Physics Teacher*, **28**, 390, 1990

Gardner, M., "Physics Trick of the Month - Transporting an Olive", *The Physics Teacher*, **29**, 51, 1991

Garstang, R. H., "How Hot Does Your Parked Car Become?", *The Physics Teacher*, **29**, 58, 1991

Gee, J. K., "The Myth of Lateral Inversion", *Physics Education,* **23**, 300, 1988

Greenler, R. G. and Mallman, A. J., "Circumscribed Halos", *Science*, **176**, 128, 1972

Haggis, G. H., Hasted, J. B. and Buchanan, T. J., "The Dielectric Properties of Water in Solutions", *Journal of Chemical Physics*, **20**, 1452, 1952

Hanson, K., "The Radiative Effectiveness of Plastic Films for Greenhouses", *Journal of Applied Meteorology*, **2**, 793, 1963

Haugland, O. A., "Hot-Air Ballooning in Physics Teaching", *The Physics Teacher*, **29**, 202, 1991

Heilmeier, G. H., "Liquid-Crystal Display Devices", *Scientific American*, **222(4)**, 100, 1970

Hewitt, P., "Figuring Physics", *The Physics Teacher*, **30**, 124, 1992

Higashiyama, A., "Anisotropic Perception of Visual Angle: Implications for the Horizontal-Vertical Illusion, Overconstancy of Size, and the Moon Illusion", *Perception & Psychophysics*, **51**, 218, 1992

Huebner, J. S. and Smith, T. L., "Multi-Ball Collisions", *The Physics Teacher*, **30**, 46, 1992

Huhn, J. K., "You Can Be Myopic and Still Survive on a Desert Island!", *The Physics Teacher*, **29**, 577, 1991

Hunt, R. G., "Bicycles in the Physics Lab", *The Physics Teacher*, **27**, 160, 1989

Hunt, R. G., "Physics of Popping Corn", *The Physics Teacher*, **29**, 230, 1991

Jacobs, D. J., "Why Don't the Earliest Sunrise and the Latest Sunset Occur on the Same Day?", *Physics Education*, **25**, 275, 1990

Jewett, J. W., "DMV Physics - The Case of the Disappearing Time Intervals", *The Physics Teacher*, **29**, 563, 1991

Jewett, J. W., "Get the LED Out - Physics Demonstrations Using Light Emitting Diodes", *The Physics Teacher*, **29**, 530, 1991

Jewett, J. W., "Giving New Meaning to 'Resonance in Pipes' ", *The Physics Teacher*, **31**, 253, 1993

Jewett, J. W., "LED's and the 'Fluttering Hearts' Phenomenon", *The Physics Teacher*, **31**, 180, 1993

Jones, E. R. and Edge, R. D., "Optics of the Rear-View Mirror: A Laboratory Experiment", *The Physics Teacher*, **24**, 221, 1986

Kaufman, L. and Rock, I., "The Moon Illusion", *Scientific American*, **207(1)**, 120, 1962

Kettler, J. E., "The Compact Disk as a Diffraction Grating", *American Journal of Physics*, **59**, 367, 1991

Kruger, C., "Some Primary Teachers' Ideas About Energy", *Physics Education*, **25**, 86, 1990

Kruglak, H., "Another Legend Debunked", *The Physics Teacher*, **28**, 491, 1990

Kruglak, H., et al, "Shattering Glass with Sound Simplified", *The Physics Teacher*, **28**, 418, 1990

Kruglak, H., "The Compact Disc as a Diffraction Grating", *Physics Education*, **25**, 255, 1990

Kuethe, D. O., "Confusion About Pressure", *The Physics Teacher*, **29**, 20, 1991

Lakhtakia, A., "Would Brewster Recognize Today's Brewster Angle?", *Optics News*, **15**, 14, 1989

Land, E. H., "Some Aspects of the Development of Sheet Polarizers", *Journal of the Optical Society of America*, **41**, 957,

Land, E. H., "The Retinex Theory of Color Vision", *Scientific American*, **237(6)**, 108, 1977

Lapp, D. R., "Wave Interference Pattern from the Wings of a Bee", *The Physics Teacher*, **28**, 335, 1990

Lee, R., "The 'Greenhouse' Effect", *Journal of Applied Meteorology*, **12**, 556, 1973

Leff, H. S., "Illuminating Physics with Light Bulbs", *The Physics Teacher*, **28**, 30, 1990

Lewis, H. W., "Ball Lightning", *Scientific American*, **208(3)**, 106, 1963

Lisensky, G. C., et al, "Periodic Properties in a Family of Common Semiconductors; Experiments with Light Emitting Diodes", *Journal of Chemical Education*, **69**, 151, 1992

Lothian, G. F., "Blue Sun and Moon", *Nature*, **168**, 1086, 1951

Lucas, A. M., "Public Knowledge of Elementary Physics", *Physics Education*, **23**, 10, 1988

MacDonald, J. E., "The Shape of Raindrops", *Scientific American*, **190(2)**, 64, 1954

MacDonald, W. M. and Hanzely, S., "The Physics of the Drive in Golf", *American Journal of Physics*, **59**, 213, 1991

Mallinckrodt, A. J. and Leff, H. S., "All About Work", *American Journal of Physics*, **60**, 356, 1992

Mamola, K. C., et al, "Light Rays in Gradient Index Media: A Laboratory Exercise", *American Journal of Physics*, **60**, 527, 1992

Marlow, A. R., "A Surprising Mechanics Demonstration", *American Journal of Physics*, **59**, 951, 1991

Matthes, G. H., "Quicksand", *Scientific American*, **188(6)**, 97, 1953

Mattila, J. O., "Physics at the Fire Station", *The Physics Teacher*, **26**, 440, 1988

McCloskey, M., "Cartoon Physics", *Psychology Today*, **18(4)**, 52, 1984

McDougall, W., "The Illusion of the 'Fluttering Heart' and the Visual Functions of the Rods in the Retina", *British Journal of Psychology*, **1**, 428, 1904

McGehee, J., "Physics Students' Day at Six Flags/Magic Mountain", *The Physics Teacher*, **26**, 13, 1988

Negret, J. P., "Boiling Water and the Height of Mountains", *Physics Education*, **24**, 290, 1986

Nicholson, T. D., "The Moon Illusion", *Natural History*, 66, August 1991

Noldeke, C., "Compact Disc Diffraction", *The Physics Teacher*, **28**, 484, 1990

Nussenzveig, H. M., "Complex Angular Momentum Theory of the Rainbow and the Glory", *Journal of the Optical Society of America*, **69**, 1068, 1979

Oberhofer, E. S., "Different Magnitude Differences", *The Physics Teacher*, **29**, 273, 1991

Otani, R. and Siegel, P., "Determining Absolute Zero in the Kitchen Sink", *The Physics Teacher*, **29**, 316, 1991

Paul, W. and Jones, R. V., "Blue Sun and Moon", *Nature*, **168**, 554, 1951

Pirie, A., "The Biochemistry of the Eye", *Nature*, **186**, 352, 1960

Porch, W. M., et al, "Blue Moon: Is This a Property of Background Aerosol?", *Applied Optics*, **12**, 34, 1973

Porch, W. M., "Blue Moons and Large Fires", *Applied Optics*, **28**, 1778, 1989

Prigo, R. B., "Liquid Beans", *The Physics Teacher*, **26**, 101, 1988

Prytulak, S., "Interdisciplinary Application", *The Physics Teacher*, **29**, 135, 1991

Ramme, G., "Colors on Soap Films - An Interference Phenomenon", *The Physics Teacher*, **28**, 479, 1990

Raybin, D. M., "The Stones of Spring and Summer", *The Physics Teacher*, **28**, 500, 1990

Reid, B., "Haidinger's Brush", *The Physics Teacher*, **28**, 598, 1990

Rinehart, J. S., "Waterfall-Generated Earth Vibrations", *Science*, **164**, 1513, 1969

Rosato, A., et al, "Why the Brazil Nuts Are on Top; Size Segregation of Particulate Matter by Shaking", *Physical Review Letters*, **58**, 1038, 1987

Rushton, W. A. H., "Effect of Humming on Vision", *Nature*, **216**, 1173, 1967

Savage, M. D. and Williams, J. S., "Centrifugal Force: Fact or Fiction?", *Physics Education*, **24**, 133, 1989

Smith, M. J., "Comment on: Shadow-Sausage Effect", *American Journal of Physics*, **36**, 912, 1968

Smith, P. A., "Let's Get Rid of 'Centripetal Force' ", *The Physics Teacher*, **30**, 316, 1992

Speers, R. S., "Physics and Roller Coasters - The Blue Streak at Cedar Point", *American Journal of Physics*, **59**, 528, 1991

Stewart, J. E., "The Collapsing Can Revisited", *The Physics Teacher*, **29**, 144, 1991

Steyn-Ross, A. and Riddell, A., "Standing Waves in a Microwave Oven", *The Physics Teacher*, **28**, 474, 1990

Strouse, W. M., "Bouncing Light Beams", *American Journal of Physics*, **40**, 913, 1972

Suzuki, K., "Moon Illusion Simulated in Complete Darkness: Planetarium Experiment Reexamined", *Perception & Psychophysics*, **49**, 349, 1991

Taylor, G. I., "Experiments with Rotating Fluids", *Proceedings of the Royal Society*, **A100**, 114, 1921

Tibbs, K. W., et al, "Helium High Pitch", *The Physics Teacher*, **27**, 230, 1989

Toepker, T. P., "Scaling Lake Erie", *The Physics Teacher*, **27**, 267, 1989

von Grünau, M. W., "The 'Fluttering Heart' and Spatio-Temporal Characteristics of Color Processing - I; Reversibility and the Influence of Luminance", *Vision Research*, **15**, 431, 1975

von Grünau, M. W., "The 'Fluttering Heart' and Spatio-Temporal Characteristics of Color Processing - II; Lateral Interactions Across the Chromatic Border", *Vision Research*, **15**, 437, 1975

von Grünau, M. W., "The 'Fluttering Heart' and Spatio-Temporal Characteristics of Color Processing - III; Interactions Between the Systems of the Rods and the Long-Wavelength Cones", *Vision Research*, **16**, 397, 1976

von Kries, J., "Über die Wirkung kurzdauernder Lichtreize auf das Sehorgan", *Psychol. Physiol. Sinnesorg.*, **12**, 81, 1896

Wagner, W. S., "Temperature and Color of Incandescent Lamps", *The Physics Teacher*, **29**, 176, 1991

Walker, J . D., "Nightmare in the Elevator", *Science World - Teacher's Edition*, **46(4)**, 6, 1989

Walker, J . D., "The Amateur Scientist - The Physics and Chemistry Underlying the Infinite Charm of a Candle Flame", *Scientific American*, **238(4)**, 154, 1978

Walker, J . D., "The Amateur Scientist - The 'Speckle ' on a Surface Lit with Laser Light Can Be Seen with Other Kinds of Illumination", *Scientific American*, **246(2)**, 162, 1982

Walker, J . D., "The Amateur Scientist - What Happens When Water Boils is a Lot More Complicated Than You Might Think", *Scientific American*, **247(6)**, 162, 1982

Walker, J . D., "Multiple Rainbows from Single Drops of Water and Other Liquids", *American Journal of Physics*, **44**, 421, 1976

Wehner, R., "Polarized-Light Navigation by Insects", *Scientific American*, **235(1)**, 106, 1976

Welford, W. T., "Laser Speckle and Surface Roughness", *Contemporary Physics*, **21**, 401, 1980

Weltner, K., "Aerodynamic Lifting Force", *The Physics Teacher*, **28**, 78, 1990

Weltner, K., "Bernoulli's Law and Aerodynamic Lifting Force", *The Physics Teacher*, **28**, 84, 1990

White, J. D., "The Role of Surface Melting in Ice Skating", *The Physics Teacher*, **30**, 495, 1992

Williams, E. R., "The Electrification of Thunderstorms", *Scientific American*, **259(5)**, 88, 1988

Williams, P. C. and Williams, T. P., "Effect of Humming on Watching Television", *Nature*, **239**, 407, 1972

Winter, R. G., "On the Difference Between Fluids and Dried Beans", *The Physics Teacher*, **28**, 104, 1990

Wolfe, J. M. and Owens, D. A., "Is Accommodation Colorblind? Focusing Chromatic Contours", *Perception*, **10**, 53, 1981

Wright, W. D., "The Rays are Not Coloured", *Nature*, **198**, 1239, 1963

Wu, J., "Are Sound Waves Isothermal or Adiabatic?", *American Journal of Physics*, **58**, 694, 1990

Zebrowski, E., "Aiming a Satellite Dish", *The Physics Teacher*, **26**, 153, 1988

Index